Springer Series in Synergetics

Editor: Hermann Haken

Synergetics, an interdisciplinary field of research, is concerned with the cooperation of individual parts of a system that produces macroscopic spatial, temporal or functional structures. It deals with deterministic as well as stochastic processes.

Volume 40 **Information and Self-Organization** A Macroscopic Approach to Complex Systems By H. Haken

Volume 41 **Propagation in Systems Far from Equilibrium** Editors: J. E. Wesfreid, H. R. Brand, P. Manneville, G. Albinet, N. Boccara

Volume 42 **Neural and Synergetic Computers** Editor: H. Haken

Volume 43 **Cooperative Dynamics in Complex Physical Systems** Editor: H. Takayama

Volume 44 **Optimal Structures in Heterogeneous Reaction Systems** Editor: P. J. Plath

Volume 45 **Synergetics of Cognition** Editors: H. Haken, M. Stadler

Volume 46 **Theories of Immune Networks** Editors: H. Atlan, I. R. Cohen

Volume 47 **Relative Information** Theories and Applications By G. Jumarie

Volume 48 **Dissipative Structures in Transport Processes and Combustion** Editor: D. Meinköhn

Volume 49 **Neuronal Cooperativity** Editor: J. Krüger

Volume 50 **Synergetic Computers and Cognition** A Top-Down Approach to Neural Nets By H. Haken

Volume 51 **Foundations of Synergetics I** Distributed Active Systems By A. S. Mikhailov

Volume 52 **Foundations of Synergetics II** Complex Patterns By A. Yu. Loskutov, A. S. Mikhailov

Volume 53 **Synergetic Economics** By W.-B. Zhang

Volume 54 **Quantum Signatures of Chaos** By F. Haake

Volume 55 **Rhythms in Physiological Systems** Editors: H. Haken, H. P. Koepchen

Volume 56 **Quantum Noise** By C. W. Gardiner

Volumes 1–39 are listed on the back inside cover

H. Haken H. P. Koepchen (Eds.)

Rhythms in Physiological Systems

Proceedings of the International Symposium at Schloß Elmau, Bavaria, October 22–25, 1990

With 208 Figures

Springer-Verlag
Berlin Heidelberg New York
London Paris Tokyo
Hong Kong Barcelona
Budapest

Professor Dr. Dr. h. c. Hermann Haken
Institut für Theoretische Physik und Synergetik der Universität Stuttgart,
Pfaffenwaldring 57/IV, W-7000 Stuttgart 80, Fed. Rep. of Germany and
Center for Complex Systems, Florida Atlantic University,
Boca Raton, FL 33431, USA

Professor Dr. Hans Peter Koepchen
Institut für Physiologie, Freie Universität Berlin, Arnimallee 22,
1000 Berlin 33, Fed. Rep. of Germany

Series Editor:

Professor Dr. Dr. h. c. Hermann Haken
Institut für Theoretische Physik und Synergetik der Universität Stuttgart,
Pfaffenwaldring 57/IV, W-7000 Stuttgart 80, Fed. Rep. of Germany and
Center for Complex Systems, Florida Atlantic University,
Boca Raton, FL 33431, USA

ISBN-13: 978-3-642-76879-8 e-ISBN-13: 978-3-642-76877-4
DOI: 10.1007/978-3-642-76877-4

Library of Congress Cataloging-in-Publication Data. Rhythms in physiological systems: proceedings of the international symposium at Schloss Elmau, Bavaria, October 22-25, 1990 / H. Haken, H. P. Koepchen (eds.). p. cm. – (Springer series in synergetics; v. 55) ISBN 3-540-54448-8 (Berlin). – ISBN 0-387-54448-8 (New York) 1. Biological rhythms–Congresses. I. Koepchen, Hans Peter. II. Haken, H. III. Series. QP84.6.R59 1992 591.1'882–dc20 91-38703

Softcover reprint of the hardcover 1st edition 1991

Typesetting: Camera ready by authors

54/3140-543210 – Printed on acid-free paper

Preface

Rhythms are a basic phenomenon in all physiological systems. They cover an enormous range of frequencies with periods from the order of milliseconds up to some years. They are described by many disciplines and are investigated usually in the context of the physiology of the respective function or organ. The importance given to the research on rhythmicity is quite different in different systems. In some cases where the functional significance is obvious rhythms are at the center of interest, as in the case of respiration or locomotion. In other fields they are considered more or less as interesting epiphenomena or at best as indicators without essential functional significance, as in the case of cardiovascular or EEG rhythms.

Recently the study of physiological rhythms has attracted growing interest in several fields, especially with respect to rhythm research in humans and its rapidly spreading applications in basic behavioral research, and as a diagnostic tool in clinical medicine.

This development was favored by two methodological and conceptual advances: on the one hand, the availability of non-invasive methods of continuous recording of physiological parameters and their computer-assisted evaluation, and on the other, the rapid development of theoretical analyses, for example, the understanding of dynamic systems, the generation of coordinated macroscopic processes in systems comprising many single elements, and the mathematical tools for treating nonlinear oscillators and their mutual coupling.

We confined the topics of this workshop to rhythms in a medium frequency range in the cardiovascular, respiratory, and motor systems, as well as the electrical cerebral activity including circadian rhythmicities. Besides the fact that there is much current interest in all these fields, these systems are quite different in their morphological and functional characteristics. Therefore the question arose whether it was actually justified to handle them in a single workshop or whether this would only lead to an accumulation of uncorrelated findings and concepts.

It was precisely the striking appearance of common principles in the broad phenomenological variety of rhythms in different systems, however, which encouraged us to try a combined effort and consider them together. This constituted the workshop on "Synergetics of Physiological Rhythms", held at Schloss Elmau, October 22–26, 1990. The contributions to the workshop are collected in this volume.

Viewing the numerous rhythmic phenomena described in the different contributions it is a fascinating experience to see common basic principles appearing in the fundamental forms of rhythm coordination. This applies not only to rhythms occurring within one single system but also to the interaction between respiration and cardiovascular functions, for example, or somatomotor and vegetative control systems.

The early discoveries of the fundamental and generally valid laws of biological rhythm coordination started with the classic work of Erich v. Holst on sliding or "relative" coordination. But in spite of the recognition of the general occurrence of these principles there was a lack of mathematical tools for a successful theoretical treatment of these phenomena. Biologists working in this field have always suffered from this lack of a thorough theoretical understanding.

In recent decades, the new science of synergetics, starting from the phenomena of interaction and cooperation between many single constituents of a system, has provided us with the conceptual understanding and the mathematical tools to cope also with the extremely complex generation and interactions of physiological rhythmicity.

Thus the time seemed ripe to bring together the hitherto mostly separate approaches to cardiorespiratory, motor, and brain rhythm research with the already highly sophisticated concepts and tools of synergetics. The cardiorespiratory rhythmicity in the medium frequency range dealt with in this workshop interacts strongly with the homeostatic feedback control, which has so far been the basis of physiological theories in this field. Thus, here we are confronted with the task of linking the aspects of vegetative autoregulation with the striking rhythmicities in all parts of these systems.

It is evident that this is an enormous challenge and that one workshop can only give some guidelines and impetus for the new way that is opened up by synergetics in this field of biology. We will be happy if this new volume of the synergetics series motivates colleagues to join us in following this new and fascinating path of interdisciplinary research.

We wish to cordially thank Ms. Ursula Funke for the traditionally perfect organization of this workshop. We owe thanks to Ms. Irmgard Möller and Ms. Deborah Hollis for their great help in editing these proceedings. This workshop was made possible by the financial support of the Volkswagen Foundation, Hannover, within its project on synergetics. The success of the interdisciplinary field of synergetics would not have been possible without the very efficient support of the Volkswagen Foundation over many years.

Stuttgart, Berlin
April 1991

H. Haken
H.P. Koepchen

Contents

Part I

Physiological Rhythmicity and Synergetics

Physiology of Rhythms and Control Systems: An Integrative Approach

H.P. Koepchen

Institute of Physiology, The Free University of Berlin,
Arnimallee 22, 1000 Berlin 33, Fed. Rep. of Germany

There is a broad spectrum of physiological rhythms in many different systems extending over a large range of frequencies. The medium ultradian frequency range is characterized by relatively great variability of frequency, intense mutual interactions and interference with the homeostatic feedback regulatory circuits and behavioral influences. In spite of the resulting multiform phenomenology, certain basic rules of "relative" sliding coordination between the subunits of one system, and between the systems themselves can be identified. They are revealed through certain kinds of mutual entrainment, the occurrence of multiple integer frequency relations, instability of phases and amplitudes during transitory states, rhythm-related excitability cycles, and through variable periods of synchronization and desynchronization. This principle is illustrated by respiratory, vascular, cardiovascular and motor rhythms and their interaction found in anaesthetized animals and in man.The fact that the same general kind of rhythm coordination is found, irrespective of the particular system and frequency range, justifies the hope of a successful analysis and understanding of rhythmicity as a basic principle of physiological self-organization. In view of their general validity, irrespective of the particular physical or living system under investigation, the concepts and analytical tools of synergetics will play a key role in future research of this whole subject.

This workshop represents a first attempt to combine the approaches of synergetics with the rhythmological aspects of neurovegetative cardiorespiratory control physiology. Moreover, it is the intention of this workshop to view these topics as one example in the broader context of other physiological rhythms, like motor and EEG rhythms.

In all these special fields, there are a large number of experimental findings and theoretical approaches. We are of the opinion that the application of concepts and methods of synergetics could give a common basis for the description, evaluation and, possibly, understanding of many different rhythmic phenomena in the organism. We will substantiate this belief by reviewing the principles of cardiorespiratory rhythmicity in a medium ultradian frequency range.

"Cardiorespiratory rhythmicity" is the rhythmicity in the respiratory and in the cardiovascular systems. The two are coupled with each other peripherally as well as centrally. This coupling is a physiological necessity because the two systems are arranged in series with respect to the transport of O_2 and CO_2. Since the demand for O_2 and CO_2 can change with varying metabolic needs by a factor of more than 20, this coupled transport has to be adapted to the actual behavioral activity from moment to moment. Cardiovascular and respiratory rhythmicity have to cope with this basic function of adaptation to the changing actual needs of the metabolism. One fundamental difference between the two systems is that the driving transport rhythm for the circulation originates in the periphery, whereas the rhythm of

Springer Series in Synergetics, Vol. 55 **Rhythms in Physiological Systems**
Eds.: H. Haken and H.P. Koepchen

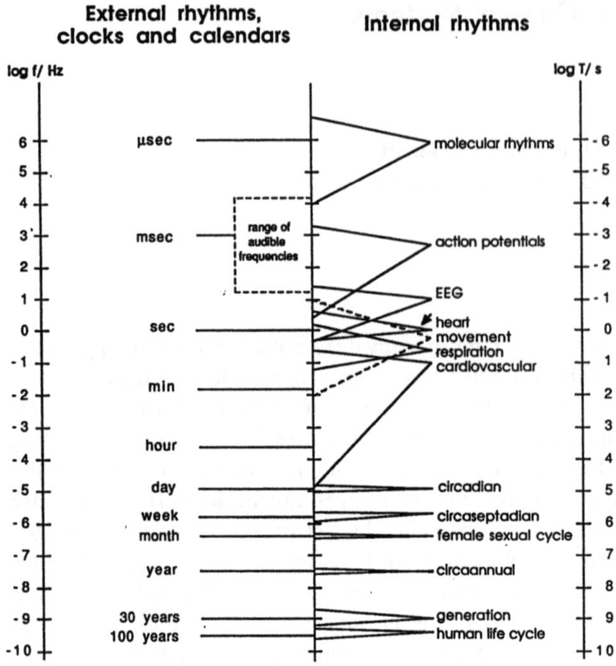

Fig. 1 Scale of frequencies and their range of variation of relevant human physiological rhythms in comparison with rhythms of the outer world.
Note the great variability and overlapping of frequencies in the cardiorespiratory and motor systems.

respiration is produced by the central nervous system. The pacemaker of the heart, in turn, is modified by central nervous influences which interfere with the central respiratory rhythm generator. As a consequence of these basic facts, behavioral activity acts strongly on both control systems, influencing the level of the basic drive as well as its rhythmic variations.

The classical physiology of cardiorespiratory control is based on the concept of homeostatic autoregulatory systems. This autoregulation is executed by feedback reflexes which have been thoroughly investigated. Here, the chemoreceptor and baroreceptor reflexes have been the center of interest for many decades. There is no doubt that in the cardiorespiratory system the homeostatic reflex control always interferes with the rhythmicity. Therefore, we have to take into consideration the relationship between rhythm and reflex control. The latter, in turn, strongly depends on the behavioral state of activity. Thus, the three aspects: rhythmicity, homeostatic reflex control, and behavioral activity cannot be separated in the context of physiological functions. Physiological rhythmicity has to be considered with this in mind.

In order to realize the peculiarities of the medium range of physiological rhythmicity, we have to localize them in the whole frequency range of biological rhythms.

Figure 1 shows the large frequency range of physiological rhythms which are important for human life. It extends over more than 15 decades of frequencies. At the upper end of the scale, we find the rhythms at the cellular or membranal level, governed mainly by physicochemical properties of the constituents: at the lower end, we see the rhythms developed in adaptation to outer cosmic rhythms with small relative frequency ranges (see the contribution of R. Wever in these proceedings). When we concentrate here on the cardiovascular and respiratory rhythms and their relation to motor rhythms we see two characteristic features: (1)

they show a relatively wide variability of frequency, (2) their frequency ranges overlap extensively. This forms the basis of mutual interactions. Moreover, these frequencies are in the same time domain as the functioning of the homeostatic feedback systems.

We will see that, irrespective of the different functional roles and the different properties of the executive peripheral organs, we can find common features in the interrelation between the systems and their single components. Within this framework, we will consider:

a) cardiovascular rhythmicity,
b) respiratory rhythmicity,
c) their mutual interaction,
d) their relationships to motor rhythmicity.

Cardiovascular Rhythms [1,2,3]

In the frequency range concerned, we find two preferential frequencies: one around 0.1 Hz, the other around 0.25 Hz. These rhythms are present in vascular movements, blood pressure and heart rate, and can easily be detected by power spectrum analysis. The interpretation of the former frequency range is controversial [2]. Here, the rhythmicity is usually explained as an oscillation in the baroreceptor homeostatic feedback system, as an expression of the autorhythmicity of the vascular smooth muscle or as the effect of primary central nervous rhythm generators projecting on the sympathetic innervation of heart and blood vessels.

The higher preferential range of cardiovascular rhythmicity roughly coincides with the frequency range of respiration. Since the classical work of Hering in 1869 [4], these oscillations of cardiovascular parameters have been interpreted as "irradiation" from the "respiratory center" to the "cardiovascular center" or, with more modern nomenclature, as a projection from central respiratory neurons onto cardiovascular efferents. That means that they would represent side effects of the respiratory rhythm, which, in this interpretation, is always considered as the primary one.

By looking at the experimental findings, we will see that the interrelations cannot be categorized by such rigid schematization. They are much more variable and follow the general rules of interaction and coordination between physiological rhythms, as in other systems too.

Vascular Rhythmicity

Vasomotor rhythms are found everywhere in the peripheral circulation. They are manifested as oscillations in blood flow, peripheral resistance to flow, vessel diameter, or other parameters. As one of many examples, Figure 2 shows the bilateral oscillations in peripheral resistance. They are recorded bilaterally using a special method of perfusing the isolated vascular bed in a skeletal muscle of both hindlimbs.

Since the oscillations are running exactly synchronously bilaterally, the first conclusion seems to be that they are caused by rhythmic vasoconstrictor innervation. Closer examination, however, shows that the synchrony is, undoubtedly, caused by bilateral symmetrical nervous activity but that the rhythm itself has a more complex origin.

Even the denervated vessels exhibit rhythmic changes in vessel diameter and, thus, resistance to flow in the same frequency domain [5,6]. This phenomenon is the expression of the inherent potential rhythmic properties of the vascular smooth muscle cells (see the contribution of G. Siegel in these proceedings). The vascular muscle forms a syncytium of electrically and mechanically coupled cells without an obligatory specialized pacemaker. That means that the synchronous rhythm in the whole muscle must result from synchronization between

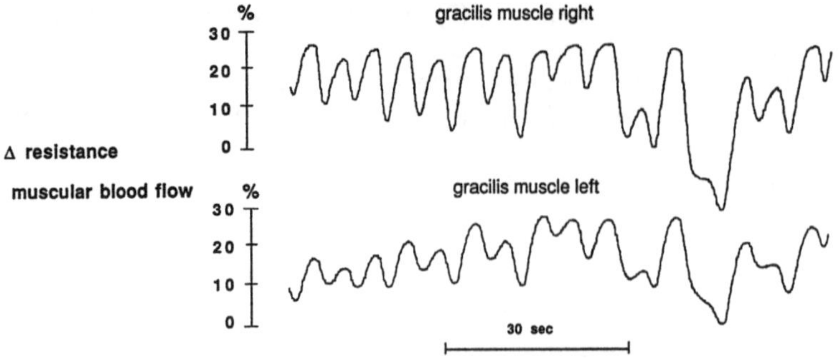

Fig. 2 Bilaterally synchronous oscillations in the resistance to flow in isolated, innervated muscular vascular beds in the hindlimbs of an anaesthetized dog. (From [35])

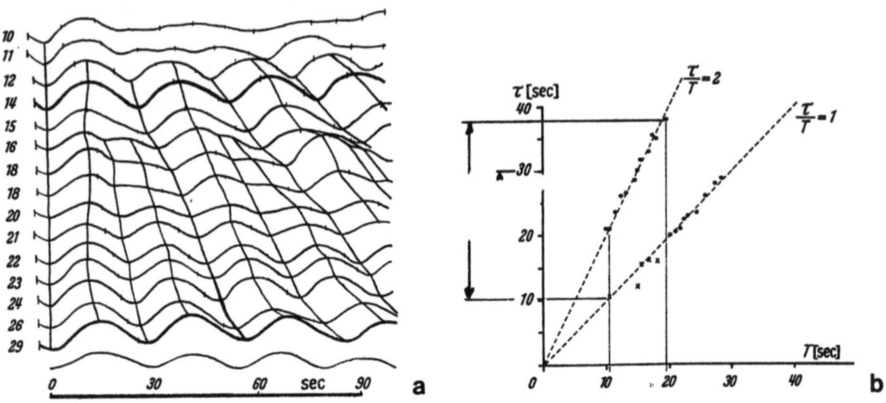

Fig. 3 a,b Spontaneous rhythmic oscillations in the resistance to skin blood flow in an anaesthetized dog and their modification by periodic stimulation of the sympathetic vasoconstrictor nerve. **(a)** Representative sections of the time courses of resistance during different kinds of rhythmic stimulation of the vasoconstrictor nerve. Numbers on the left indicate the intervals between the stimulations, small vertical bars on the individual traces indicate the moment of the single stimulation. Lowest trace represents the time course of resistance without stimulation. Note nearly identical time courses when the stimulation frequency is halved (stimulation intervals 14 sec and 29 sec, thick traces). **(b)** Dependence of the period t of vascular rhythm on the period T of stimulation. The periods are strictly grouped around the 1:1 and 2:1 lines. Ranges of entrainment between 10 and 30 sec stimulation periods for 1:1 relation and between 10 and 20 sec for 2:1 relation. (Modified from [7])

the single cells in a self-organizing process. The studies hitherto have been performed either at the single cell level or in the whole vessel, so that this process of intercellular self-organization in smooth muscular organs is still widely unexplored.

We were able to study, however, at the next higher level of organization, where the rhythmic vasoconstrictor nervous activity interferes with the endogenous rhythmicity of the vascular muscle [7]. In a preparation using a perfused isolated skin area of anaesthetized dogs, it was possible to superimpose artificial rhythmic nervous activity on the spontaneous oscillations of smooth muscle activity.

In this way, various kinds of "relative coordination", according to v. Holst's definition [8], could be detected between the rhythm of vasoconstrictor innervation and the response of vascular smooth muscle as manifested in the rhythmic oscillations of resistance to flow. Figure

3 shows the various forms of coordination between the rhythm of vasoconstrictor stimulation and the endogenous rhythmic constrictions of the vascular muscle.

Characteristic phenomena which are known from other mutual interactions can be observed between biological rhythms or between artificial phase-related stimulation and the response of a physiological rhythm. There is a certain frequency range of entrainment around the endogenous frequency, in which the rhythm can be synchronized by the superimposed rhythm of stimulation. Outside this range, more irregular or alternating forms of coordination are observed, switching over to more stable multiple integer relationships in faster or slower ranges of rhythmic stimulation. These relations are based on an excitability cycle of vascular smooth muscle which we demonstrated in the same preparation. Another consequence is a resonance-like enhancement of the amplitude of the oscillations when the stimulating rhythm approaches the endogenous smooth muscular frequency. Analogous phenomena can be found in other mutual interactions between biological rhythms or between artificial phase-related stimulation and the response of a physiological rhythm (see the contribution of F.L. Eldridge in these proceedings).

Looking for the rhythmic circulatory phenomena at the next higher level of integration, i.e. circulation at the level of whole organs such as skin and muscle, we have to state that the known fluctuations of blood flow, for example, neither reflect exactly the central vaso-motor innervation nor solely the autorhythmicity of vascular muscle. They are the result of interference between two rhythmic processes. The stimulation experiments, as in the case of the respiratory rhythm, are an experimental model of what is going on naturally in the spontaneous coordination between leading and dependent biological rhythms in the individual organism.

Blood Pressure Rhythms

The most easily accessible and clinically relevant circulatory parameters are blood pressure and heart rate. The first continuous registrations of blood pressure in the last century revealed blood pressure "waves" [9] and the rhythmic fluctuations of heart rate were described, for the first time, by A. v. Haller in 1760 [10]. Blood pressure is a resultant of many cardiovascular parameters. Since the first detection, a great number of studies on the causes of these phenomena and many different theories about them have been published.

The majority of these theories are based on more or less monocausal and morphologically oriented explanations. Our consideration of the hierarchy of cardiovascular rhythms, from the single cell level up to the whole circulatory control system, illustrates that we have to look for the rules of self-organization between the many single components. In the case of blood pressure "waves", we must realize that many oscillatory systems interfere with each other [2], as is depicted in Figure 4. This simplified scheme already shows that the classical question of cause and effect cannot lead to an understanding of this complex system.

On the contrary, in our opinion the actually occurring rhythm in the blood pressure is the result of the self-organization of many rhythmic subunits. This concept concurs with the variable shape of blood pressure waves. Every change in the properties of one of the single components can change the resulting rhythm in the entire system. This is a typical synergetic process. Therefore, we expect that quantitative treatment using the methods developed in the science of synergetics [11] will be the next important step, surpassing a qualitative visualization of the interrelations, like that of Figure 4. Our observation that the general rules of rhythm coordination, known from quite different systems, also apply to cardiovascular rhythms as well encourages such an approach.

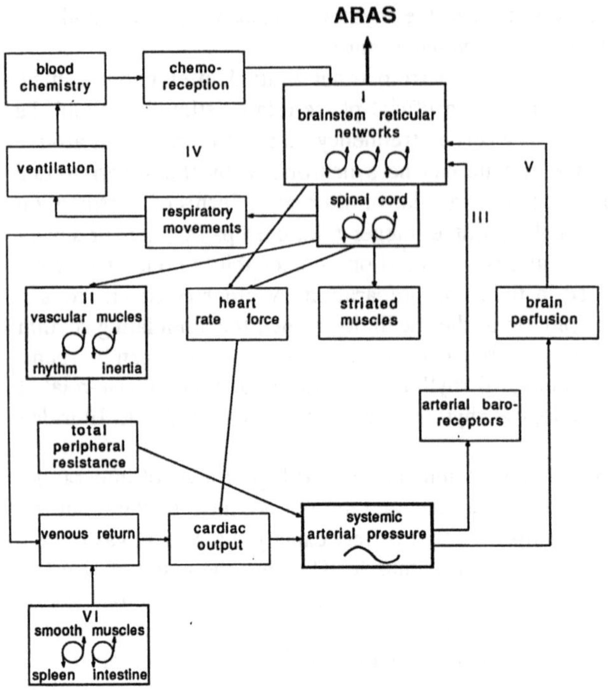

Fig. 4 Scheme of the main hitherto known oscillatory systems potentially influencing the arterial blood pressure. Circles symbolize local oscillators in the central nervous system or in smooth muscle tissue. Lines and arrows represent feedback systems which can oscillate under special conditions. III: Baroreceptor circuit. IV: Chemoreceptor circuit. V: Brain ischemic circuit ("Cushing reflex"). (Modified from [2])

Respiratory Rhythm

Research on respiratory rhythmogenesis has been performed at two different levels: studies of the behavior and the reactions of the whole system and neurophysiological research at the single neuron level. We have followed a combined approach by studying the behavior of groups of identified central respiratory neurons in relation to systemic respiratory events [12]. In the present context, the following example clearly demonstrates the excitability cycle of this rhythm. Stimulation in the center of a group of caudal retro-ambigual expiratory neurons causes a "resetting" of the respiratory rhythm strongly dependent on the time of stimulation relative to the respiratory cycle (Figure 5).

The same stimulus at the same localization produces prolongation, shortening, or no change of respiratory cycle duration depending on the phase of stimulation. As a consequence, repetitive stimuli can lead to acceleration or deceleration of the respiratory rhythm. Irrespective of the actual neurophysiological mechanism (in this case probably based on antidromic stimulation of descending neural pathways), it is striking that this phase dependence closely resembles the phase dependent resetting effects produced by stimulation of pulmonary afferents or at other sites within the brain stem (see the contribution of F.L. Eldridge in these proceedings). This means that the response is determined not so much by the particular kind of interference as by the basic properties of the respiratory rhythm shared by other biological rhythmic systems, such as the vascular smooth muscle.

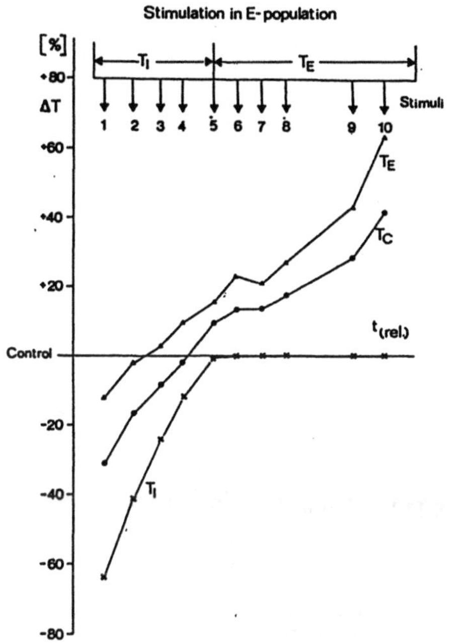

Fig. 5 Effect of single stimuli in the center of the para-ambigual population of expiratory neurons on the duration of the actual respiratory cycle. Positive values: percentage prolongation, negative values: percentage shortening in comparison with the mean duration of preceding control respiratory cycles. TZ: Total cycle duration (i.e. reciprocal value of instantaneous respiratory frequency). TI: inspiratory time. (Modified from [36])

Another observation demonstrates the astonishing self-reorganizing ability of central respiratory rhythmogenesis. Based on neurophysiological studies, it is generally agreed that respiratory rhythm originates from the interplay between several partial central mechanisms, each of them governing a certain phase of the respiratory cycle [13,14]. By local mechanical destruction or cooling at a defined site in the brain stem, it was possible to impair the partial mechanism which normally terminates the inspiration [15,16]. This interference resulted in extreme prolongation and irregularity of the inspiratory phase with a decrease in respiratory frequency. The most interesting result, however, was the spontaneous restoration of the normal respiratory rhythm in the course of some hours, without any change in the external conditions or the afferent information (Figure 6).

This demonstrates that the special partial function of respiratory rhythmogenesis is not necessarily bound to a certain anatomical substrate. Conversely, after the loss of part of the tissue, the respiratory rhythm is no longer as stable as it was before: a small dose of anaesthetic, which was ineffective under control conditions, now caused the reappearance of the alterations in rhythmogenesis.

Thus, we can conclude that the tissue destroyed contributed to the stability of rhythmogenesis together with other parts of the brain stem. This finding shows that the old and still continuing search for the primary precise anatomical site of the respiratory rhythm generator starts from an erroneous presumption. It would be highly interesting to follow with the methods of synergetics the time course of restoration from the disturbed to the normalized respiratory rhythm.

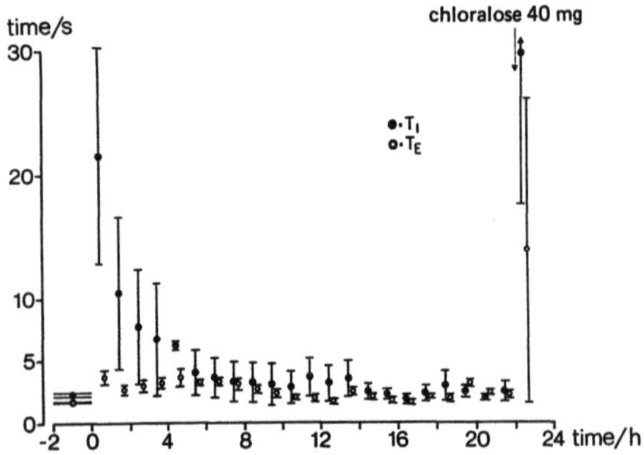

Fig. 6 Severe disturbance and spontaneous restoration of respiratory rhythm after bilateral localized electrolytic lesion in the region of the ventrolateral solitary tract nucleus (vlNTS) in an anaesthetized, vagotomized, immobilized, artificially ventilated cat. This special lesion causes extreme prolongation and increase in variability of inspiratory time TI without essential alteration of expiratory time TE. After more than 11 hours under continued constant artificial ventilation and maintained identical conditions of anaesthesia and blood chemistry the respiratory pattern is spontaneously normalized. Note the severe disturbance of the respiratory rhythm, now involving also expiratory timing, following the application of a small additional dose of chloralose, which is never observed with intact brain stem tissue. (From experiments by Lazar H, Hukuhara T, Klüßendorf D and Koepchen HP)

Relationship Between Cardiovascular and Respiratory Rhythms [1,17,18]

"Respiratory" rhythm is found in many cardiovascular parameters and in the discharge of numerous nerves innervating heart and blood vessels. Unfortunately, in this field the term "respiratory", for rhythms appearing in the cardiovascular system, has biased the thinking. It starts from the presumption that all rhythmicity in the frequency range of respiration is a secondary consequence of respiratory rhythmogenesis, destined for respiratory muscles but "irradiating" to other systems.

This is more than a semantic problem. This presumption has led to the practice of taking the respiratory rhythm as a fixed reference, for example in the evaluation of power spectra of heart rate fluctuations (see the contribution of R.Stark in these proceedings), or of sympathetic discharge. In this way, the variable coordination between respiratory and cardiovascular rhythmicity is widely overlooked. One important hint in this direction is the observation that the phase-relation between respiratory and sympathetic rhythms are species-dependent and can change with the functional state (see the contribution of A.Trzebski in these proceedings).

One of the first indications of the relationship between respiratory and cardiovascular rhythms detected was the observations of blood pressure "waves" with the frequency of respiration. Other waves occur in the lower frequency range around 0.1 Hz. The description was followed by a classification into two categories: "respiratory" waves and "Mayer waves". Unfortunately, even today many authors follow this misleading categorization, mostly with the suggestion that the former waves are caused by "irradiation" from a "respiratory center" and the latter ones by an endogenous rhythm of a "vasomotor center". Even in publications where these classical terms are not used, the concept is still the same.

As a consequence of this classification, the variable nature of the coordination was frequently overlooked. One crucial point is the fact that the respiratory rhythm cannot be used as

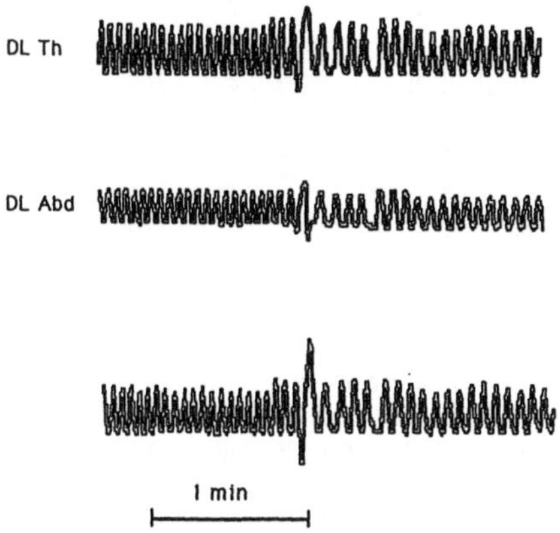

Fig. 7 Spontaneous change of respiratory pattern in a quiet sitting human subject, comparable with a change in gait in locomotion. The change starts with a deep inspiration ("sigh"). Recording of respiration by means of a thoracic (DL Th) and an abdominal (DL Abd) belt, together with the on-line computed sum signal of both traces. (From experiments by Abel HH, Klüßendorf D and Koepchen HP)

a fixed reference but can change from breath to breath and can switch over into the lower frequency range. Figure 7 shows one example of a recording from a human subject. In the middle of this recording period the respiratory pattern suddenly changes spontaneously, comparable to a change of gait in locomotion.

We have critically revised Hering's famous recording from 1869, from which the theory of "irradiation" took its origin, taking into account the different latencies between the central events and their peripheral manifestations [19]. It revealed that, in this case, the respiratory rhythm running in synchrony with the rhythm of blood pressure waves did not follow the respiratory rhythm, but preceded it. In other experiments, similar respiratory rhythm-related blood pressure waves continued when the respiratory neuronal output in the phrenic nerve was abolished by artificial hyperventilation [1].

Including many other observations from animal experiments and recordings from human subjects, we come to the general result that respiratory and cardiovascular rhythmicity is not simply a case of cause and effect with an obligatory primacy of the respiratory rhythm. Rather, they are potentially independent rhythms. As a consequence, they can exhibit many different kinds of mutual coordination, as found in other examples of coordination between other biological rhythms.

One famous example is the forms of coordination between the movements of individual fins in fish which has been analyzed extensively in v. Holst's fundamental work on "relative coordination" [8]. It is striking to see the exact parallelism between the coordination among these motor rhythms and that between the single components of cardiorespiratory neurovegetative rhythmicity.

One of these characteristic features is the appearance of integer multiple frequency relations, often in a relation of 2:1 between respiratory and cardiovascular rhythms [1]. One example is shown in Figure 8 taken from an experiment on an anaesthetized rabbit. There is an exact 2:1 relation between the respiratory rhythm and the concomitant blood pressure rhythm. At the same time, this example shows that this kind of coordination is not rigidly fixed; it is temporarily lost in the middle of the trace. Shortly afterwards, the 2:1 relation is restored but now with the inverse phase-relation.

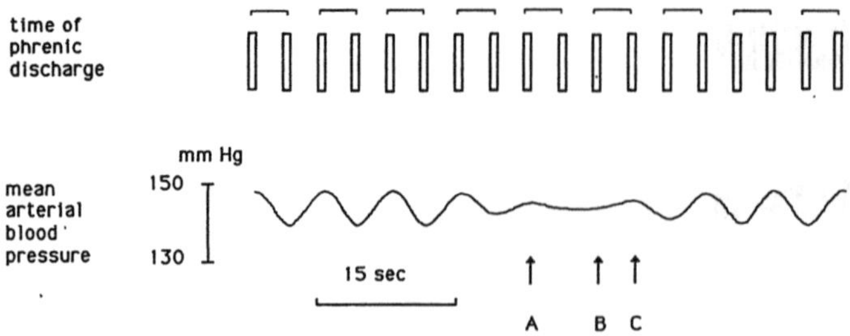

Fig. 8 2:1 relationship between respiratory and blood pressure rhythms in an immobilized artificially ventilated rabbit. Bars in the upper trace: time of phrenic nerve discharge. Lower trace: time course of mean arterial blood pressure. Note the spontaneous change in phase coordination between A and C. Before this transition, the first of each second inspiratory discharge is coordinated with a peak of a blood pressure wave. After the transition this relation is reversed. (From [1])

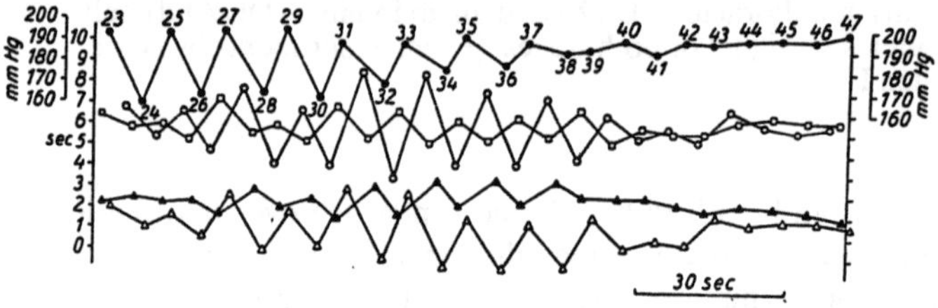

Fig. 9 Change in the coordination between respiratory rhythm, revealed by phrenic nerve activity, and blood pressure waves in an anaesthetized, artificially ventilated dog. Instability of phases and amplitudes during transition from 2:1 to 1:1 coordination. Closed circles: time and amplitude of peaks of blood pressure waves. Open circles: period of blood pressure rhythm. Closed triangles: phase lag between respiratory and blood pressure rhythm. Open triangles: phase lag betwen blood pressure rhythm and respiratory rhythm. (From [1])

The more stable 1:1 and 2:1 relations are frequently found in anaesthetized animals when no outer influences are acting on the systems. In addition, however, there are short transitory periods between these more stable states. Here we find changing periodicities with alternating amplitudes and periods. Figure 9 shows an example of a transition from a 2:1 to a 1:1 coordination between respiratory and blood pressure rhythms in an anaesthetized dog. The irregularities during this phase are reminiscent of similar irregularities in the transition between different kinds of motor coordination [20].

Heart Rate Rhythmicity

Taking heart rate as an indicator of cardiorespiratory rhythm coordination, we can start from the very well known and frequently studied phenomenon of the so-called "respiratory sinus arrhythmia". Recent development of medical engineering has made computer-assisted analysis of heart rate fluctuations for clinical and psychophysiological applications possible. Most of these studies use power spectrum analysis.

Since the first publication in this field [21], there is general agreement that in the time domain concerned, there are two prominent peaks in the spectrum, one around 0.1 Hz, the other one around 0.25 Hz [22,23]. These ranges are often defined as the "low frequency band" and the "high frequency band", respectively. The latter corresponds to the preferential frequency of the resting respiratory rhythm. Therefore, this peak is often called "respiratory". With this classification, we again find the old classification of blood pressure waves into "respiratory" and "vasomotor" in another form. Also in the case of heart rate rhythmicity such a rigid scheme cannot cope with the complex and variable interactions between the different rhythms. This applies even more if the "low frequency band" around 0.1 Hz is simply used as a convenient measure of sympathetic "tone" and the "high frequency band" of vagal tone. So, we have found that the rhythmicity in both frequency bands goes down with increasing mental or physical activity [24]. Moreover, the power in both these frequency domains increases with decreasing respiratory frequency. This applies also to the "high frequency" rhythm when the respiratory frequency has changed into the lower frequency range, so that now this oscillation is no longer "respiratory". We hope that an analysis using synergetic methods will lead to a deeper understanding, surpassing the rigid classical categories and the first orientation given by the power spectrum.

Further details of the power spectrum analysis of heart rate and its applications are dealt with in other parts of this workshop (see the contributions of R. Stark, J.P. Saul, M. Pagani and N. Honzikova).

Here we will deal with another aspect of respiration-related heart rate oscillation as a paradigm of the interrelations between physiological rhythmicity, homeostatic feedback control and behavioral control. There is a well known relation between blood pressure and heart beat interval: the size of the blood pressure stimulus at the arterial baroreceptors determines the successive heart beat intervals or momentaneous heart rate. This is part of the potent baroreceptor reflex self-regulation of arterial blood pressure.

Precise phase-related measurement of the effects of baroreceptor stimulation in the anaesthetized dog revealed that the efficiency of this reflex control fluctuates within the respiratory cycle; it is much smaller during the inspiratory than during the expiratory phase [25] (Figure 10a). This has been confirmed in man [26] and applies also to the reflex inhibition of sympathetic activity induced by the baroreceptors [27] (Figure 10b).

It is noteworthy that periodic changes in the baroreflex efficiency in this frequency range continue without respiration [25]. This means that this fluctuation of reflex efficiency is

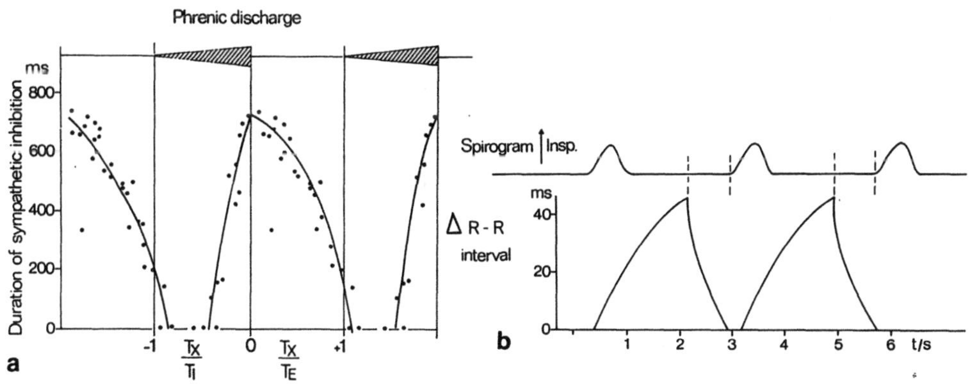

Fig. 10 (a) Rhythmic changes of baroreceptor heart rate reflex efficiency, correlated to the respiratory rhythm, in an anaesthetized dog. Upper trace: respiration, lower trace: prolongation of the succeeding heart beat interval above the control value after electrical stimulation of both sinus nerves. (Modified from [25]. **(b)** Rhythmic changes of baroreceptor-sympathetic reflex efficiency, measured by the duration of the pause in renal sympathetic nerve discharge following stimulation of both sinus nerves. (Modified from [27]). Note the decrease in reflex efficiency, starting before the beginning of inspiration.

not a simple "irradiation" of a "respiratory" rhythm generator but rather the expression of an independent rhythm. In most cases the latter is synchronized with the central rhythm which projects onto the respiratory muscles. We can presume that the mechanisms of interrelation between physiological rhythms will also be valid for the interaction between these two rhythms. This question requires further analysis.

Central Nervous Neuronal Substrates

When considering the physiology of the central network processing, the baroreceptor reflexes open a more general view of the physiological meaning of the interferences mentioned between the respiratory and the cardiovascular rhythms. The decrease in baroreceptor efficiency is not restricted to the respiration-related changes. Every mental, emotional or physical activation leads to such a decrease. Thus, we can state that there is a general antagonism between behavioral activation and the baroreceptor homeostatic feedback system. Baroreceptor afferents exert a general dampening effect on the central nervous system, including behavioral effects. Conversely, behavioral activation diminishes the inactivating baroreceptor influences on the central nervous system.

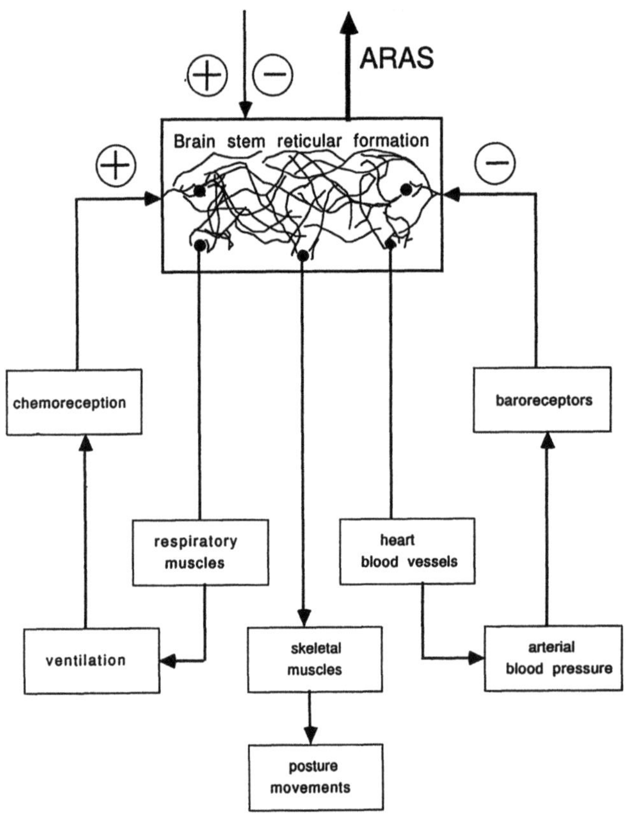

Fig. 11 Scheme of the functional relationships between the main homeostatic feedback systems for circulation and respiration mediated by the network of the brain stem reticular formation. These functional connections form the basis for the variable common cardiorespiratory and motor rhythmicity with the involvement of reflex efficiency and the relation to behavioral activation.

The other great cardiorespiratory feedback system, the chemoreceptors mediating the excitatory effects of CO_2 and hypoxia, act in a similar general fashion but in the reverse direction. Thus, the baroreceptor and the chemoreceptor feedback systems are stabilizers of the general state of central nervous activity in a wider sense. Figure 11 outlines these interrelations in a schematic manner.

For our considerations upon vegetative rhythms, this general functional structure has two consequences:

a) Through their action on the activating reticular ascending system the main cardiorespiratory afferents provide a functional relation between the control systems of circulation and respiration and the rhythmicity of the EEG, dealt with in the second part of the workshop.

Since these functional relations show distinct rhythmicity in various time domains, the feedback control of vigilance is embedded in this rhythmicity; the state of central nervous activity in turn severely influences the peripheral cardiorespiratory rhythms.

Examples of this bidirectional interaction are the respiration-related changes in behavioral response time, and the respiratory rhythm in the EEG found in animal experiments.

b) The scheme of Figure 11 shows the dual functional role of baroreceptor and chemoreceptor afferents. They act as an afferent limb of a classical autoregulatory feedback system, but they also determine the excitatory state in the reticular network [28]. This network, in turn, determines the efficiency of the reflex transmission. In this way, these afferents influence their own central processing. One can assume it causes non-linearity in the reflex control. Such a non-linearity, indeed, was demonstrated many years ago in the processing of cardioinhibitory reflexes [29]. Recently, this phenomenon has been found in the form of the dependence of heart rate rhythmicity on mean heart rate, especially in the vagally mediated "high frequency band" of the power spectrum [30].

Relationship Between Cardiorespiratory and Motor Rhythms

As we have seen, the general rules of biological rhythm organization also apply to the relations between respiratory and cardiovascular rhythms and between their subsystems. Formally, these are the same phenomena of mutual coordination as in the realm of motor rhythms (see the contribution of J.A.S. Kelso in these proceedings). Moreover, there are numerous studies on the relationship between respiration and movements (see the contribution of F. Raschke in these proceedings).

We looked for a possible coordination between vegetative and somatomotor rhythms in the resting state. Indeed, their existence could be demonstrated using the same method in principle as in the above-mentioned studies on baroreflex efficiency, i.e. the search for subthreshold variations in excitability by repeated elicitation of reflexes [31,32]. We investigated the so-called patellar tendon reflex, measuring myographically the amplitude of muscular reflex excitation. We found a more or less clear respiration-related rhythm in the reflex responses. At first sight, this could again be interpreted as a pure side effect of respiration in the sense of "irradiation".

Closer analysis, with the aid of autocorrelation functions, however, revealed sliding kinds of coordination, limited ranges of entrainment, multiple integer frequency relations (Figure 12), and periods of transition between different kinds of coordination. From these observations, we can draw the general conclusion that there is no basic difference between the mechanisms of coordination between different motor rhythms, different vegetative rhythms, and, finally, between vegetative and somatomotor rhythms. This recognition encourages the

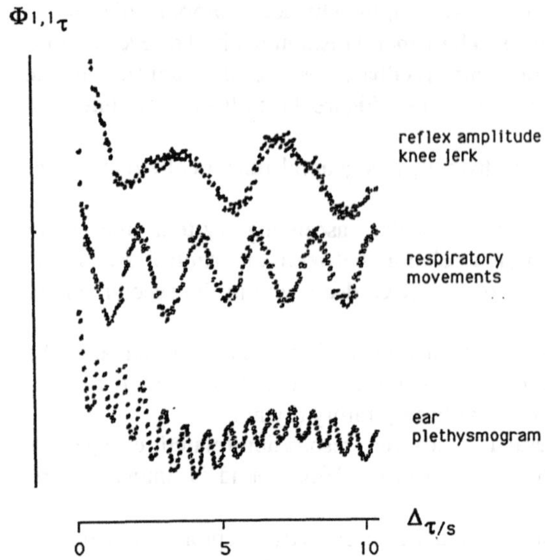

Fig. 12 Autocorrelation functions of the amplitudes of rhythmically elicited knee jerk reflexes, respiration and ear plethysmogram in a quietly sitting subject. Note the approximate 1:2:4 relation of periods between these three parameters. (From [31])

application of synergetic methods to the wide and scarcely investigated field of interaction between the vegetative and the somatomotor rhythmicity.

From the neurophysiological point of view, these common features are not so surprising. Microrecordings within the brain stem have shown that brain stem neurons controlling respiration, cardiovascular functions and motor tone cannot be definitively discriminated within the central network. This fact and the observations upon the coordination between the peripheral parameters have led to the assumption that some of the neuronal substrates must be responsible for the control of more than one peripheral organ or organ system [33,34].

Thus, there are concrete hints about the existence of a neurophysiological background for that kind of variable coordination between the peripheral systems and subsystems which we have called "common central rhythmicity". After all, the connections within this network have to be assumed to be variable and can be reorganized according to the actual functional requirements (see the contribution of A. Trzebski in these proceedings for example). Thus, it is unlikely that the search for the detailed site and identification of control substrates for a certain peripheral function will be successful. It is another exciting challenge to treat the behavior of such a variable multifunctional network with the synergetic approach.

General Conclusions

By considering the rhythmicities in the various systems and subsystems and their mutual interactions, we can gain some insight into the common features of physiological rhythmicity. First we see that rhythms are present in every function and in every frequency range. They interact with each other in a very variable manner without a fixed relation of cause and effect. One rhythm can lead, entrain, or influence the other in a more or less strong way.

The role of the leading or dependent rhythm can change according to the actual conditions. So, for example, a strong drive from the chemoreceptors can intensify the respiratory rhythm so much that it spreads also to other systems, especially the cardiovascular system. On the other hand, closer analysis has shown that we have to give up the dogma of the obligatory leading role of the respiratory rhythm like a rigid metronome. The respiratory rhythm is embedded into the whole concert of physiological rhythmicity (see the contribution of C.L. Webber in these proceedings) and can be governed and entrained by rhythms appearing in the sympathetic system, by pulmonary mechanical feedback afferents, by motor and locomotor rhythms and even by rhythmic acoustic stimuli (see the contribution of F. Raschke in these proceedings).

There are many kinds of possible transitions between full independence and strict synchronization. Without outer interferences, certain states of coordination are preferred. Besides synchronism, we find preferential states of integer multiple frequency relations where the low whole number relations like 2:1 or 3:1 are more stable than the more complicated features.

In the analysis of the rhythms, it is easier to work with an experimental stimulus which can exactly be timed and graduated by the investigator as a first step (see the contribution of F.L. Eldridge in these proceedings). Since the response of the experimentally influenced rhythm is basically the same as its response to another, internal leading rhythm (acting as a periodic stimulus), the results gained by the stimulation method can also be applied for an understanding of internal rhythm coordination.

In the realm of cardiovascular, respiratory and motor rhythmicity, the rhythms interact with each other and with the homeostatic feedback control. In addition, they are under the strong influence of behavioral factors. Therefore, it is not surprising that complicated patterns appear when we record and analyze a single parameter. On the other hand, we have seen that the same basic rules of interaction are found at all hierarchical levels, from the single organ, such as the isolated blood vessel, to the interaction between whole control systems, such as respiratory and cardiovascular control or respiration and locomotion.

This recognition opens the chance for a rational analysis of the complex appearance of the individual variable. The more we are able to analyze the individual rhythmic variable and correlate it with the other rhythmic events in the rich spectrum of physiological rhythms, the more we will understand the time course of the respective parameter. In this context, in the hierarchy of rhythms (see Figure 1), a slower rhythm can formally appear as a "trend" in the time series of a faster one but can also act as an order parameter for the latter.

Under these aspects, the complex physiological rhythmicity will appear not as a confusing phenomenon but as a general principle of organization, allowing a compromise between stability and flexibility of adaptation. We are optimistic that the concepts and analytical tools of synergetics will open up a new way for a comprehensive gradual analysis and understanding of this biological principle. Within the framework of this task, the greatest gap of knowledge is still in the field of interaction between the different systems.

The physiological rhythmicity by itself is a typical example of the principle of synergetics, according to which many individual elements are coordinated by a process of self-organization in which the resulting rhythmic order, in turn, "enslaves" the individual elements.

It is noteworthy that Goethe already recognized and formulated this general biological principle and emphasized its importance for physiological science.

Goethe's original text reads as follows:

" Allein noch wäre zu wünschen, daß zu einem schnelleren Fortschritte der Physiologie im Ganzen die Wechselwirkung aller Teile eines lebendigen Körpers sich niemals aus den

Augen verlöre; denn bloß allein der Begriff, daß in einem organischen Körper alle Teile auf einen Teil hinwirken und jeder wieder auf alle seinen Einfluß ausübe, können wir nach und nach die Lücken der Physiologie auszufüllen hoffen."

In English:
"It would be highly desirable, in order to accelerate the progress of the whole of physiology, not to lose sight of the interaction of all parts of a living organism. Only by recognizing that in an organic body all parts together act on the single part and, in turn, each single part influences all others, can we hope to fill in step by step the gaps of physiology."

In other words: physiology needs synergetics.

From the point of view of the history of science, we are observing a remarkable evolution: the integrative and intuitive thinking of Goethe in the field of science remained without major influence in the further development of established physiological sciences and scientific medicine in the 19th and 20th centuries. The latter was mainly based on a very successful morphological and physico-chemical reductionistic approach. The lack of acceptance of Goethe's ideas in the natural sciences may partly be due to his peculiar aversion for mathematical approaches in life sciences.

It is a fascinating aspect today that integrative ideas are beginning to reappear in biology, starting just from mathematical concepts. The junction between synergetics and the physiology of rhythms intended by this workshop can be considered as one step in this direction.

References

1. Koepchen HP (1962) Die Blutdruckrhythmik. Dr Dietrich Steinkopff Verlag, Darmstadt

2. Koepchen HP (1984) History of Studies and Concepts of Blood Pressure Waves. In: Miyakawa K, Koepchen HP, Polosa C (eds) Mechanisms of Blood Pressure Waves. Japan Scientific Societies Press, Tokyo. Springer, Berlin Heidelberg New York Tokyo, pp 3-23

3. Koepchen HP, Abel HH, Klüßendorf D, Lazar H (1986) Respiratory and Cardiovascular Rhythmicity. In: Von Euler C, Lagercrantz H (eds) Neurobiology of the Control of Breathing. Raven Press, New York, pp 303-312

4. Hering E (1869) Über Atembewegungen des Gefäßsystems. Sber Akad Wiss, Wien, Math naturwiss Kl 2. Abt. 60:829

5. Siegel G, Hofer HW, Schnalke F, Adler A, Walter A, Koepchen HP (1989) Membrane Physiological Basis of Vascular Autorhythmicity. Prog Appl Microcirc 15:10-31

6. Golenhofen K (1970) Slow Rhythms in Smooth Muscle (Minute-Rhythm). In: Bülbring E, Shuba MF (eds) Smooth Muscle. Edward Arnold, London

7. Seller H, Langhorst P, Polster J, Koepchen HP (1967) Zeitliche Eigenschaften der Vasomotorik. II. Erscheinungsformen und Entstehung spontaner und nervös induzierter Gefäßrhythmen. Pfluegers Arch 296:110-132

8. Von Holst E (1939) Die relative Koordination als Phänomen und als Methode zentralnervöser Funktionsanalyse. Erg Physiol 42:228-306

9. Ludwig C (1847) Beiträge zur Kenntnis des Einflusses der Respirationsbewegungen auf den Blutumlauf im Aortensystem. Arch Anat Physiol, pp 242-302

10. von Haller A (1760) Elementa physiologica. T II Lit VI, Lausanne, p 330

11. Haken H (1983) Synergetics: An Introduction. 3rd ed. Springer, Berlin Heidelberg New York

12. Koepchen HP (1976) Quantitative Approach to Neural Control of Ventilation. In: Loeschcke HH (ed) Acid Base Homeostasis of the Brain Extracellular Fluid and the Respiratory Control System. Thieme, Stuttgart, pp 164-186

13. von Euler C (1986) Brain Stem Mechanisms for Generation and Control of Breathing Pattern. In: Fishman AP, Cherniak NS, Widdicombe JG, Geiger SR (eds) Handbook of Physiology, section 3: The Respiratory System, vol II: Control of Breathing, part 1, Bethesda, Am Physiol Soc, pp 1-67

14. Richter DW (1982) Generation and Maintenance of the Respiratory Rhythm. J Exp Biol 100: 93-107

15. Koepchen HP, Klüßendorf D, Lazar H, Hukuhara T, Abel HH (1985) Conclusions on Respiratory Rhythmogenesis Drawn from Lesion and Cooling Experiments Predominantly in the Region of Ventrolateral Nucleus of Solitary Tract (vlNTS). In: Bianchi AI, Denavit-Saubié M (eds) Neurogenesis of Central Respiratory Rhythm. MTP Press Ltd, Lancaster Boston The Hague Dordrecht, pp 77-80

16. Koepchen HP, Lazar H, Klüßendorf D, Hukuhara T (1986) 'Medullary Apneusis' by Lesions and Cooling in the Ventrolateral Solitary Tract Region and Genesis of Respiratory Rhythm. J Autonom Nerv Syst, Suppl 63-69

17. Koepchen HP, Klüßendorf D, Sommer D (1981) Neurophysiological Background of Central Neural Cardiovascular-Respiratory Coordination: Basic Remarks and Experimental Approach. J Autonom Nerv Syst 3:335-368

18. Koepchen HP (1983) Respiratory and Cardiovascular "Centres": Functional Entirety of Separate Structures? In: Schläfke ME, Koepchen HP, See WR (eds) Central Neurone Environment and the Control Systems of Breathing and Circulation. Springer, Berlin Heidelberg New York, pp 221-237

19. Koepchen HP, Abel HH, Klüßendorf D (1987) Integrative Neurovegetative and Motor Control: Phenomena and Theory. Funct Neurol 2, 4:389-406

20. Haken H, Kelso JAS, Bunz H (1985) A Theoretical Model of Phase Transitions in Human Hand Movements. Biol Cybern 51:347-356

21. Penaz J, Roukens J, van der Waal HJ (1968) Spectral Analysis of Some Spontaneous Rhythms in the Circulation. Biokybernetic I: 233-236

22. Pagani M, Lombardi F, Guzzetti S, Rimoldo O, Furlan R, Pizinelli P, Sondrone G, Malfatto G, Dell'Orto S, Piccaluga E, Turiel M, Baselli G, Cerutti S, Malliani A (1986) Power Spectral Analysis of Heart Rate and Arterial Pressure Variabilities as a Marker of Sympatho-Vagal Interaction in Man and Conscious Dog. Circ Res 59:178-193

23. Akselrod S, Gordon D, Ubel FA, Shannon DC, Barger AC, Cohen RJ (1981) Power Spectrum Analysis of Heart Rate Fluctuation: A Quantitative Probe of Beat-to-Beat Cardiovascular Control. Science 213:220-222

24. Abel HH, Klüßendorf D, Droh R, Koepchen HP (1991) Cardiorespiratory Relations in Human Heart Rate Pattern. In: Koepchen HP, Huopaniemi T (eds) Cardiorespiratory and Motor Coordination. Springer, Berlin Heidelberg New York (in press)

25. Koepchen HP, Wagner PH, Lux HD (1961) Über die Zusammenhänge zwischen zentraler Erregbarkeit, reflektorischem Tonus und Atemrhythmus bei der nervösen Steuerung der Herzfrequenz. Pfluegers Arch 273:443-465

26. Eckberg DL, Kifle YT, Roberts VL (1980) Phase Relationship between Normal Human Respiration and Baroreflex Responsiveness J Physiol 304:489-502

27. Seller H, Langhorst P, Richter D, Koepchen HP (1968) Über die Abhängigkeit der pressorezeptorischen Hemmung des Sympathikus von der Atemphase und ihre Auswirkung in der Vasomotorik. Pfluegers Arch 302:300-314

28. Langhorst P, Schulz B, Schulz G, Lambertz M (1983) Reticular Formation of the Lower Brainstem. A Common System for Cardiorespiratory and Somatomotor Functions: Discharge Pattern of Neighbouring Neurons Influenced by Cardiovascular and Respiratory Afferents. J Aut Nerv Syst 9:411-432

29. Koepchen HP, Lux HD, Wagner PH (1961) Untersuchungen über Zeitbedarf und zentrale Verarbeitung des pressorezeptorischen Herzreflexes. Pfluegers Arch 273:413-430

30. Abel HH, Klüßendorf D, Koepchen HP (1989) Relation between Tone and Rhythmicity of Cardiac Chronotropic Innervation. Pfluegers Arch 423 (Suppl 1): R 11 [Abstract]

31. Schmidt-Vanderheyden W, Heinich L, Koepchen HP (1970) Investigations into the Fluctuations of Proprioceptive Reflexes in Man. I. Fluctuations of the Patellar Tendon Reflex and Their Relation to the Vegetative Rhythms during Spontaneous Respiration. Pfluegers Arch 317:56-71

32. Schmidt-Vanderheyden W, Koepchen HP (1970) Investigations into the Fluctuations of Proprioceptive Reflexes in Man. II. Fluctuations of the Patellar Tendon Reflex and Their Relation to Vegetative Rhythms during Controlled Respiration. Pfluegers Arch 317:72-83

33. Koepchen HP (1969) Vegetative-Somatic Relationships in Single Neuron Activity in the Lower Brain Stem. In: Evans CR, Mulholland TB (eds) Attention in Neurophysiology, Butterworth, London, pp 83-99

34. Koepchen HP, Langhorst P, Seller H (1975) The Problem of Identification of Autonomic Neurons in the Lower Brain Stem. Brain Res 87:375-393

35. Koepchen HP, Seller H, Polster J, Langhorst P (1968) Über die Feinvasomotorik der Muskelstrombahn und ihre Beziehung zur Ateminnervation. Pfluegers Arch 302:285-299

36. Koepchen HP, Sommer D, Frank Ch, Klüßendorf D, Forstreuter K (1979) Characteristics and Functional Significance of the Expiratory Bulbar Neuron Pools. In: Von Euler C, Lagercrantz H (eds) Central Nervous Control Mechanisms in Breathing. Pergamon Press, Oxford New York, pp 217-232

Synergetics – Can It Help Physiology?

H. Haken

Institut für Theoretische Physik und Synergetik,
Pfaffenwaldring 57/4, W-7000 Stuttgart 80, Fed. Rep. of Germany

Abstract. This paper discusses how the interdisciplinary field of synergetics may enable us to interpret physiological data on heart beat, blood circulation, breathing, EEGs, etc. and to model the underlying processes. It is argued that physiological systems are close to instability points, which allows the systems to adapt to new situations rapidly. At those instability points even the complex dynamics of multicomponent systems may be described by just a few variables, the so-called order parameters. These order parameters either move on a low-dimensional attractor, perform transients towards an attractor or undergo transitions between attractor states. These observations cast new light on dimension analysis used in chaos theory and give hints about how one may model these physiological processes. Emphasis is laid on the possibility of various couplings between oscillators. Such couplings may combine stability in amplitude with adaptability in phase.

1. Introduction

Both traditional and more advanced methods allow physiologists and medical doctors to collect an enormous amount of physiological data [1]. This is particularly pronounced in intensive care. On the other hand, the question of how to interpret this enormous amount of information becomes more and more pressing. Quite clearly, an adequate understanding of these data will be of utmost importance for fundamental research, diagnostics and therapy. On the other hand we must be aware of the fact that the human body is an incredibly complex system. This leads us quite naturally to the question whether there are theoretical concepts or approaches which allow us to understand such complex systems. In this author's opinion, synergetics [2], [3] is the most advanced theoretical tool which is presently at our disposal for that purpose. Nevertheless, there is still an enormous gap between the experimental data and their theoretical interpretation. But I hope that synergetics will help us to narrow this gap in the years to come.

Over the past years synergetics, which originated from physics, has found applications in biology and medicine. Among the problems treated are the following: Models of morphogenesis [4], modelling movement coordination [5], EEG analysis [6], and computer models for visual perception including its oscillations and fluctuations [7].

My present article may be seen in the context of my previous articles on similar occasions, namely in the Springer Proceedings on "Synergetics of the Brain" [8] and "Synergetics of Cognition" [9]. A number of

Springer Series in Synergetics, Vol. 55 **Rhythms in Physiological Systems**
Eds.: H. Haken and H.P. Koepchen

statements made in the present article may be found there to some extent though some crucial points will be added here, too.

2. Relevant Concepts of Synergetics

According to its definition, synergetics is concerned with systems that are composed of many parts that interact with each other and in this way may produce spatial, temporal, or functional structures. Quite clearly, any biological system is such a system. Biological systems show an enormous degree of coordination or coherence. We may just recall heart beat, breathing, blood circulation, or locomotion of animals, or, at a still higher level, perception and thinking of humans. In all these processes literally billions of individual cells participate in a highly regulated fashion. In the past, a number of trends in physiology based on specific schools of thought developed in order to cope with these phenomena. One school is certainly related to teleology, another one is strongly based on cybernetics. Cybernetics deals with regulation or steering by means of a steering center. But in any biological system the question of what steers the steering center immediately arises. Thus one can arrive at the suspicion that still other principles are needed. Among them are the principles of self-organization, which are the central point of study by synergetics.

Indeed, over the past twenty years a rather comprehensive mathematical theory of self-organization has been developed. In the context of this volume it will be more appropriate, however, to elucidate the basic concepts and ideas of synergetics by means of model systems.

Comparatively simple model systems have been provided by physics in which, in particular, transitions from disordered states to ordered, i.e. highly coordinated, states have been studied. Typical examples are the light source laser, fluid dynamics and flames as physical or chemical systems. My favorite example, the laser, allows us to understand basic concepts of synergetics rather simply [10]. Consider first a lamp consisting of a gas in a gas tube, where the individual gas atoms are excited by an electric current sent through this tube. After each act of excitation, an atom may emit a light wave train. The individual atoms emit their waves independently of each other so that a microscopically chaotic light field emerges (Fig. 1). It is as if we are throwing a handful of pebbles into water, whose surface then becomes excited quite irregularly. On the other hand, in the laser the atoms emit light in a well-regulated fashion so that a single well-ordered coherent light wave emerges (Fig. 1). We shall not dwell on the detailed mechanism of this process but rather mention a few important concepts of synergetics.

The laser light wave acts as a so-called order parameter. It describes the state of the laser and at the same time it gives orders to the individual parts of the system, i.e. the individual atoms, so that they support this laser wave in a well-regulated fashion. In technical terms, the order parameter enslaves the atoms. On the other hand, the individual atoms support the order parameter, i.e. the laser light wave, by their emission processes. In this example the order parameter is a material quantity. Liquids may perform specific movements, for instance a liquid heated from below can spontaneously form macroscopic rolls as indicated in Fig. 2. This roll pattern can be described mathe-

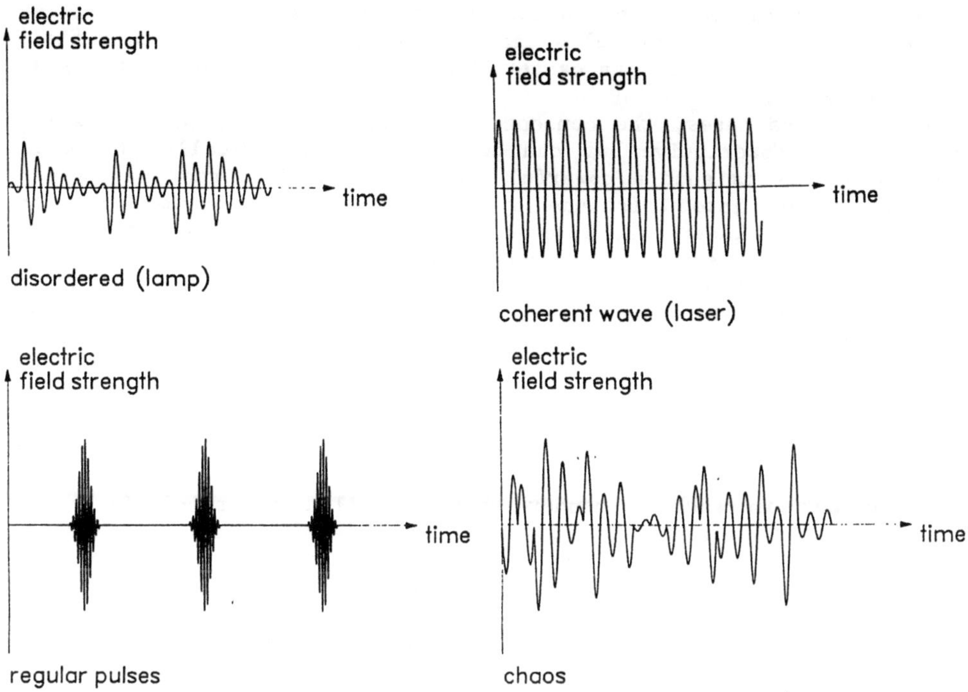

Fig.1. These figures illustrate how a simple system that is composed of many individual parts of the same type can produce qualitatively different kinds of behavior. Each figure shows the variation of of the electric field strength of the laser in the course of time.
Upper left: When the power input into the laser is small, microscopically chaotic light composed of individual wave trains is generated.
Upper right: At an increased pump rate a well-ordered coherent wave emerges.
Lower left: At a still higher pump rate ultrashort regular pulses of laser light are produced.
Lower right: Deterministic chaos of laser light under somewhat different pump and loss conditions.

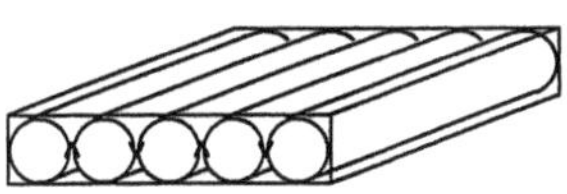

Fig. 2. Illustration of the emergence of a roll pattern in a fluid heated from below.

matically by means of an order parameter, which in this case is just a concept (which has a well-defined mathematical meaning, however). But this order parameter enslaves the individual parts of the liquid to follow the macroscopic roll motion. A further example is provided by biology in the case of slime mold, which is a little mushroom a few millimeters high. Usually its cells live separately, but if their food supply becomes scarce, they start to emit cAMP molecules that diffuse, may hit other cells and cause them to emit further cAMP molecules. Quite in analogy to the laser, by these processes a well-ordered concentration field (in the form of spirals) of these molecules is formed. The individual cells can measure the gradient of this field and then migrate towards its center. The concentration field acts as an order parameter.

In all these cases, the concepts of order parameters and enslaving are not only descriptive but can be derived mathematically. (This is why synergetics is such a powerful theoretical tool.) Order parameters can be brought into existence by changing control parameters. For instance, when a laser is energetically excited only weakly, it acts as a lamp and its order parameter vanishes. On the other hand, when the control parameter exceeds a critical value, one or several order parameters can form spontaneously. A system can form several order parameters, which may compete so that only one order parameter wins, or they may cooperate so that a new structure evolves. Order parameters can be described by their amplitudes and phases. If order parameters describe oscillators, specific relations, called phase-locking, can occur. As a consequence, in the laser well-ordered regular pulses emerge (Fig. 1). Under different circumstances a few order parameters can interact in such a manner that so-called deterministic chaos occurs. It is important to note that in deterministic chaos only a few order parameters are involved, and this kind of chaos must be clearly distinguished from microscopic chaos, where an enormous number of individual actions take place in a highly uncorrelated fashion. During a transition from the disordered to the ordered state specific phenomena may occur, such as pronounced fluctuations of the order parameters and so-called critical slowing down, i.e. when an order parameter is disturbed in the transition region, it relaxes very slowly to its equilibrium value.

Synergetics sheds new light on the notion of attractors originally introduced in dynamic systems theory. Namely instead of considering an extremely high-dimensional space of all the variables, it suffices to consider the low-dimensional space of order parameters and to discuss the kinds of attractors available in such a low-dimensional space. When a system relaxes towards an equilibrium state, one may say that it relaxes to a so-called stable fixed point. When it oscillates coherently, one speaks of a limit cycle attractor, which is described in particular by a specific frequency. There are motions on an attractor which are described by several frequencies; in technical terms one speaks of a motion on a torus.

Finally, there are the so-called chaotic attractors. In this case the variables describing the system relax towards a specific region of "phase space", but in this region they perform quite an irregular motion. Modern theory tries to characterize this chaotic motion by specific quantities, such as Lyapunov exponents and fractal dimensions [11]. Figures 3 - 6 show an analysis performed by Schanz et al. [12] of several time series representing EEG, ECG, etc. They show the power spectrum, correlation functions, and a dimensional analysis according to the standard Grassberger/Procaccia method [13]. Several difficulties in doing such an analysis will be discussed below. Microscopic fluctuations representing microscopic chaos may be superimposed on the pure attractor states. When experimental data are analyzed, it is not always easy to separate the macroscopic order parameters and microscopic fluctuations clearly from each other.

One of the main achievements of synergetics is the insight into the mechanism that produces macroscopic structures, be it temporal, spatial,

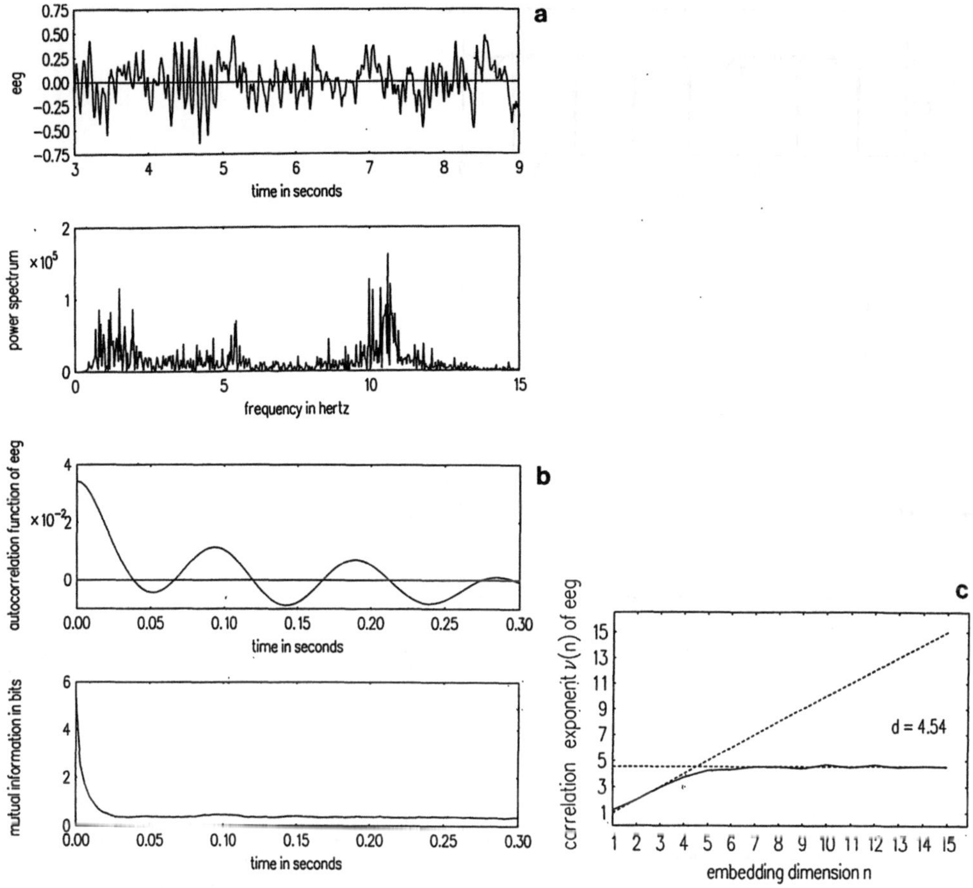

Fig. 3. Analysis of an EEG time series. (The experimental data of this figure and the following ones were provided by Prof. Koepchen.)
a) Upper part: The time series
   Lower Part: Power spectrum
b) Upper part: Autocorrelation function
   Lower part: Mutual information in bits
c) Determination of dimension according to the Procaccia/Grassberger method

or functional. When certain parameters of a system, the so-called control parameters, are changed, the system which had formerly been on a specific attractor becomes destabilized and the attractor is replaced by a new one. For instance, a fixed point attractor may be replaced by two fixed point attractors or several of them. In such a case, a specific spatial pattern or homogeneous state is replaced by new spatial patterns. In other cases, a fixed point attractor can be replaced by a limit cycle attractor or a torus. In turn, a limit cycle attractor or a torus can become destabilized and may be replaced by a chaotic attractor. In each case, the transition of the system can be described by changes of a few order parameters. Indeed, close to transition points, the systems are capable of forming a very small number of macroscopic

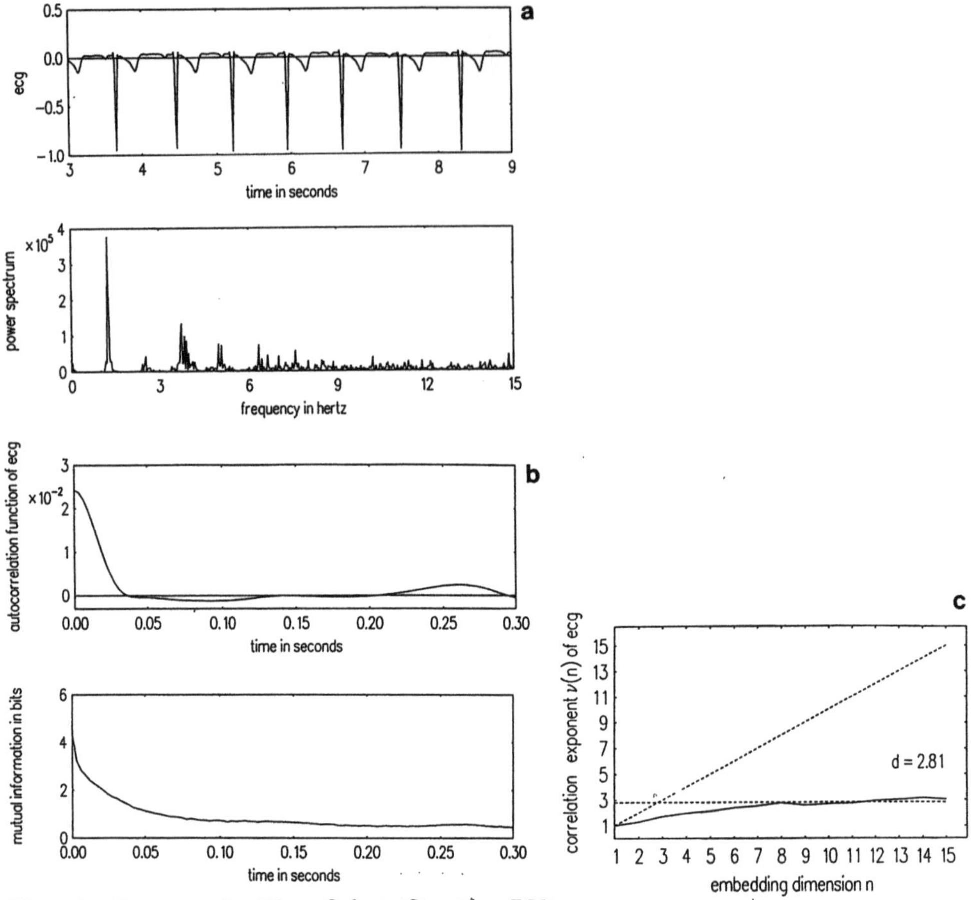

Fig. 4. Same as in Fig. 3 but for the ECG.

order parameters that determine the dynamics of the whole system. In other words, the originally high-dimensional problem, i.e. a problem with very many variables, is reduced to one of rather low dimension, i.e. a problem with few variables. This means especially that very many cells of a body are determined in their behavior by just a few quantities, namely the order parameters. In such cases it is sufficient to study the dynamics in such a low-dimensional space and to classify the attractors.

In some cases the identification of the order parameters and their attractor states is rather simple. For instance, in measurements of the coordination between the movement of fingers, the relative phase was easily identified as the appropriate order parameter and the order parameter dynamics were established, which allowed us incidentally to identify the attractor states [5]. In other cases a crucial task will be the identification of the appropriate order parameters and their equations of motion. As we have mentioned above, the transition from a disordered state to an ordered state, or, more generally, from one type of ordered state into another type of ordered state, is induced by a change of control parameters. In biological systems a rather complicated

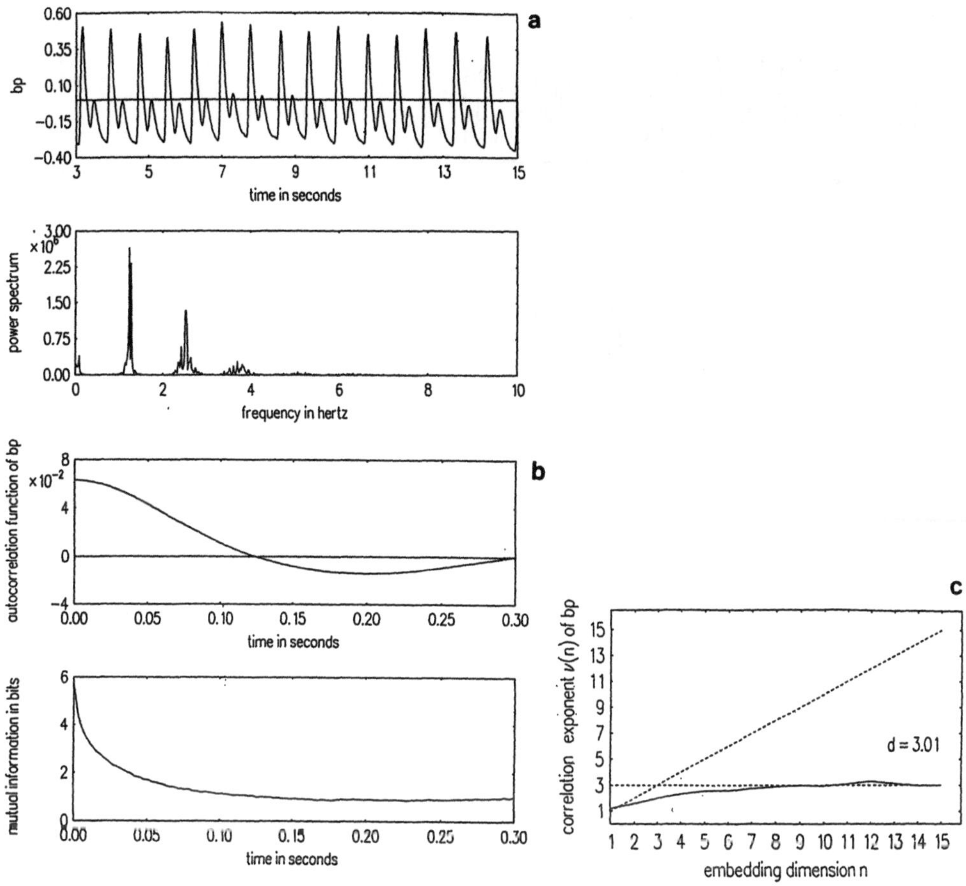

Fig. 5. Same as in Fig. 3 but for blood pressure.

network is established in which control parameters of one subsystem may be order parameters of another system or vice versa. This leads to a phenomenon in which a system does not stay in a specific attractor all the time but may change between one attractor and another. Instead of attractors one has rather to speak of <u>quasi attractors</u>, which exist for a while but then disappear or are replaced by new attractors. A difficulty may arise here. A typical example is provided by the so-called Lorenz attractor of fluid dynamics and laser physics (Fig. 7). In this case, a point describing the macroscopic state of the system cycles around a center for a while, then jumps to another region in phase space, where it cycles around another center for a certain while and jumps back again, and so on. This process takes place in an irregular fashion ad infinitum. Thus, if we studied the motion of the point for a while, we would come to the conclusion that it is on something like a limit cycle. But as soon as we extend the time series to a longer time, we may identify the second part of the Lorenz attractor. This example may become a new paradigm in the study of physiological systems. While it is certainly useful to use the concept of an attractor with respect to specific order parameters for a certain

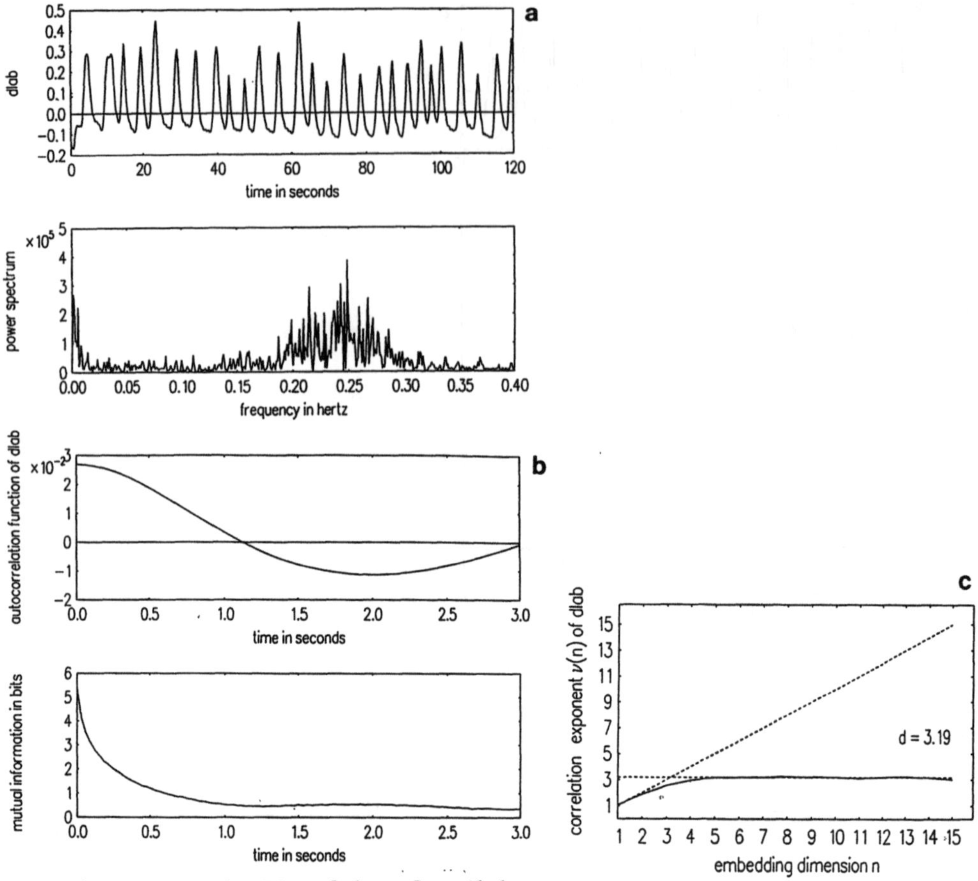

Fig. 6. Same as in Fig. 3 but for dlab.

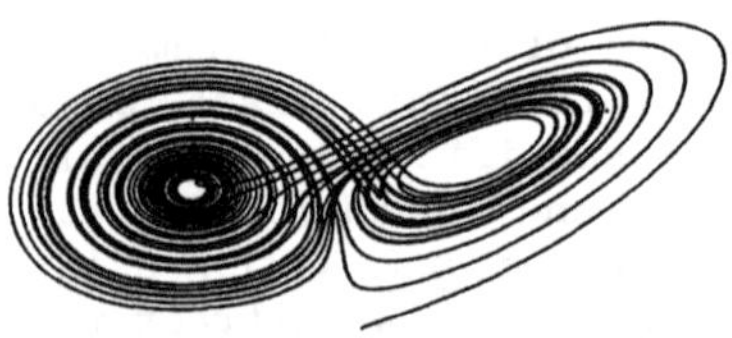

Fig. 7. The Lorenz attractor as an example in which a point in phase-space circles in one part for a while and then jumps to another part where it starts circling for a while until it returns to the first part of phase-space, and so on.

period of time, one has rather to think in terms of quasi attractors in which the system stays at one for only a limited amount of time, then goes to another attractor or another part of it. It is also possible for the attractor to be replaced by a new one. In this way the system may migrate between different parts of the state space, and this may lead to a new difficulty, namely in defining the state space appropriately. Thinking of the Lorenz attractor again, we see that it may be useful to study the behavior of the sytem within a part of the state space and then to study the jumps of the system between different parts of the state space. In this way it can happen that the dimensions of state space must be enlarged again and again.

So one may ask the question, why do physiological systems show this kind of migration? Though we would like to avoid teleological reasoning, we may state nevertheless that in the course of evolution physiological systems have certainly developed in such a fashion to be best adapted to their surroundings and to the requirements of the total body. As has been revealed by synergetics quite generally, there is some kind of "exclusion principle" between stability and adaptability. So any biological system has to find a compromise between these two limiting cases. A possible answer to this problem is to resort to the use of oscillators. One can keep them at a rather stable amplitude, but nevertheless one has a rather high adaptability in making adequate choices with respect to their phases. Our point of view is underlined by experimental findings of Kelso. When we interpreted his results on movement coordinations of fingers in terms of synergetics, we were able to identify specific attractor states. As Kelso found, these attractor states play a fundamental role when a person is learning or when a system is disturbed [14]. An example of biological quasi attractors is provided by speech. It is composed of well-defined, reproducible building blocks - the phonemes. One may speculate that the muscle movements engaged in the formation of phonemes are governed by one or a few order parameters.

Our remarks shed new light on attempts to apply chaos theory to the identification of attractors. An important method is to analyze a time series in order to reconstruct the attractor. An example is provided in the contribution by Friedrich et al. [15], where EEGs are analyzed. Generally speaking, a number of typical difficulties may occur in a time-series analysis:

1) Microscopic chaos is superimposed on the chaotic attractor. This may make it difficult to determine the dimension of the attractor.
2) The number of data points may be insufficient. It was shown by Eckmann and Ruelle [16] that in order to find the correct dimension of a chaotic attractor it is necessary to collect enough data points. If one tries to increase the number of data points by collecting them very quickly within a given frequency band, one gets into trouble with the sampling theorem, which tells us that the number of data points within a sampling interval is limited by the band width. Thus, the only way out is to collect data on a longer time scale. But then it is an every-days experience in medicine that the system is not stationary,so that the dimensions collected over different time scales can vary quite considerably. In this situation the concept of quasi attractors may become useful, where one allows the system to stay in an attractor for a while and determines the corresponding dimension during this limited time interval (at the risk of underestimating the "true" dimension).
3) In specific tasks, for instance seizing of objects, transients towards an attractor rather than motion on an attractor may occur.

To overcome the above-mentioned difficulties at least partly, one may consider time series referring to the same quantity but taken at different locations of the body, or time series of different quantities. An example of the former approach is given in the above-mentioned paper by Friedrich et al., where EEG-potentials are used that were derived from different positions on the scalp. As transpires from this example, an ultimate goal is to establish equations for the order parameters instead of deriving a few numbers such as Lyapunov exponents and dimensions. The coefficients of the order parameter equations contain control parameters so that by a careful theoretical study of the behavior of the solutions as a function of the control parameters we obtain a new insight into physiological mechanisms.

A wide field of further experimental and theoretical study will be the phenomena of mode-locking between two or several physiological oscillators. Oscillators with somewhat different frequencies ω_1 and ω_2 may acquire the same frequency ω and develop a fixed phase relationship once they are coupled together. As was shown experimentally by Kelso and modelled by Haken et al. [5], this phase relationship may change spontaneously in coordinated finger movements. The nonlinear terms we had to introduce into the equations describing the coupling between finger movements lead to terms describing specific kinds of frequency locking in which originally different frequencies ω_1, ω_2 are locked within ratios of natural numbers, such as 2:3, etc. Phenomena of this sort are well known in physiology, e.g. between breathing and heart beat (for instance 1:4 or 1:3). One may argue that nature uses the same principles time and again, which would mean that the same coupling terms occur in quite different physiological subsystems. Such coupling terms were introduced in the above-mentioned paper by Haken et al., and they can give rise in discrete maps to the so-called Arnold tongues, as shown by Kelso.

Both from our above arguments and from our previous publications, the following strategy for modelling physiological systems emerges: In a first step we may search for appropriate order parameters describing collective behavior of many individual cells within specific subsystems, e.g. the heart, blood circulation, and breathing. In other words, these subsystems of the body are described by a few variables. Then we may attempt to derive equations describing the coupling between these systems at the level of order parameters, where the resulting order parameter equations will depend on a few control parameters. This kind of approach differs considerably from a traditional one in which it is assumed that the coupling between different physiological subsystems is caused by specific neurones. In fact, couplings may be caused at quite different levels between the different systems, and it is probably a better strategy to model these couplings in a global manner than to try to locate them in specific regions. The picture that emerges in this way is one of indirect controls via control parameters rather than by a direct steering in which each individual subsystem is directly steered by means of neurones, or where in other cases one subsystem, like breathing, enslaves another one, like heart beat.

3. Concluding Remarks

In view of the jungle of data with which physiology provides us, we need ideas about how to cut a trail through this jungle. Though we are just at the beginning I believe that synergetics may prove to be a trail-blazer by providing us with appropriate concepts. They are as follows: It is argued that physiological systems are kept close to instability points. This allows us to use the concept of macroscopic variables, namely the order parameters, and the concept of quasi attractors, and it suggests that we look for situations in which mode-locking and mode-delocking occur. When one browses through the more recent literature on physiology and its theoretical discussions, one comes upon the terminology of dynamic systems theory more and more often quoting attractors and other notions from chaos theory, such as dimensions and Lyapunov exponents. I do not believe that qualitative discussions offered there using the terminology of dynamic systems theory will be sufficient. What we really need are precise measurements and definitions of order parameters and their equations of motion. This may be a hard job, but as we have shown in previous publications, it can be achieved in a number of well-defined experimental settings, and I am convinced that this may be the only reliable way in which we may proceed towards a deeper understanding of physiological processes.

I wish to thank Prof. H.P. Koepchen for many stimulating discussions.

References

[1] H.P. Koepchen, these proceedings, with many further references

[2] H. Haken, Synergetics, An Introduction, 3rd ed., Springer, Berlin (1983)

[3] H. Haken, Advanced Synergetics, Instability Hierarchies of Self-Organizing Systems and Devices. Springer, Berlin (1983)

[4] H.Haken and H. Olbrich, J. Math. Biol. 6, 317 (1978)
C. Berding and H. Haken, J. Math. Biol. 14, 133 (1982)

[5] H. Haken, J.A.S. Kelso, H. Bunz, Biol. Cybern. 51, 347-356 (1985)
G. Schöner, H. Haken, J.A.S. Kelso, Biol. Cybern. 53, 247 (1986)

[6] A. Fuchs, R. Friedrich, H. Haken, and D. Lehmann, in Computational Systems - Natural and Artificial, ed. by H. Haken, Springer, Berlin (1987)
R. Friedrich, A. Fuchs, and H. Haken, Modelling of Spatio-Temporal EEG Patterns, Proceedings of the workshop "Mathematical Approaches to Brain Functioning Diagnostics", ed. by A. Holden, to appear in: Manchester University Press Series; Proceedings in Nonlinear Science

[7] T. Ditzinger, H. Haken, Biol. Cybern. 61, 279-287 (1989); 63, 453-456 (1990)

[8] E. Basar, H. Flohr, H. Haken, and A.J. Mandell, (eds.), Synergetics of the Brain, Springer, Berlin (1983)

[9] H. Haken and M. Stadler, (eds.), Synergetics of Cognition, Springer, Berlin (1990)

[10] H. Haken, Synergetics, the Science of Structure, Van Nostrand Reinhold, New York (1984)

[11] H.G. Schuster, Deterministic Chaos, 2nd ed., VCH Verlag, Weinheim (1989)

[12] M. Schanz, H. Haken, H. P. Koepchen, to be published

[13] P. Grassberger, I. Procaccia, Physica 9D, 189 (1983)

[14] J.A.S. Kelso, these proceedings, with further references

[15] R. Friedrich, A. Fuchs, and H. Haken, these proceedings, with further references

[16] J.P. Eckmann and D. Ruelle, preprint IHES/P/90/11

Part II

Cardiovascular Rhythms

Autorhythmicity in Blood Vessels: Its Biophysical and Biochemical Bases

G. Siegel[1], *H.W. Hofer*[2], *A. Walter*[1], *K. Rückborn*[3], *F. Schnalke*[1], *and H.P. Koepchen*[1]

[1]Institute of Physiology, The Free University of Berlin, Arnimallee 22, 1000 Berlin 33, Fed. Rep. of Germany
[2]Department of Biology, The University of Konstanz, Universitätsstr. 10, W-7750 Konstanz, Fed. Rep. of Germany
[3]Institute of Physiology, The University of Rostock, O-2500 Rostock 1, Fed. Rep. of Germany

Abstract. The rhythmogenic properties of vascular smooth muscle are closely linked to the intact circulation. The electrical and mechanical oscillations, which can be traced back to rhythmic activity of the active, electrogenic Na^+/K^+ pump, could originate in the allosteric qualities of the enzyme phosphofructokinase. Thus, phosphofructokinase represents a rhythmogenic enzyme which may well serve as an example of the connection between biological properties on a molecular level and the system's behaviour in space and time.

1. Introduction

Rhythms and oscillations are an expression of the time-organization of living systems. We are acquainted with countless rhythmical phenomena in the living world: the yearly and monthly rhythms, the day-and-night rhythms, the respiratory and circulatory rhythms, the rhythms of rapid-firing neurones - to all of which man is subject [1,17]. For a few years it has been known that glycolysis, cell respiration and cellular transport processes can be oscillatory. The cause of the oscillations is the periodically varying effectiveness of regulatory enzymes and transport proteins [1,2,7,15]. Oscillation is observed when enzymes show nonlinear kinetics, when the biochemical or biophysical systems are simply or multiply feedback-coupled, and when they are far from equilibrium and thermodynamically open [7, 10,15]. Consequently these periodical events are limit cycle oscillations and oscillatory domains must be assumed to exist [37].

Synergetics, a comparatively new field of interdisciplinary research, deals with the cooperation of the individual parts of a system which creates macroscopically spatial, temporal or functional structures [11]. It is concerned with not only deterministic but also stochastic processes. The systems may belong to physics, chemistry, biology, computer science or economics. Synergetics has shown that large classes of systems exist in which the formation of spatial, temporal or functional structures is governed by the same principles. Typical examples of macroscopic structures in biology and chemistry are: evolution models, population dynamics, models of morphogenesis, movement coordination, perception, all-encompassing patterns, oscillations [11]. This study presents the system "vascular smooth musculature", which tends towards electro-mechanical oscillation, in its temporal and functional organization. With our current knowledge, we are not yet in the position to include diffusion processes which contribute to the spatial organization of the system in our consideration [cf. 10]. Much less can we define the order parameters of the system and their numbers to gain a unified theory of the spontaneous formation of structures using synergetic means [12].

2. Materials and Methods

Experimental Preparations and Solutions

Intracellular potential recordings as well as measurements of tension and transmembrane ion fluxes were performed on the vascular smooth muscle of the A. carotis communis. The carotid arteries were surgically removed from sacrificed dogs within 3-4 min and equilibrated in a modified Krebs solution having the following composition: Na^+ 151.16; K^+ 4.69; Ca^{2+} 2.52; Mg^{2+} 1.1; Cl^- 145.4; HCO_3^- 16.31; $H_2PO_4^-$ 1.38; glucose 7.77 mmol/l (temperature 37°C; pH 7.2-7.4). The solution was aerated with a 95% O_2-5% CO_2 gas mixture. The experiments were performed on small carotid segments (5-20 mg) after a period of not less than 2 h equilibration, when the preparation is in equilibrium regarding ion concentrations and fluxes [33,38,41].

For the biochemical measurements the following organs, or parts thereof, were taken from sacrificed dogs within 5 min as representative of obligate and facultative spontaneously rhythmic muscle types: carotid arteries (smooth muscle); sinus node, left atrium and left ventricle myocardium (heart muscle). The preparations were very rapidly frozen in liquid nitrogen, stored for short periods in dry ice, and prepared as required for chemical analysis. Before freezing, superficial connective tissue was cut off the carotis; from atrium and ventricle myocardium, pieces of organ which had minimum foreign tissue were cut out. Sinus node tissue was subjected to several histological controls by light and electron microscopy to guarantee its identity. Serial selections for light microscopy showed that no myocardial foreign tissue was present in excised sinus nodes.

Mechanical and Electrical Recording

4-5 mm long cylinders from the carotid arteries were cut lengthwise. The folded-out vessel ring, 10-15 mm long, was attached at the cut ends to an isometrically measuring tension device. The specimens were equilibrated for 1 h in Krebs solution at an initial tension of 2 g. Then solutions with different noradrenaline concentrations (10^{-8} to 10^{-3} mol/l) were applied. Propranolol was added in concentrations between 10^{-5} and 10^{-3} mol/l.

Simultaneous intracellular recordings of membrane potential were made with glass microelectrodes filled with 3 mol/l KCl. The resistances ranged from 60 to 100 MΩ and the tip potentials from -40 to -80 mV. The electrodes were shielded to just below the tips; otherwise conventional recording techniques were used [32,36,38,41].

Na^+-K^+ Exchange Measurements

The preparations were loaded until saturation for 2 h in a radioactive bath containing ${}^{24}_{11}Na^+$ and ${}^{42}_{19}K^+$. Washout was then performed over 40 min at intervals of 10 s down to 1 s, rinsing with inactive solutions of the same composition and pH. VIP series coaxial Ge(Li) detector systems with a ratio of peak height to Compton plateau height of 30:1 were used, which permitted adequate discrimination of the γ-lines (system resolution 2.2 keV using 1.333 MeV photons) of ^{24}Na at 1.37 MeV or of ^{42}K at 1.52 MeV by means of stabilized single channel analyzers [29,31-33,35]. The combined γ-spectra were measured continuously in the preparation, accumulated at variable intervals and transferred to a PDP 11/23 Plus. During the measurements, a strictly constant and laminar flow of the washout medium was maintained. The fluid in the measurement chamber was completely exchanged 1.2 times per second (chamber volume 2.3 ml, flow rate 160 ml/min). Experiments with dye indicators confirmed the favourable flow conditions. Thus, the accumulation of tracer ions in the preparation chamber, interstitium, or on the surface of the specimen was avoided. After each experiment, the content of Na, K, Mg, Ca and Cl in the preparation and in the incubation and washout solution was determined by atomic absorption spectroscopy [33].

Fluorimetric Measurements

Measurements for the investigation of NADH oscillations in the vascular smooth musculature were carried out on a fluorescence spectrophotometer (Perkin-Elmer, model MPF3). A 1 cm long segment of a canine carotid artery was mounted on a plastic frame and pushed into a quartz cuvette continuously perfused with an equilibrated Krebs solution. Temperature was kept constant at 37°C. The spectral band width was adjusted to 6 nm by slit control for the excitation and emission beams. The wavelength was driven through with a velocity of 50 nm/min; the chart speed was 25 mm/min. The maxima for the excitation and emission spectrum of NADH were determined to be λ_{max} = 380 nm and λ_{max} = 450 nm, respectively [37]. The corresponding maximum of the excitation spectrum of pure NADH is at approximately 345 nm.

For the time-dependent spectrum, the monochromators for excitation and emission wavelength, respectively, were adjusted to the observed maxima, and the spectral band width was restricted to 4.5 nm in order to obtain a better resolution. All other adjustments were maintained.

Fructose-2,6-Bisphosphate Determinations and Phosphofructokinase Activity Assays

Determination of fructose-2,6-bisphosphate (Fru-2,6-P_2) concentrations was performed according to *Van Schaftingen and Hers* [45]. Phosphofructokinase-1 activities were determined as described previously [16], protein concentrations were measured by the method of *Bradford* [3].

Mathematical Evaluation Procedure

The calculation of significant oscillations in membrane potential, tone, transmembrane ion exchange and NADH fluorescence of the vascular smooth muscle cells was performed by auto-covariance and cross-covariance functions and analyses of the power density spectrum. These evaluation procedures were represented in detail in earlier publications [29,30,32,34,35].

Model of Glycolytic Oscillations

For further investigations of the glycolytic energy supply in vascular smooth muscle we designed a simple model which reflects the main mechanisms known to be responsible for glycolytic oscillations and which can be based upon experimental data. We restricted our computations to the upper part of the glycolytic pathway, where the allosteric enzyme PFK converts fructose-6-phosphate (F6P) and adenosine triphosphate (ATP) to fructose-1,6-bisphosphate (FBP) and adenosine diphosphate (ADP), respectively. For an analysis of the regulatory properties of PFK the enzyme system of Fig. 1 is proposed, where two effectors were considered: FBP as an allosteric activator and ATP as inhibitor of the enzyme [32,34,35].

The sum of ADP and ATP concentrations was kept constant. For the reconstitution of ATP from ADP in the lower part of the glycolytic pathway and in other biochemical cell reactions, an overall rate representing these different reactions was used and chosen to be equal to the rate of removal of the product FBP. The input rate for F6P was constant and corresponded to physiological values in vascular smooth muscle. A computer programme [8] designed for solving kinetic problems in chemistry was utilized for studying the PFK system with the two substrates rate law [25] formulated for the concerted model of *Monod, Wyman, and Changeux* [24].

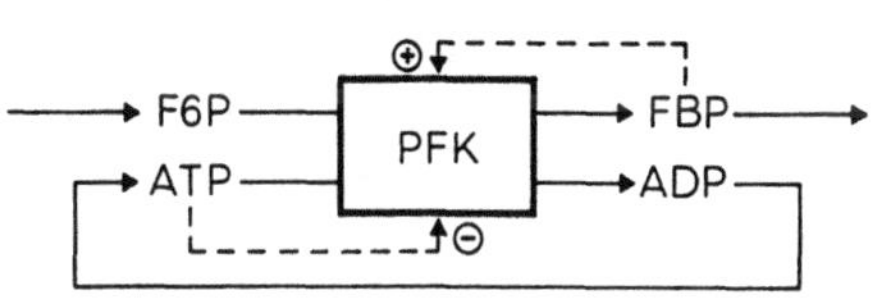

Fig. 1. Enzyme model for phosphofructokinase activity and glycolytic oscillations.

$$\alpha_i = \frac{S_i}{K_{R,S_i}}; \quad \gamma_i = \frac{P_i}{K_{R,P_i}}; \quad q = \frac{V_{T,max}}{V_{R,max}}; \quad c_i = \frac{K_{R,i}}{K_{T,i}}; \quad (i = 1,2)$$

$$V = V_{R,max} \cdot v\,(\alpha_1, \alpha_2, \gamma_1, \gamma_2)$$

$$v = (g_R \cdot \alpha_1 \cdot \alpha_2 \cdot R^{n-1} + q \cdot L \cdot g_T \cdot c_1 \cdot \alpha_1 \cdot c_2 \cdot \alpha_2 \cdot T^{n-1}) \cdot (R^n + L \cdot T^n)^{-1}$$

$$R = 1 + \alpha_1 + \alpha_2 + g_R \cdot \alpha_1 \cdot \alpha_2$$

$$T = 1 + c_1 \cdot \alpha_1 + c_2 \cdot \alpha_2 + g_T \cdot c_1 \cdot \alpha_1 \cdot c_2 \cdot \alpha_2$$

$$L = L_o \cdot \prod_i \left(\frac{1 + c_{M,i} \cdot \mu_i}{1 + \mu_i}\right)^m; \quad \mu_i = \frac{M_i}{K_{R,M_i}}; \quad c_{M,i} = \frac{K_{R,M_i}}{K_{T,M_i}}$$

S_i, P_i, M_i denote substrates, products and modifiers, respectively; g_R, g_T are the affinity coefficients for the two possible states of enzyme configuration.

It has been shown for PFK isolated from vascular smooth and cardiac muscle of the dog that the enzyme exists as a tetramer. Furthermore, a distinction between the number n of functional subunits converting substrates to products and the number m of subunits with identical binding sites for effectors seems to be reasonable, since it is known from X-ray diffraction studies on a bacterial PFK that two protomers are necessary to bind one molecule of substrate, while on the other hand each of the four subunits seems to be able to bind the effectors [9]. This led to the assumption $n = 2$ and $m = 4$ for our calculations [34]. The influence of effectors on v was formulated in the term L under the simplifying hypothesis that the effectors ATP and FBP bind to different allosteric centres and act independently. This assumption is justified, as the exact mechanism is not yet known.

The coefficients g_R and g_T and the ratio q were evaluated by a computer fit from curves as shown in Fig. 6 using an algorithm for least squares estimates of non-linear parameters [23]. The results of two independent sets of experimental data on PFK activity were averaged. For the subsequent simulations the mean values $g_R = 0.98$, $g_T = 36.97$, $q = 0.145$ were applied. The rest of the parameters were determined by experiments:

	L_o	$= 2.6 \cdot 10^5$	$V_{R,max}$	= 0.050 mmol/l/s
for the substrates:	$K_{R(F6P)}$	= 0.08 mmol/l	$K_{R(ATP)}$	= 0.005 mmol/l
	$K_{T(F6P)}$	=15.00 mmol/l	$K_{T(ATP)}$	= 0.500 mmol/l
and for the effectors:	$K_{R(FBP)}$	= 0.03 mmol/l	$K_{R(ATP)}$	= 4.000 mmol/l
	$K_{T(FBP)}$	= 6.00 mmol/l	$K_{T(ATP)}$	= 0.600 mmol/l

The input rate $v_{in(F6P)}$ and the rate $v_{out(FBP)}$ for the removal of the product were varied in the different computer runs.

In the R state the binding of one ligand has almost no effect on the binding of the other ($g_R = 0.98 \approx 1$); in the T state the affinity for the second ligand is decreased by the binding of the first one ($g_T = 36.97 > 1$). From the dissociation constants $K^d_{R(F6P)} = 0.08$ mmol/l and $K^d_{T(F6P)} = 15$ mmol/l the association constants $K'_{R(F6P)} = 12.76$ l/mmol and $K'_{T(F6P)} = 0.002$ l/mmol are obtained. Therefore, in the R state the association with the ternary complex is significantly favoured. The maximum turnover rate $V_{R,max}$ is 6.9 times as high as $V_{T,max}$ ($q = 0.145$).

3. Results and Discussion

At the beginning of this communication there is a flow chart of experimental and theoretical steps, which have been followed during recent years, for a better understanding of vascular rhythm and its significance for a functioning circulation (Fig. 2). These steps will now be referred to so that the underlying questions and answers are ever present. The flow chart should be read from bottom to top rather than the other way around. Based upon the findings from the single cell element, conclusions may be drawn about the properties of the whole structure of "vascular smooth muscle" and its organization.

Topics of discussion are the molecular origins and the functional consequences of rhythmogenesis in smooth muscle, particularly that in the blood vessels. Smooth musculature shows spontaneous-rhythmic cell potentials just as does cardiac muscle, whereby the influence of the autonomic nervous system is merely a modifying one. In Fig. 3 one can see slow fluctuations of the membrane potential in intestinal smooth muscle cells, on which action potentials may be superimposed during the phases of depolarization [21]. These spike potentials cannot be found in arterial vessels. In the latter, each slow potential wave ("slow wave") is followed by a muscle contraction [37]. In the following we would like to talk about the origin of autorhythmicity starting with the single element, the vascular smooth muscle cell, in order to reach a deeper knowledge of the reactions of the whole organ to different circulatory demands.

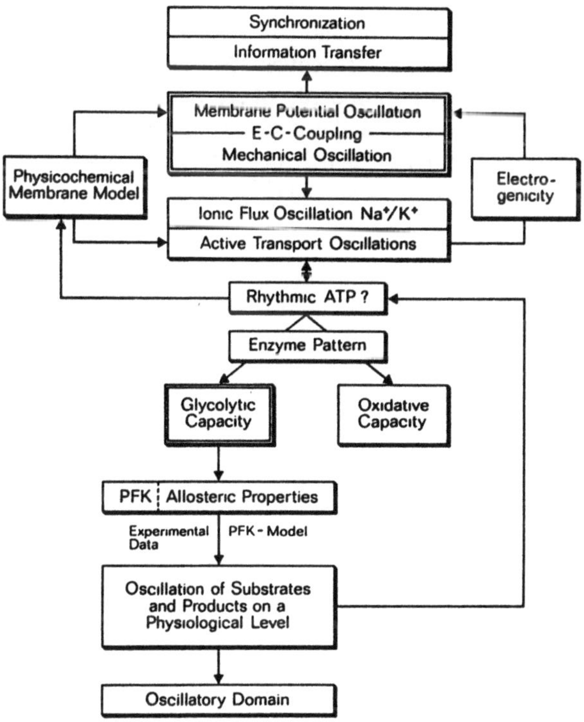

Fig. 2. Flow chart of experimental and theoretical findings to illustrate the concept of the search for the origin of rhythmogenesis.

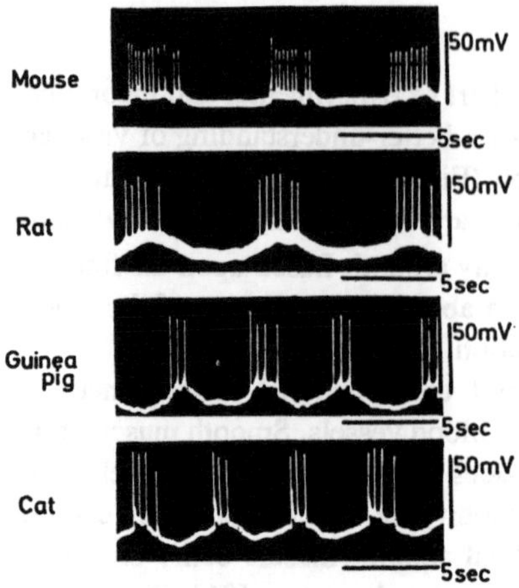

Fig. 3. Spontaneous discharges recorded from the longitudinal smooth muscle of jejunum in various species. From *Kuriyama* [21].

Already in the fifties, Professor *Edith Bülbring* [4] discovered the dependence of the electromechanical properties of intestinal smooth muscle on the glycolytic metabolism. Professor *Benno Hess* [2,15] described the phosphofructokinase (PFK) as being the most important allosteric-cooperative enzyme in glycolysis, which is capable of oscillations because of its nonlinear reaction kinetics. We took up this idea and used the following 2-substrate/2-effector model as a starting point for our investigations (cf. Fig. 1). We restricted our computations to the upper part of the glycolytic pathway, where the allosteric enzyme PFK converts fructose-6-phosphate (F6P) and adenosine trisphosphate (ATP) to fructose-1,6-bisphosphate (FBP) and adenosine diphosphate (ADP), respectively. For an analysis of the regulatory properties of PFK this enzyme system is proposed, where two effectors were considered: FBP as an allosteric activator and ATP as inhibitor of the enzyme.

By inserting the normal biochemical data into this model, one obtains the oscillatory domain covering the range of undamped glycolytic oscillations (Fig. 4). It was calculated

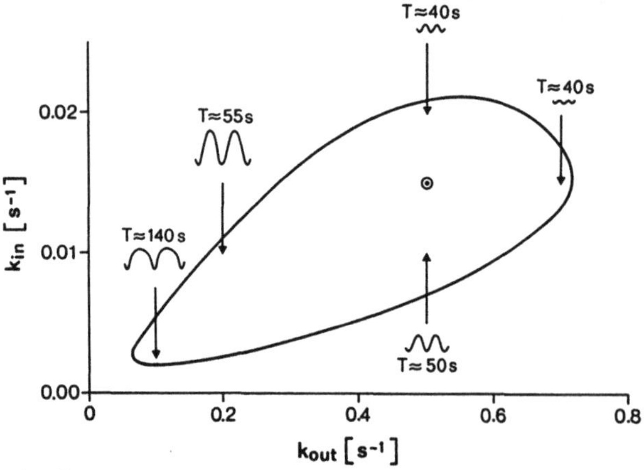

Fig. 4. Oscillatory domain of spontaneous glycolytic oscillations for the system of Fig. 1. The wave-like lines illustrate the time course of ATP concentration at the points indicated by the arrows. The period duration (T) of the oscillations is also given. The marked point within the limit cycle represents the $k_{in(F6P)}$-$k_{out(FBP)}$-constellation used for Fig. 5.

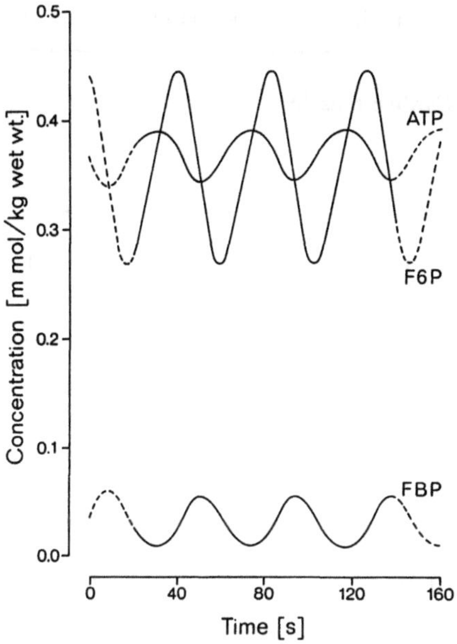

Fig. 5. Simulation of stable glycolytic oscillations of substrate and product concentrations applying the enzyme model of Fig. 1. Parameters used as given in the section on "Materials and Methods".

with systematic variations of input (F6P) and output (FBP) rate. Period durations range from 35 s up to 140 s. The wavelike lines illustrate the time course of ATP. One sees that via the glycolytic turnover rate, the frequency, amplitude and form of oscillations in metabolite and effector concentrations can be altered and regulated within relatively wide boundaries [26, 29,31,34]. An example of the effects of ATP oscillation and its dependence on glycolytic flow rate can be perceived from the recordings of Fig. 13.

Since direct measurement of glycolysis-linked ATP-oscillations *in vivo* has not yet been possible, we tried to simulate the oscillations using the described enzyme model (cf. Fig. 1). If glycolysis is driven by an input rate $k_{in(F6P)}$ and rate of removal $k_{out(FBP)}$ corresponding to that of vascular smooth muscle *in vivo*, then there appear undamped oscillations of substrate and product concentrations on a level verified by our experiments with a period of 43 s (Fig. 5). This is in good agreement with the corresponding frequencies found in the electrophysiological, mechanical and tracer measurements [34,37,40]. The phase shift found between F6P and ATP is 275°. The amplitude of the ATP oscillation is 0.045 mmol/kg wet wt. and thus about 12% that of the total ATP level.

After these computations, in which a rhythmic ATP production in cell glycolysis has been proven, the question arises as to the ratio of glycolytic to oxidative ATP. Enzyme pattern studies showed that the oxidative metabolic capacity of spontaneously rhythmic organs of the cardiovascular system increases in the order carotid vascular smooth muscle, sinus node, atrium and ventricle myocardium (Table 1). For example, this can be clearly seen in the ratio triosephosphate dehydrogenase/3-hydroxyacyl-CoA dehydrogenase (glycolysis/β-oxidation of fatty acids). According to this quotient, the oxidative capacity of the ventricle myocardium is about three times as high as that of carotid smooth muscle. Obviously, among these selected muscles of the "vascular muscle tube" the intrinsically rhythmic organs have a higher glycolytic capacity [29,35].

Table 1. Ratios of enzyme activities in vascular smooth muscle and cardiac muscle

	Vascular smooth muscle	Cardiac muscle		
	Carotid artery	Sinus node	Atrium	Ventricle
triosephosphate dehydrogenase / β-hydroxyacyl-CoA dehydrogenase	14.04	11.05	7.22	4.76
hexokinase / triosephosphate dehydrogenase	$1.21 \cdot 10^{-2}$	$2.19 \cdot 10^{-2}$	$0.38 \cdot 10^{-2}$	$0.36 \cdot 10^{2}$
phosphorylase / hexokinase	1.10	0.21	1.61	5.67
m-glycerolphosphate dehydrogenase / triosephosphate dehydrogenase	$5.10 \cdot 10^{-3}$	$15.70 \cdot 10^{-3}$	$6.00 \cdot 10^{-3}$	$3.00 \cdot 10^{3}$

As was already mentioned, the control of the glycolytic pathway of energy-supplying metabolism is exerted at the level of the enzyme PFK. Oscillations of glycolysis are created by this allosteric enzyme. The prerequisites to rhythmic changes in the ATP levels are i) the negative feedback of ATP to the regulatory control point of the glycolytic system, i.e. the PFK, and ii) the allosteric response of the enzyme to fluctuations in the concentrations of substrates and effectors; both are fulfilled for the PFK enzymes from blood vessels and the sinus node of the heart [32,34,40]. These enzymes show cooperative saturation for their second substrate fructose-6-phosphate (F6P). Fig. 6 demonstrates substrate saturation curves of PFK from the carotid artery of the dog for F6P in the absence and presence of effectors. In the absence of effectors, the enzyme shows distinctly sigmoid saturation behaviour in the presence of F6P concentrations lower than 1 mmol/l. This is in the range of the "over-all" substrate concentration in intact cells.

As is evident from the oscillatory domain derived from the 2-substrate/2-effector model, the simulation of glycolytic ATP oscillations resulted in higher period durations than those observed under *in vivo* conditions [29,37]. This problem could be solved by the discovery of PFK-2 and fructose-2,6-bisphosphate (Fru-2,6-P_2) also in vascular smooth musculature. Fru-2,6-P_2 has recently been determined to be the strongest activator of PFK-1. Fig. 7 shows the dependence of the activity of PFK-1 on the concentrations of Fru-2,6-P_2 in the presence of three different concentrations of the substrate F6P. Under the conditions of the experiments, which were performed with a physiological ATP concentration and a physiological pH, the enzyme was inactive in the absence of the activator. The increase of activity in the presence of Fru-2,6-P_2 followed a more or less distinct sigmoidal curve depending on the concentrations of F6P in the assays [37]. The curves were fitted to the Hill equation. The parameters of the fit are given in Table 2. The estimated V_{max} values decreased

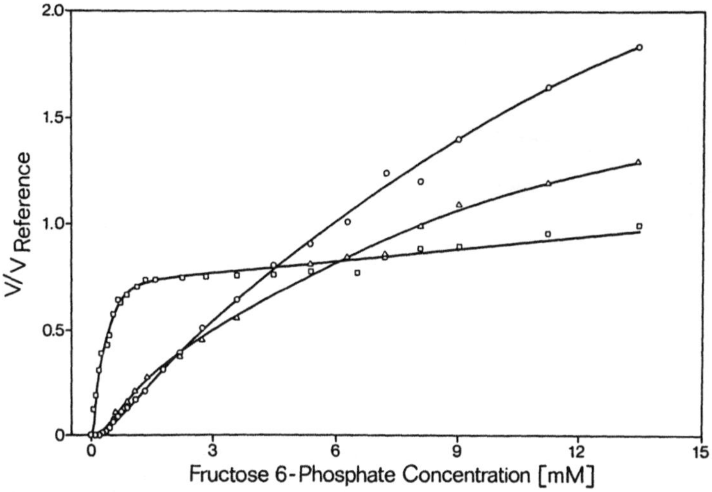

Fig. 6. Substrate saturation curves of phosphofructokinase from vascular smooth muscle of the canine carotid artery for F6P in the absence and presence of effectors. The PFK activity was determined using an optical assay system as follows: imidazole-HCl 0.1 mol/l, F6P 0.5 mmol/l, ATP 5 mmol/l, $MgCl_2$ 5 mmol/l, NADH 0.25 mmol/l, FBP aldolase 10 μg/ml, triosephosphate isomerase 4 μg/ml, glycerol-3-phosphate dehydrogenase 8 μg/ml (pH 6.9, temp.=25°C). The reaction rates V were relative to a reference activity ($V_{reference}$) determined at pH 7.6 in the presence of 3 mmol/l F6P and 1.5 mmol/l ATP. The extracts were extensively dialysed against the extraction medium and diluted to about the same reference activity as before the experiment. ○ without additions; □ +0.9 mmol/l glucose-1,6-bisphosphate; Δ +0.85 mmol/l 5'-AMP.

Table 2. Maximal reaction velocity (V_{max}), activator concentration at half-maximal reaction velocity ($A_{0.5}$), Hill coefficient (n_H) and quotient 90% V_{max} to 10% V_{max} (Q 90/10) depending on the substrate concentration in response to activation of PFK-1 from canine carotid artery with Fru-2,6-P_2: calculations were based upon the substrate saturation curves of PFK-1 in Fig. 7.

	Substrate concentration		
	0.5 mmol/l F6P	0.75 mmol/l F6P	1.0 mmol/l F6P
V_{max}	0.023	0.034	0.046
$A_{0.5}$	0.985	1.379	1.261
n_H	1.760	1.181	1.100
Q-90/10	12	41	54

The values summarized in the table were computed according to the equation:

$$v = \frac{V_{max} \cdot A^{n_H}}{A_{0.5}^{n_H} + A^{n_H}}$$

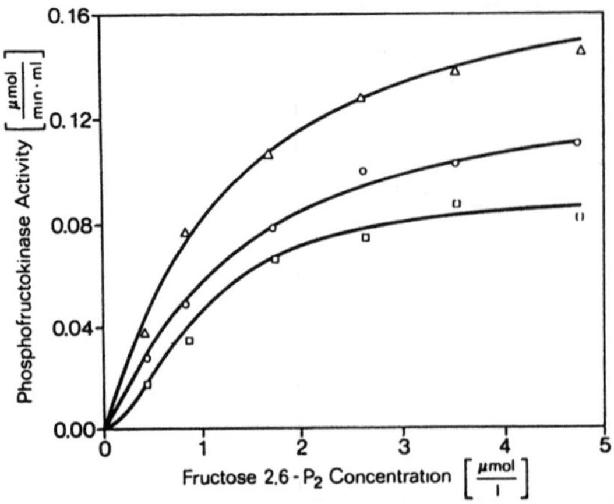

Fig. 7. Activation by Fru-2,6-P_2 of phosphofructokinase-1 from the smooth muscle of the carotid artery in the presence of 1.0 mmol/l (Δ), 0.75 mmol/l (○), and 0.5 mmol/l (□) of F6P. The optical assays were performed at pH 7.0 in the presence of ATP (3 mmol/l) and Mg^{2+} (5 mmol/l).

when the F6P concentrations were reduced, while the values of the Hill coefficients increased, indicating enhanced cooperativity. Extrapolating to a physiological substrate concentration of about 0.1 mmol/l F6P, the n_H would increase even more. The high cooperativity implies that within a certain range of activator concentration small changes in the concentration of Fru-2,6-P_2 cause large changes in the rate of substrate turnover. This fact is also illustrated by the values of Q-90/10 listed in the bottom line of Table 2. In the presence of 0.5 mmol/l F6P, a 12-fold increase in the concentration of Fru-2,6-P_2 is sufficient to stimulate PFK activity from 10 to 90% of V_{max}. The $A_{0.5}$ values listed in Table 2 are in the same order of magnitude as the Fru-2,6-P_2 concentrations found in the cardiovascular tissues.

In order to achieve a better simulation of the physiological situation, the original 2-substrate/2-effector model had to be extended to a 2-substrate/3-effector model including the strongest activator Fru-2,6-P_2, which causes an increase in the calculated activity of PFK-1 and therefore glycolytic flux. Using this flux rate for the recalculation of the oscillatory domain leads to an enlargement of the domain to higher frequencies, thus allowing a frequency of oscillation as high as 2 Hz.

Since ATP oscillation can be predicted on account of the measured allosteric qualities of the enzyme PFK, we strived to obtain more direct experimental evidence. In fact, we succeeded in proving oscillations in the NADH fluorescence of isolated vascular strips (Fig. 8). Part b represents an excerpt from an original recording (16 min) examined for significant oscillations. Not only in the autocovariance function (part c) but also in the power spectrum (part d) a main period of 40 s duration is found [43].

After we proved rhythmic ATP production at the cellular level, primarily in the glycolytic intermediary metabolism, we had to examine which consequences such oscillations of the ubiquitous energy carrier have for the outer cell membrane and for the higher organized cell formation "vascular muscle". The anisotropic transmembrane ion distribution, especially of Na^+ ions, is maintained by the consumption of metabolic energy (Na^+ outward, K^+ inward pump). Fluctuations in the concentration of the chemical energy carrier should be detectable in the active transport component of the transmembrane ion current. Therefore, we carried

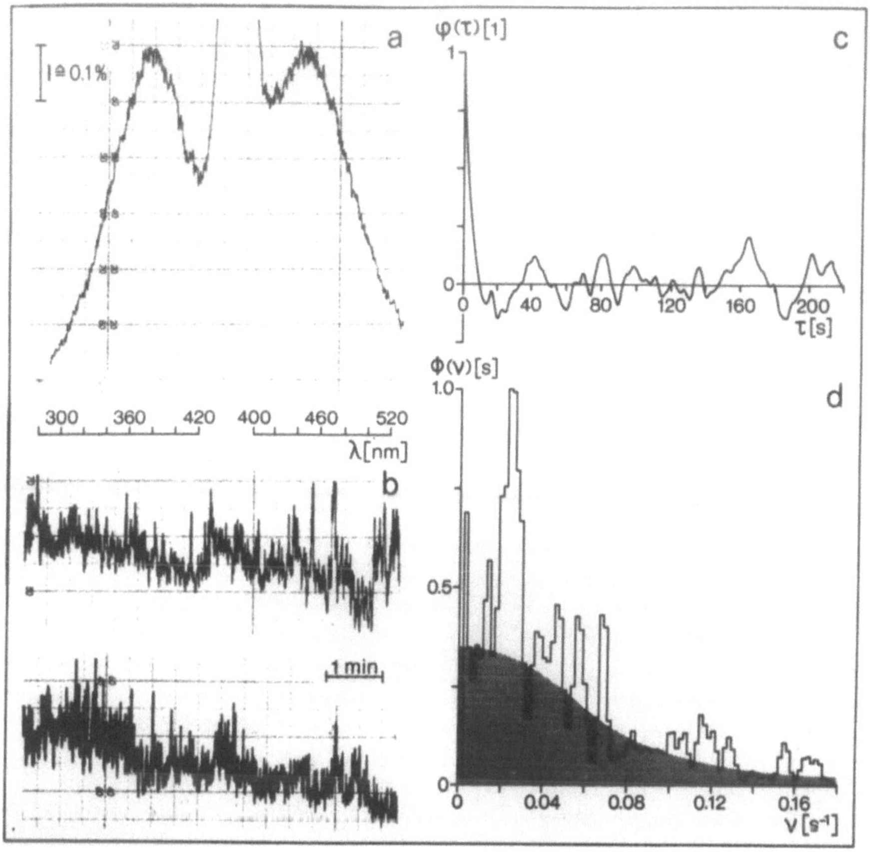

Fig. 8. Oscillations of the NADH fluorescence in an isolated vessel strip of the carotid artery. *a* Excitation and emission spectrum. *b* Time-dependent recording of the NADH emission (excerpt from an original recording). *c* Autocovariance function of section *b*. *d* Power spectrum with significant oscillation frequencies.

out continuous Na^+ and K^+ exchange measurements by use of double tracer techniques. A characteristic example of a simultaneous $^{24}Na^+$-$^{42}K^+$ efflux experiment in normal Krebs solution is shown in parts A of Fig. 9. Time is indicated along the abscissa; instantaneous exchange ratio along the ordinate. Every increase of the instantaneous exchange quotient corresponds to an increase of flux of equivalent magnitude; every decrease to an equivalent reduction of flux. The ideal time course of the Na^+ and K^+ efflux is given by the thinly drawn, computer-fitted curves. In the later, flat part of the washout curves, the recorded flux fluctuations are much larger than in the initial, steep trace [34,38-40]. This indicates that they are related to the cellular exchange. This behaviour is clarified in the insets labelled B, in which the percent deviation of the measured values from the function values of the fitted flux curves are given. Increase and decrease rates for the Na^+ and K^+ effluxes of 20 to 100% are obtained in the later part of the cellular exchange phases. Parts C and D result from calculation of the autocovariance functions and the power spectra using the relative deviations in part B. In part D, main oscillations with period durations of 50 and 100 s for Na^+ and 45 and 135 s for K^+ can be seen.

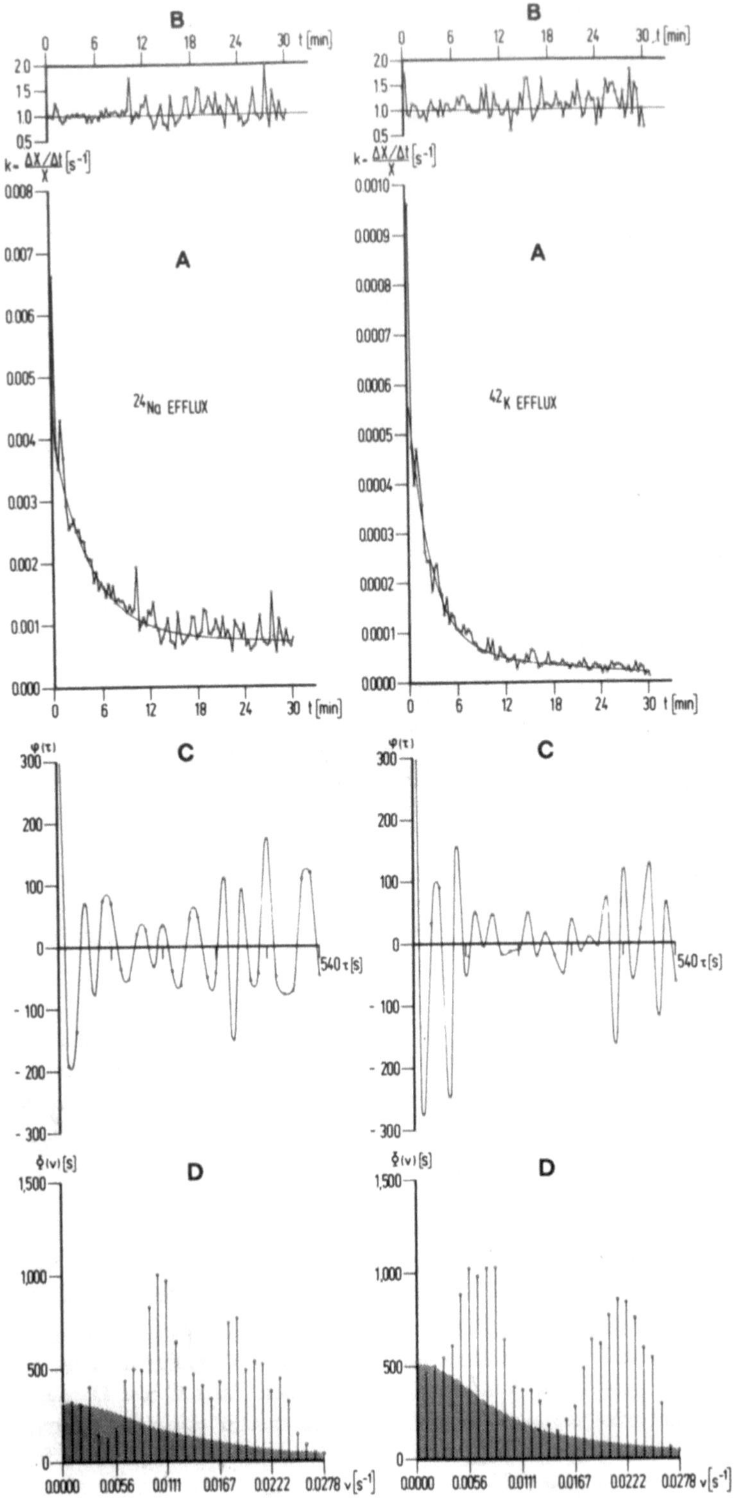
B
0
6
12
18
24
30
t [min]
2.0
1.5
1.0
0.5
k = (ΔX/Δt)/X [s⁻¹]
A
²⁴Na EFFLUX
⁴²K EFFLUX
0.008
0.007
0.006
0.005
0.004
0.003
0.002
0.001
0.000
0.0010
0.0009
0.0008
0.0007
0.0006
0.0005
0.0004
0.0003
0.0002
0.0001
0.0000
C
φ(τ)
300
200
100
0
- 100
- 200
- 300
540 τ [s]
D
Φ(ν) [s]
1,500
1,000
500
0.0000
0.0056
0.0111
0.0167
0.0222
0.0278
ν [s⁻¹]

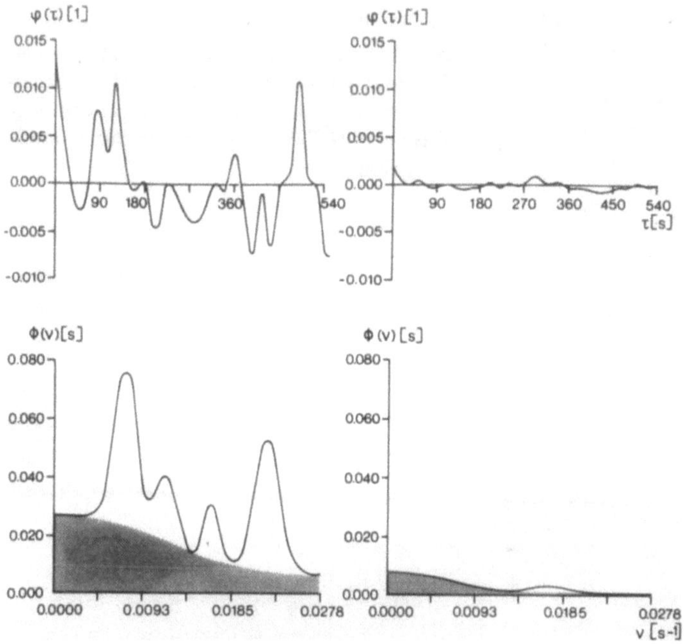

Fig. 10. Autocovariance functions and power spectra of a simultaneous $^{24}Na^{+}$-$^{42}K^{+}$ efflux experiment in the carotid artery of the dog. The left half of the figure is derived from the Na^{+} efflux, the right half from the K^{+} efflux. Oscillations with period durations of 45, 62, 87 and 134 s can be seen in the Na^{+} efflux.

Fig. 10 indicates autocovariances and power spectra of a simultaneous Na^{+}-K^{+} efflux experiment, in which the oscillations appear only in the Na^{+} outward transport. This experiment clearly proves that the rhythmicity is not caused by a mechanical squeezing out of radioactivity from the preparation or turbulences in the perfusion system. Fig. 11 shows the radioactivity of Na^{+} in the washout fluid which was recorded each second. A clear oscillation with a period duration of about two minutes can be demonstrated in the cellular exchange component.

Fluctuations in the Na^{+} efflux are probably due to periodic changes in active transport since in this tissue this component amounts to about 97% of the total Na^{+} efflux [33,42]. Only if the active Na^{+} outward transport does not balance the active K^{+} inward transport, in other words, if an electrogenic ion net transport is shifted across the membrane, is an

◀ **Fig. 9.** *A* Instantaneous exchange quotient k(t) of $^{24}_{11}Na^{+}$-$^{42}_{19}K^{+}$ effluxes determined simultaneously in the carotid artery of the dog by the indirect method. The radioactivity of washout fluid was continuously monitored in one experiment and printed out in intervals of 18 s. The thin-traced line represents the optimal triple-exponential function found by computer fitting. *B* Relative deviation of the measured values from the functional values of the optimal curve. Clear flux fluctuations appear first in the later phase of cellular exchange. *C* Autocovariance functions of the percent deviations given in part *B*. Periods of distinct oscillations can be read on the abscissa. *D* Power spectra of the autocovariance functions. The frequencies of the oscillations shown in *C* may be directly observed on the abscissa of the power spectra. The shaded areas indicate the power noise. Peak:noise relations of 4:1 to 11:1 are the general rule. Maximum lag interval m = 30.

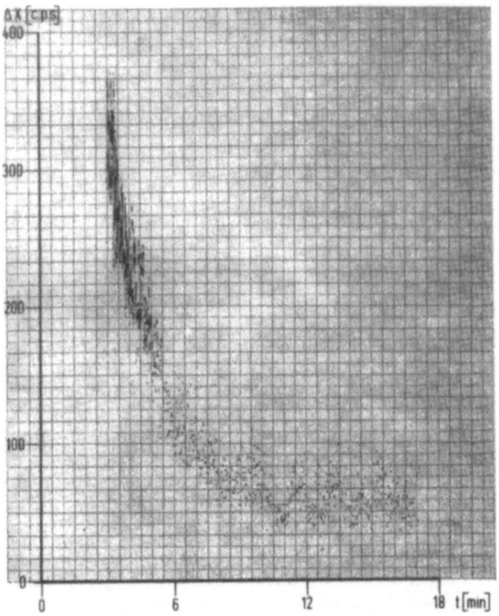

Fig. 11. Decrease in $^{24}Na^{+}$ in an indirect efflux experiment on a canine carotid artery segment. The γ-ray level in the washout fluid was recorded each second. During the cellular exchange phase a periodicity in the Na^{+} efflux is easily recognizable.

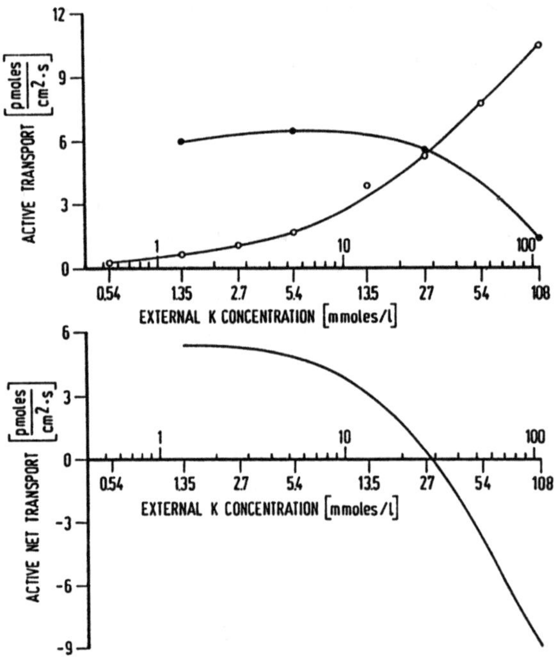

Fig. 12. *Upper part:* K^{+} and Na^{+} transport as a function of the K^{+} and Na^{+} concentration of the external solution. $[K^{+}]_o$ and $[Na^{+}]_o$ vary oppositely such that their sum remains 118.2 mmol/l. (●) Na^{+} transport (corresponding to the Na^{+} efflux deficit under DNP); (○) K^{+} transport (corresponding to the K^{+} influx deficit under DNP). *Lower part:* Stationary, active net transport as a function of the external K^{+} concentration. The curve was obtained by subtraction of the active K^{+} inward transport from the active Na^{+} outward transport shown in the upper half of the figure.

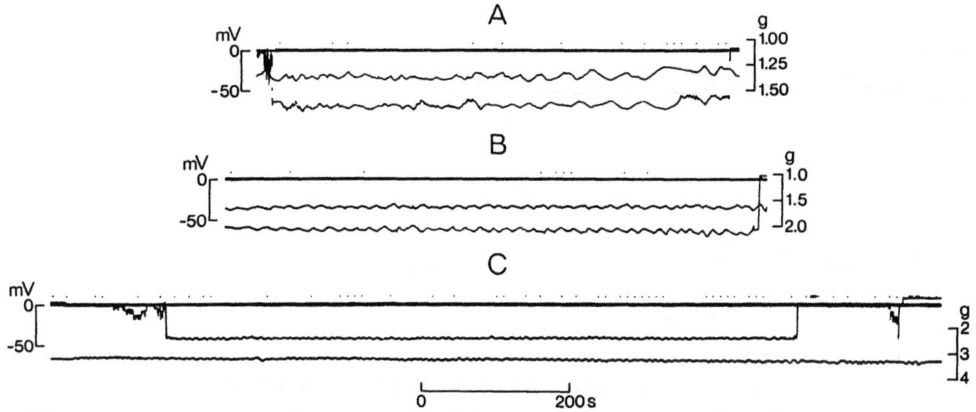

Fig. 13. Intracellular recording of membrane potential and recording of mechanical force development in segments of the canine carotid artery (slow wave activity). *A:* Normal Krebs solution; membrane potential (lower trace) 69.7 ± 3.3 mV; tension (upper trace) 1.360 ± 0.051 g; mean period duration 24 s. *B:* Noradrenaline concentration 10^{-7} mol/l; membrane potential (lower trace) 60.6 ± 2.1 mV; tension (upper trace) 1.632 ± 0.030 g; mean period duration 19 s. *C:* Noradrenaline concentration 10^{-4} mol/l; membrane potential (upper trace) 41.9 ± 1.1 mV; tension (lower trace) 3.296 ± 0.032 g; mean period duration 9 s. Time scale throughout: 10 s (uppermost trace in *A*, *B* and *C*).

influence on the actual membrane potential of the cells conceivable. To test whether this is indeed so, we measured the active Na^+ outward transport rate and the K^+ inward rate within a wide range of external Na^+ and K^+ concentrations [29,41]. In Fig. 12, on the abscissa the external K^+ concentration is plotted, on the ordinate the active Na^+ and K^+ transport in the Na^+ efflux or K^+ influx, respectively. Both transport curves cannot be brought to congruence, so that an electroneutral pump can be excluded. In the lower part of the figure the active net transport is plotted. Under physiological conditions, with 4 mmol/l $[K^+]_o$, the active, electrogenic Na^+ outward transport, and thus metabolism, may play a decisive role in the generation of the membrane potential and its oscillations. The electrogenic share of the membrane potential has been estimated to be some 5 to 20 mV [14].

Because of these measurements we can predict that not only is the potential genesis partly due to active transport processes, but also that oscillations in the intermediary metabolism must be mediated to the membrane potential through the electrogenic Na^+ transport. Indeed, Fig. 13 demonstrates electrical and mechanical auto-oscillations in normal Krebs solution (A) and at two noradrenaline concentrations, 10^{-7} mol/l (B) and 10^{-4} mol/l (C), in the vascular smooth muscle of the canine carotid artery [32,37]. Such slow waves can be recorded frequently under noradrenaline, but also in normal Krebs solution. First of all, the perfect electromechanical coupling is striking: every depolarization is followed by constriction, every hyperpolarization by relaxation. With the same time scale in A, B and C, it can be seen further that with a more depolarized membrane and increased tension the period duration as well as the relative oscillation amplitude decrease. Upon a further increase in noradrenaline concentration, the oscillation again ceases. The effects described can easily be explained by the knowledge of the oscillatory domain (cf. Fig. 4). As the domain of spontaneous glycolytic oscillations induced by canine PFK is very small in comparison to that of other structures [2,7,10], it is evident that a slight imbalance in the metabolic state leads to cessation of the oscillations. Such a perturbing effect is exerted by removal of the preparation

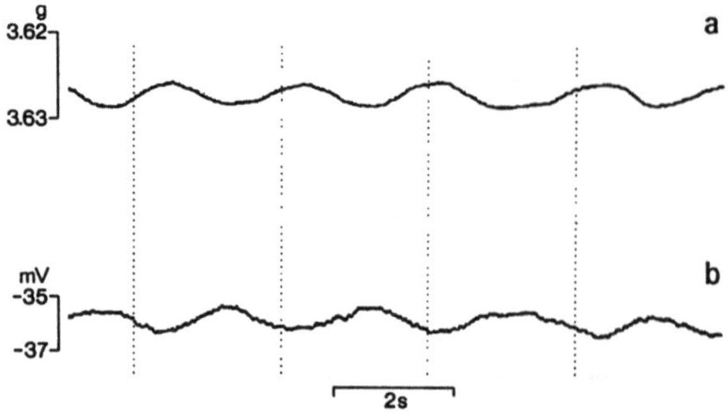

Fig. 14. Slow wave activity of membrane potential recorded intracellularly, and of mechanical force development in a segment of the canine carotid artery under 10^{-4} mol/l noradrenaline and 10^{-5} mol/l propranolol. *a:* Tension 3.627 ± 0.016 g. *b:* Membrane potential 35.9 ± 0.6 mV; mean period duration 2.3 s. Observing the recordings thoroughly, one recognizes that the electrical wave precedes the mechanical wave.

from *in vivo* conditions, by its equilibration in Krebs solution without adequate catecholamine and thyroid hormone concentration [20], or finally by a too high noradrenaline concentration. Often the addition of very low concentrations of catecholamines or superfusion/perfusion with the animal's own serum immediately leads to slow wave activity in formerly inert specimens [28]. The observation of electrically and mechanically quiescent preparations kept in artificial blood substitute solutions, whose glycolytic flow rates are obviously outside the oscillatory domain, is probably due to the experimental conditions which are not comparable to *in vivo* conditions [6,22,29].

High-frequency waves under noradrenaline are also shown in Fig. 14, where these oscillations are represented with a strong amplification. A statistical approach, the calculation of autocovariance functions and power density spectra, was used to prove and to analyse the frequency of the rhythmic processes in the membrane potential and ion exchange [29]. In Fig. 15, the power spectrum of the noradrenaline oscillations is represented, showing a

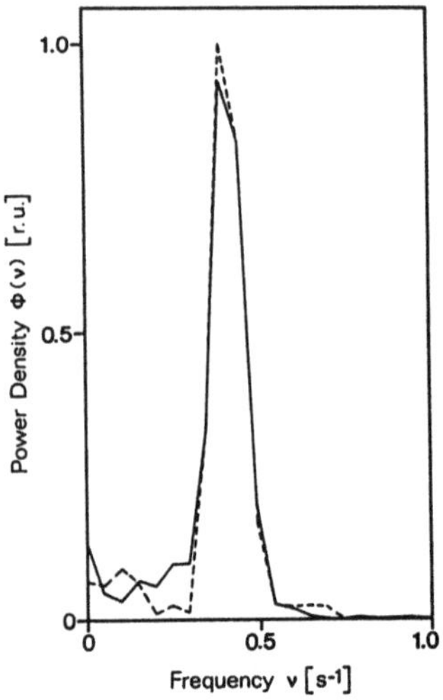

Fig. 15. Power density spectra of the fast membrane potential (dashed curve) and tension oscillations (solid curve) of the recordings illustrated in Fig. 14. The main frequency for both electrical and mechanical slow waves is 0.39 Hz.

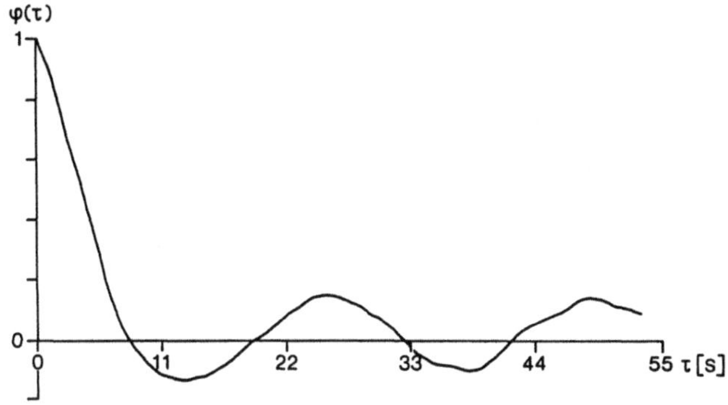

Fig. 16. Autocovariance function of slow membrane potential oscillations of an intracellular recording from the carotid artery of the dog.

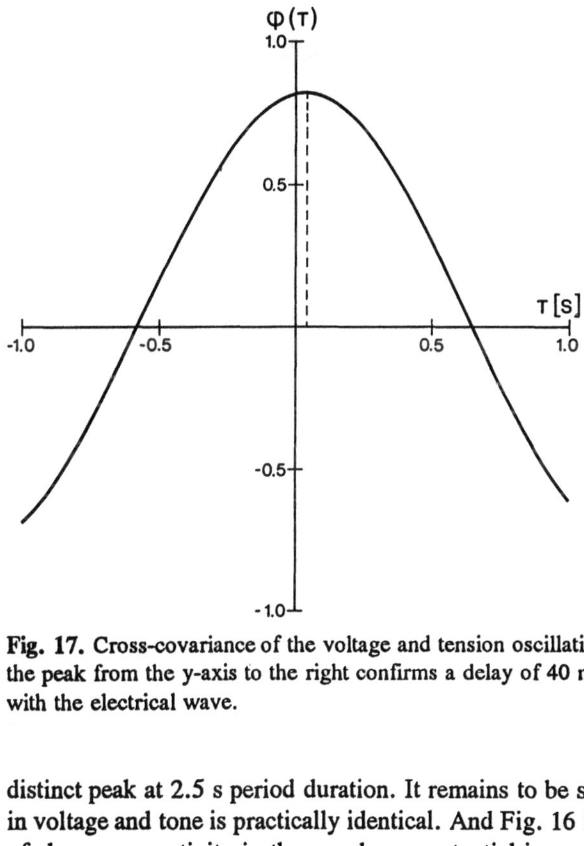

Fig. 17. Cross-covariance of the voltage and tension oscillations shown in Fig. 14. The deviation of the peak from the y-axis to the right confirms a delay of 40 ms of the mechanical wave as compared with the electrical wave.

distinct peak at 2.5 s period duration. It remains to be stated that the frequency distribution in voltage and tone is practically identical. And Fig. 16 indicates an autocovariance function of slow wave activity in the membrane potential in a normal Krebs solution with a period duration of about 25 s, typical of this graded depolarization type of smooth muscle [28,29, 32,35].

To be able to answer the question of electromechanical coupling of the potential to the tension wave, one would like to know whether the oscillation in voltage precedes the oscillation in force. Fig. 17 represents the result of a cross-covariance analysis of the fast oscillations with 2.3 s periodicity depicted in Fig. 14. One recognizes that the peak is shifted

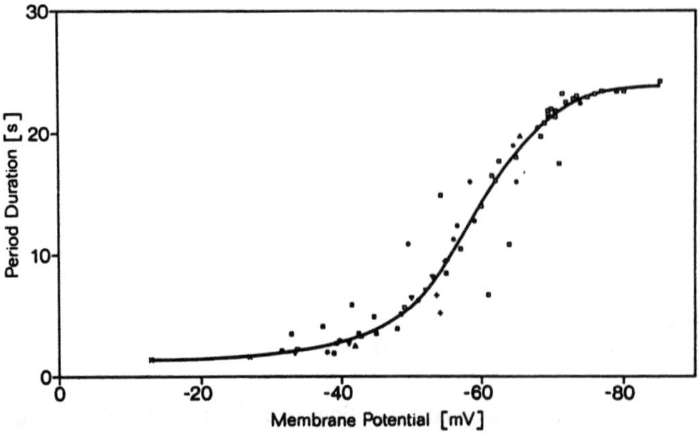

Fig. 18. Period duration of slow waves in voltage and tension as a function of the mean membrane potential of the vascular smooth muscle cells. The mean membrane potential varied by changes in the composition of the Krebs solution: □, Krebs solution; Krebs solution + varying noradrenaline concentrations: △, 10^{-8} mol/l; ○, 10^{-7} mol/l; +, 10^{-6} mol/l; ▽, $3 \cdot 10^{-6}$ mol/l; ■, 10^{-5} mol/l; ▲, 10^{-4} mol/l; ●, 10^{-3} mol/l; ▼, 10^{-4} mol/l + 10^{-5} mol/l propranolol; ×, 10^{-3} mol/l + 10^{-5} mol/l propranolol.

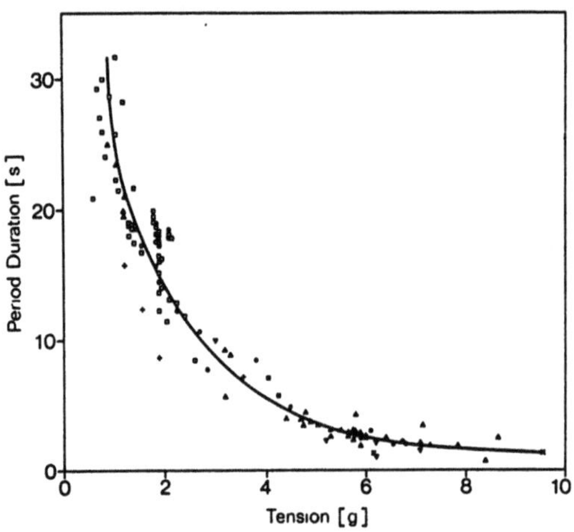

Fig. 19. Period duration of slow waves in membrane potential and tone as a function of the mean tension of vascular segments of the canine carotid artery. The mean tension varied by changes in the composition of the Krebs solution. Symbols as used in Fig. 18.

from the y-axis by 40 ms to the right, which means the mechanical wave was delayed by 40 ms compared with the electrical wave. *Cannell and Lederer* [5] found very similar values in the cardiac muscle, and *Hermsmeyer* [13] in the spiking vascular smooth muscle. This establishes the existence of a tight electromechanical coupling even under these nonstationary conditions.

Now that we know the frequencies of the PFK-dependent ATP oscillations and the tight electromechanical coupling, further frequency analysis of the potential and tension oscillations found under *in vivo* conditions is advisable. Based on a great number of experiments carried out under different conditions, Fig. 18 shows the variation of period

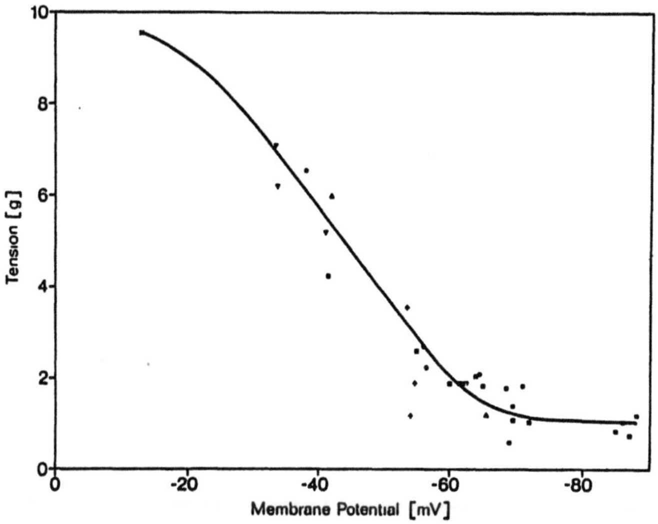

Fig. 20. Dependence of mechanical tension on the membrane potential in isolated carotid arteries. The stationary activation curve was derived from the relations in Figs. 5 and 6 by elimination of the parameter "period duration". Symbols as used in Fig. 18.

duration in dependence on the mean membrane potential. With membrane depolarization the period duration decreases in the form of a sigmoid function. With strong depolarization (V > -30 mV) or hyperpolarization (V < -75 mV) the limit values of 1.5 and 23 s, respectively, hardly change any more [37]. Similar dependences result when the relationship "period duration-tension" is plotted (Fig. 19). Since depolarization is coupled to an increase in tension, very short period durations correspond to strong mean tensions, and long period durations are related to weak mean tensions. However, the period duration varies with tension not in a sigmoid, but in a hyperbolic manner. For that reason, the product of period duration and mean tension remains constant. High tension means a small period duration and vice versa. A physical interpretation of this statement is not yet possible.

Since in Figs. 18 and 19 the period duration is the dependent variable, this parameter can be eliminated and the developed force plotted as a function of the membrane potential (Fig. 20). The relation thus derived represents the stationary, sigmoid activation curve. It is identical to the previously measured activation curve, which was recorded under static conditions [31,44]. Even the slope of 5.3 mV/g for the linear portion of the activation curve is in good agreement with the formerly determined value of 5.1 mV/g. Thus, the activation curve obtained from data on dynamic processes supplies a further strong proof of an electromechanical coupling also in the vascular smooth muscle.

In the total regulatory system of the organ "vascular smooth muscle", another basic characteristic of rhythmic systems should be mentioned, and that is the phase dependence of effector actions. Spontaneous mechanical rhythmicity can be observed especially well in a peripheral, functioning vascular system [27,28]. In Fig. 21, fluctuations in resistance are illustrated in an isolated, perfused skin area of a dog. Part A shows spontaneous oscillations with a period duration of 25 s (cf. Fig. 16). In part B, various phases of changing resistance were modified by a single stimulation of the vasoconstrictor nerve to the area. The excitability, as measured by the constricting effect of the same nerve stimulation, depends

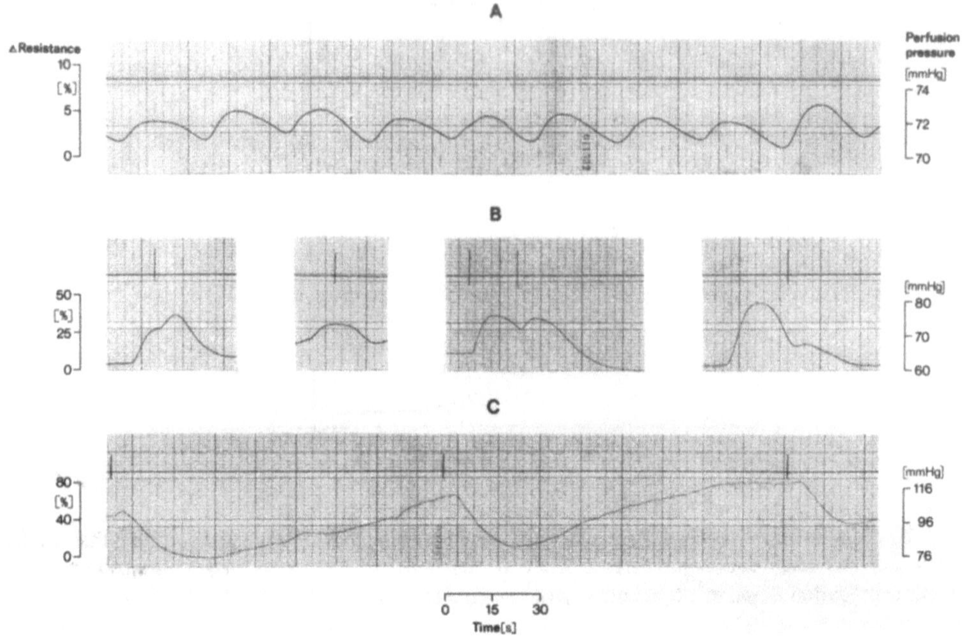

Fig. 21. Oscillations of resistance in an isolated perfused skin area of a dog. *A* Spontaneous fluctuations. *B* During single stimulation of the vasoconstrictor nerve at various phases of resistance increase. *C* During a short impulse series to the vasoconstrictor nerve of a spontaneously rhythmic vessel.

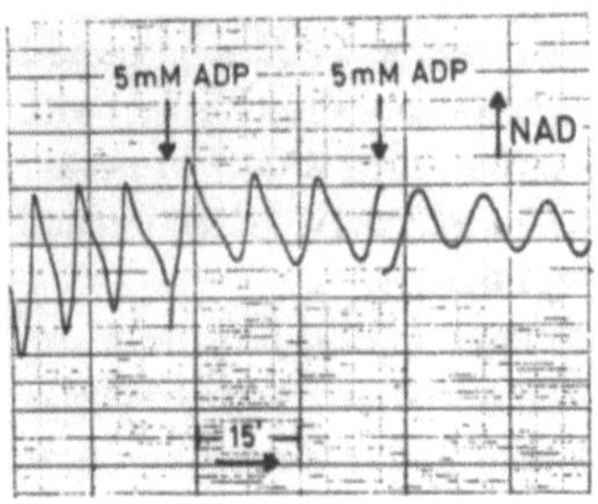

Fig. 22. Titration of glycolytic oscillations with ADP. Recording of NADH fluorescence in an extract of *S. carlsbergensis*. From *Hess* [15].

on the phase of the spontaneous constriction-dilatation cycle. Thus, for example, a stimulus applied at the peak of the constriction phase has no effect. By contrast, if the stimulus is applied later, in the dilatory phase, the vascular resistance suddenly changes in the opposite direction and a vasoconstriction follows.

Hess [15] observed similar phase jumps by titration of glycolytic oscillations with ADP (Fig. 22). ADP is a dynamic control variable and exerts a positive feedback on the PFK activity. The concentration of NADH as a function of time was registered. From this picture, one sees that titration with ADP can reverse the phase of the oscillation without essentially changing the frequency and amplitude. However, this only occurs in the quadrant of the oscillation phase in which the metabolite ADP controls the state of glycolysis via an interaction with a primary element of the oscillating system. A coupling of the two

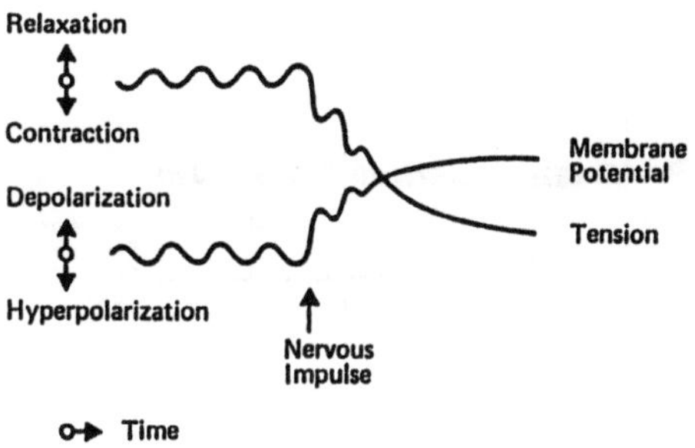

Fig. 23. Schematic representation of the reaction of a blood vessel oscillating in membrane potential and tone to an effector intervention (nervous impulse). Under decrease of period duration and oscillation amplitude, membrane depolarization and increase in tension occur.

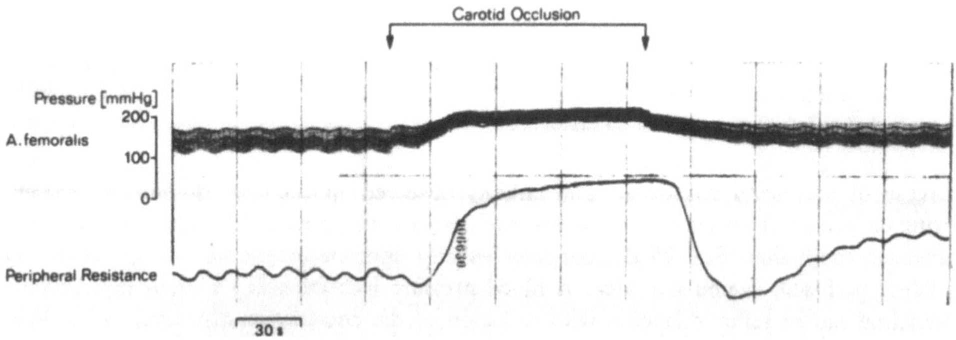

Fig. 24. Recording of peripheral resistance in the vascular bed of an isolated, perfused *m. gracilis* of the dog. The preparation is connected with the rest of the animal by the intact saphenous nerve. Increase of sympathetic activity by carotid occlusion.

systems - titration of NADH oscillations with ADP, and sympathetic nerve stimulation on spontaneous rhythms of vascular resistance - is of course speculative, but a sudden change in the energy metabolism, as indicated by a phase jump, could alter the activity of electrogenic ion pumps quickly and drastically.

Thus, cell metabolism exerts at least a double influence on vascular regulation: on the one hand, via the electrogenic Na^+ pump, which gives a variable contribution to the membrane potential. On the other hand, the electrogenic ion pump can become periodically effective and thus induce slow wave activity in membrane potential and tone. This rhythmicity is a prerequisite to a very fast vascular reactivity, e.g. with an alteration of sympathetic tone [28]. Consequently, the reaction of a vessel oscillating in membrane potential and tension can be conceived as outlined in Fig. 23. The electro-mechanical oscillation conforms to a distinct metabolic state as characterized by a defined constellation of input (F6P) and output rate (FBP) within the oscillatory domain [31]. A practical example of this is given in the recordings of Fig. 13. As the oscillatory domain of spontaneous glycolytic oscillations induced by canine PFK is very small in comparison to that of other structures [2,7,10], it is already evident that a slight imbalance in the metabolic state leads to cessation of the oscillations. Therefore, with increasing noradrenaline concentrations the oscillatory domain can be traversed (nervous impulse) or a previously quiescent preparation started to oscillate.

The former case is illustrated in Fig. 24. Fluctuations in resistance were registered in a small, isolated, perfused muscular vessel of the dog, which was still innervated [18,19]. Clamping of the carotid arteries leads to a strong sympathetic drive and thus to a step-like

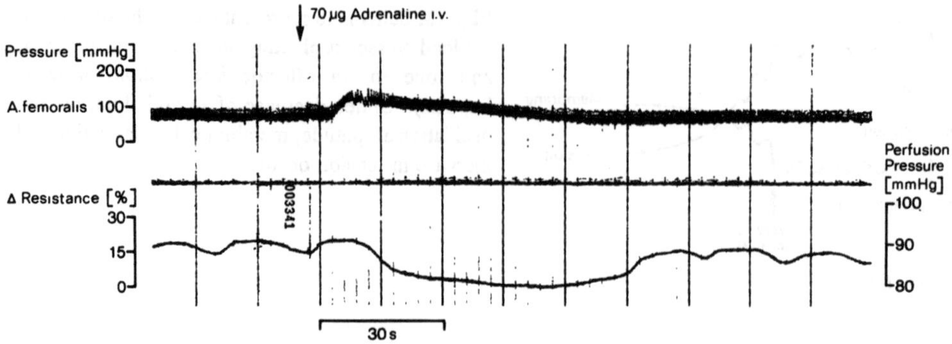

Fig. 25. Recording of peripheral resistance in an isolated perfused skin area of the hindleg of a dog. Blood pressure increase by bolus injection of 70 μg adrenaline i.v. causes reflex dilatation with disappearance of rhythmic vascular contractions.

increase of peripheral resistance. The initially observed spontaneous rhythmicity ceases during the jump in resistance. The reduction of the perfusion pressure can also terminate the resistance oscillation. Fig. 25 demonstrates exactly opposite conditions to Fig. 24 in an isolated, perfused, canine skin area. A blood pressure increase using a bolus injection of adrenaline causes reflex dilatation with cessation of the constriction-dilatation cycle. The oscillatory domain lies between the maximal (Fig. 24) and minimal peripheral resistance (Fig. 25). Note the higher oscillation frequencies in muscular as compared to skin vascular bed. The long latency between blood pressure increase and reactive dilatation in Fig. 25 is probably due to the "wrong" phase of the constriction cycle [28] in which reflex withdrawal of sympathetic activity occurs (cf. Figs. 21B and 22).

4. Acknowledgements

The authors thank Mrs. *Ch. Fuhrmann* for her excellent technical assistance and Mr. *H. Ewald* from the precision workshop for his help in the development of new analytical methods. We are grateful to Mrs. *M. Krawczynski* for her outstanding work in preparing the illustrations and to Mrs. *A. Scheuermann* for the translation, editorial elaboration and writing of the manuscript.

5. References

1. Aschoff, J.: Circadian clocks. Amsterdam: North-Holland Publ. Comp. 1965.

2. Boiteux, A., Goldbeter, A., Hess, B.: Control of oscillating glycolysis of yeast by stochastic, periodic, and steady source of substrate: a model and experimental study. Proc. Natl. Acad. Sci. U.S.A. **72**, 3829-3833 (1975).

3. Bradford, M.M.: A rapid and sensitive method for the quantitation of microgram quantities of protein utilizing the principle of protein-dye binding. Anal. Biochem. **72**, 248-254 (1976).

4. Bülbring, Edith, Lüllmann, H.: The effect of metabolic inhibitors on the electrical and mechanical activity of the smooth muscle of the guinea-pig's taenia coli. J. Physiol. (Lond.) **136**, 310-323 (1957).

5. Cannell, M.B., Lederer, W.J.: The arrhythmogenic current I_{TI} in the absence of electrogenic sodium-calcium exchange in sheep cardiac Purkinje fibres. J. Physiol. (Lond.) **374**, 201-219 (1986).

6. Casteels, R., Kitamura, K., Kuriyama, H., Suzuki, H.: The membrane properties of the smooth muscle cells of the rabbit main pulmonary artery. J. Physiol. (Lond.) **271**, 41-61 (1977).

7. Chance, B., Pye, E.K., Ghosh, A.K., Hess, B.: Biological and biochemical oscillators. New York, London: Academic Press 1973.

8. Chance, E.M., Curtis, A.R., Jones, I.P., Kirby, C.R.: FACSIMILE: a computer program for flow and chemistry simulation, and general initial value problems. Harwell: UK Atomic Energy Authority 1977.

9. Evans, P.R., Hudson, P.J.: Three dimensional structure of phosphofructokinase from *Bacillus stearothermophilus*. In: Protein: structure, function and industrial applications, edited by Hofmann, E., Pfeil, W., Aurich, H., pp. 349-357. Oxford: Pergamon Press 1979.

10. Goldbeter, A.: Patterns of spatiotemporal organization in an allosteric enzyme model. Proc. Natl. Acad. Sci. U.S.A. **70**, 3255-3259 (1973).

11. Haken, H.: Entwicklungslinien der Synergetik, I. Naturwissenschaften **75**, 163-172 (1988).

12. Haken, H.: Entwicklungslinien der Synergetik, II. Naturwissenschaften **75**, 225-234 (1988).

13. Hermsmeyer, K.: High shortening velocity of isolated single arterial muscle cells. Experientia **35**, 1599-1602 (1979).

14. Hermsmeyer, K., Harder, D.: Membrane ATPase mechanism of K^+-return relaxation in arterial muscles of stroke-prone SHR and WKY. Am. J. Physiol. **250**, C557-C562 (1986).

15. Hess, B.: Modelle enzymatischer Prozesse. In: Biologische Modelle, Band 33, edited by Scharf, J.-H., Bruns, G., pp. 195-230. Leipzig: Johann Ambrosius Barth 1968.

16. Hofer, H.W., Pette, D.: Verfahren einer standardisierten Extraktion und Reinigung der Phosphofructokinase aus Kaninchen-Skelettmuskel. Hoppe-Seyler's Z. physiol. Chem. **349**, 995-1012 (1968).

17. Koepchen, H.P.: Die Blutdruckrhythmik. Darmstadt: Dr. Dietrich Steinkopff Verlag 1962.

18. Koepchen, H.P., Seller, H., Polster, J.: Hochempfindliche Widerstandsregistrierung bei kleinen Flüssen. Pflügers Arch. ges. Physiol. **294**, 72-78 (1967).

19. Koepchen, H.P., Seller, H., Polster, J., Langhorst, P.: Über die Fein-Vasomotorik der Muskelstrombahn und ihre Beziehung zur Ateminnervation. Pflügers Arch. ges. Physiol. **302**, 285-299 (1968).

20. Kubista, V., Kubistova, J., Pette, D.: Thyroid hormone induced changes in the enzyme activity pattern of energy-supplying metabolism of fast (white), slow (red), and heart muscle of the rat. Eur. J. Biochem. **18**, 553-560 (1971).

21. Kuriyama, H.: Ionic basis of smooth muscle action potentials. In: Handbook of Physiology, Section 6: Alimentary Canal, vol. IV, Motility, edited by Code, Ch.F., pp. 1767-1791. Washington, D.C.: American Physiological Society 1968.

22. Kuriyama, H., Suyama, A.: Multiple actions of cocaine on neuromuscular transmission and smooth muscle cells of the guinea-pig mesenteric artery. J. Physiol. (Lond.) **337**, 631-654 (1983).

23. Marquardt, D.L.: An algorithm for least squares estimates of non-linear parameters. J. Siam **11**, 431-441 (1963).

24. Monod, J., Wyman, J., Changeux, J.-P.: On the nature of allosteric transitions: a plausible model. J. mol. Biol. **12**, 88-118 (1965).

25. Plesser, Th.: Dynamic states of allosteric enzymes. In: VII. Internationale Konferenz über nichtlineare Schwingungen, vol. II/2, pp. 273-280. Berlin: Akademie-Verlag 1977.

26. Plesser, Th., Lamprecht, I.: Chemical turnover and the rate of heat production in complex reaction systems. Springer Series in Synergetics **39**, 182-192 (1988).

27. Polster, J., Seller, H., Langhorst, P., Koepchen, H.P.: Zeitliche Eigenschaften der Vasomotorik. I. Über den Verlauf von Widerstandsänderungen an Hautgefäßen bei indirekter Reizung. Pflügers Arch. ges. Physiol. **296**, 95-109 (1967).

28. Seller, H., Langhorst, P., Polster, J., Koepchen, H.-P.: Zeitliche Eigenschaften der Vasomotorik. II. Erscheinungsformen und Entstehung spontaner und nervös induzierter Gefäßrhythmen. Pflügers Arch. ges. Physiol. **296**, 110-132 (1967).

29. Siegel, G.: Principles of vascular rhythmogenesis. Prog. appl. Microcirc. **3**, 40-62 (1983).

30. Siegel, G.: Membranphysiologische Grundlagen der Gefäßeigenrhythmik. In: Vasomotion und quantitative Kapillaroskopie, hrsg. von Meßmer, K., Hammersen, F., pp. 42-70. Basel, München, Paris, London, New York, Tokyo, Sydney: S. Karger 1983.

31. Siegel, G.: Membranphysiologische Grundlagen der peripheren Gefäßregulation. Physiol. akt. **1**, 31-52 (1986).

32. Siegel, G., Adler, A., Ebeling, B.J., Roedel, H., Hofer, H.W., Nolte, J.: Temporal behaviour of transmembrane ion exchange in vascular smooth muscle. Angéiologie **36**, 261-285 (1984).

33. Siegel, G., Carl, A., Adler, A., Stock, G.: Effect of the prostacyclin analogue iloprost on K^+ permeability in the smooth muscle cells of the canine carotid artery. Eicosanoids **2**, 213-222 (1989).

34. Siegel, G., Ebeling, B.J., Hofer, H.W.: Foundations of vascular rhythm. Ber. Bunsenges. Phys. Chem. **84**, 403-406 (1980).

35. Siegel, G., Ebeling, B.J., Hofer, H.W., Nolte, J., Roedel, H., Klüßendorf, D.: Vascular smooth muscle rhythmicity. In: Mechanisms of blood pressure waves, edited by Miyakawa, K., Koepchen, H.P., Polosa, C., pp. 319-340. Tokyo, Berlin, Heidelberg, New York: Japan Sci. Soc. Press and Springer-Verlag 1984.

36. Siegel, G., Ehehalt, R., Koepchen, H.P.: Membrane potential and relaxation in vascular smooth muscle. In: Mechanisms of vasodilatation, edited by Vanhoutte, P.M., Leusen, I., pp. 56-72. Basel, München, Paris, London, New York, Sydney: S. Karger 1978.

37. Siegel, G., Hofer, H.W., Schnalke, F., Adler, A., Walter, A., Koepchen, H.P.: Membrane physiological basis of vascular autorhythmicity. Prog. appl. Microcirc. **15**, 10-31 (1989).

38. Siegel, G., Jäger, R., Nolte, J., Bertsche, O., Roedel, H., Schröter, R.: Ionic concentrations and membrane potential in cerebral and extracerebral arteries. In: Pathology of cerebral microcirculation, edited by Cervós-Navarro, J., pp. 96-120. Berlin, New York: Walter de Gruyter 1974.

39. Siegel, G., Koepchen, H.P., Roedel, H.: Slow oscillations of transmembrane Na and K fluxes in vascular smooth muscle. In: Vascular smooth muscle, edited by Betz, E., pp. 3-6. Berlin, Heidelberg, New York: Springer-Verlag 1972.

40. Siegel, G., Roedel, H., Hofer, H.W.: Basic rhythms in vascular smooth muscle. INSERM Coll. **50**, 215-232 (1976).

41. Siegel, G., Roedel, H., Nolte, J., Hofer, H.W., Bertsche, O.: Ionic composition and ion exchange in vascular smooth muscle. In: Physiology of smooth muscle, edited by Bülbring, E., Shuba, M.F., pp. 19-39. New York: Raven Press 1976.

42. Siegel, G., Schnalke, F., Stock, G.: Vasorelaxation in prostacyclin-hyperpolarized arterial smooth musculature. Prog. Clin. Biol. Res. **301**, 441-447 (1989).

43. Siegel, G., Thiel, M., Walter, A., Adler, A.: Membranphysiologische Grundlagen der Vasomotorik. In: Berichtsband der 5. Gemeinsamen Jahrestagung der Angiologischen Gesellschaften der Bundesrepublik Deutschland, Österreichs und der Schweiz, hrsg. von Häring, R., pp. 147-153. Gräfelfing: Demeter Verlag 1986.

44. Siegel, G., Walter, A., Bostanjoglo, M., Jans, A.W.H., Kinne, R., Piculell, L., Lindman, B.: Ion transport and cation-polyanion interactions in vascular biomembranes. J. Membrane Sci. **41**, 353-375 (1989).

45. Van Schaftingen, E., Hers, H.-G.: Formation of fructose 2,6-bisphosphate from fructose 1,6-bisphosphate by intramolecular cyclisation followed by alkaline hydrolysis. Eur. J. Biochem. **117**, 319-323 (1981).

Role of the Rostroventrolateral Medulla in the Generation of Rhythmicities of the Sympathetic Activity

A. Trzebski and S. Baradziej
Medical Academy, Department of Physiology,
Krakowskie Przedmiescie 26/28, PL-00-325 Warsaw, Poland

Abstract. In anaesthetized, paralyzed, bilaterally vagotomized, and artificially ventilated rats, the ventral medullary surface was exposed. Spontaneous sympathetic activity (SA) of the preganglionic cervical and postganglionic lumbar nerves was recorded and integrated. For the evaluation of respiratory-related synchronization in the time domain, phrenic nerve activity and SA were integrated and averaged. In the frequency domain, the analysis of the periodicities was determined by Fast Fourier Transform (FFT). Ipsi-, contra- or bilateral microinjections of calcium channel blockers, $CoCl_2$ or $MgCl_2$ into the rostroventrolateral medulla (RVLM) abolished respiratory frequency and cardiac frequency related peaks in the power density spectrum (PDS), and significantly reduced the power of other periodicities. After bilateral microinjections of Ca-channel blockers into the RVLM, the magnitude of PDS peaks was reduced to those which remained in spinal animals. A similar effect was produced by kynurenate. An increase in blood pressure and an augmented desynchronized sympathetic discharge accompanied this central effect produced by Ca-channel blockers. SA desynchronization was reversible, lasted for 15-25 minutes, and could be reproduced by repetitive microinjections. It is concluded that in the time domain a restricted neuronal pool within the rostral para-ambigual area of the RVLM determines the respiratory-related rhythmical SA pattern occurring in each respiratory cycle. In the frequency domain, the RVLM neurons are critical for supraspinal rhythmogenesis of SA.

1. Introduction

In recent years the importance of oscillations of the spontaneous sympathetic activity (SA) has attracted the attention of those studying cardiovascular control, because a release of the vasoconstrictor peptide co-transmitter (NPY) requires an intermittent, burst-like rhythmical, high frequency discharge rather than low frequency continuous firing /1/. The respiratory-related sympathetic discharge pattern in spontaneously hypertensive rats differs from that typical of normotensive animals, and may contribute to exaggerated sympathetic activity in primary arterial hypertension /2/. Although rhythmical, respiratory and cardiac frequency related discharges of sympathetic nerves have been known for almost 60 years since the original SA recordings /3/, their central mechanism is still obscure. The range of frequencies of spontaneous SA rhythmicities varies between very slow, below respiratory frequency, up to higher frequencies of 2-6 Hz regarded as a specific rhythm characterizing the central sympathetic oscillator /4/. Several concepts have been proposed to explain SA periodicities: 1. The earliest one, the so-called "irradiation" hypothesis: - a rhythmical input from respiratory neurons to central sympathetic neurons /5/. 2. Common brain stem system entraining both central respiratory and sympathetic rhythms according to the principle of sliding coordination of different neuronal rhythmicities /6,7/. 3. Specific neuronal network, antecedent to central sympathetic premotoneurons, responsible for the generation of the high frequency (2-6 Hz) sympathetic rhythm in the manner of a probabilistic neuronal oscillator /4/. 4. Central pacemaker neurons within the rostroventrolateral medulla generating intrinsic rhythmicities, and projecting to the spinal cord /8/.

The rostroventrolateral medulla (RVLM) contains sympathoexcitatory neurons antecedent to spinal preganglionic neurons (for references see /9/). The free-running 22 Hz intrinsic activity reported in sympathetic pacemaker premotoneurons within the RVLM /8,10/ is, however, too high to account for much slower spontaneous rhythmicities recorded in the efferent sympathetic fibers in resting conditions. The purpose of this study was to test the following hypothesis: are efferent SA periodicities dependent on synaptic circuitry within the RVLM? Could local synaptic connections within the RVLM synchronize high frequency intrinsic activities of single pacemaker neurons and transform them into the low frequency periodicities typical of the efferent sympathetic output?

2. Material and Methods

40 Wistar rats were anaesthetized with pentobarbital, paralyzed by Pentacuronium, vagotomized and artificially ventilated. In some experiments carotid sinus nerves were cut bilaterally. Preganglionic cervical and mainly postganglionic lumbar nerves efferent activity was recorded and integrated. Hexametonium was used for blockade of sympathetic ganglionic transmission. SA was averaged over 16 respiratory cycles, with phrenic burst used as a trigger, to analyze respiratory rhythm related discharges within the respiratory cycle in the time domain. Power density spectra computed by the FFT method on 16 sequential time windows each of 10 s were averaged. The signal was sampled at 100 Hz by a Cambridge Electronic Design (CED) real-time computer. The ventral medulla was exposed and 150 nl 0.5 M kynurenate, and non-specific glutamate receptor blocker, was microinjected into the RVLM in order to interrupt excitatory glutaminergic transmission. To block reversibly Ca-channels, 150-300 nl of 4.0 mM $CoCl_2$ or $MgCl_2$ were microinjected into the RVLM. The effectiveness of synaptic blockade in the RVLM was checked and confirmed by the disappearance of sympathoinhibitory baroreceptor reflex induced by an i.v. pressor dose of phenylephrine.

3. Results

Kynurenate, $CoCl_2$ and $MgCl_2$ microinjections into the RVLM abolished respiratory-related SA rhythm within the respiratory cycle (Figure 1) Effects of Co^{2+} and Mg^{2+} ions were reversible, lasted for 20-25 min. and could be reproduced repeated microinjections. In the frequency domain analysis, control PDS exhibited frequencies in the range up to 6 Hz with peaks of the power corresponding to the respiratory, cardiac and non-respiratory and non-cardiac broad frequency range between 2 and 6 Hz. All synchronized activities were significantly reduced following ipsi- or contralateral microinjections of drugs (Figure 2A). Bilateral microinjections desynchronized SA significantly. The low level of power in the PDS was identical to that observed after spinal cord transection (Figure 2B). Microinjections of smaller volumes (30-50 nl) of blockers into restricted areas of the RVLM eliminated only some of the peaks in the PDS. Respiratory frequency-related peak in the PDS was selectively abolished by blocking agents applied to the region localized in the area situated at the ventral and medial border of the nucleus ambiguus (para-ambigual area, PA, Figure 3).

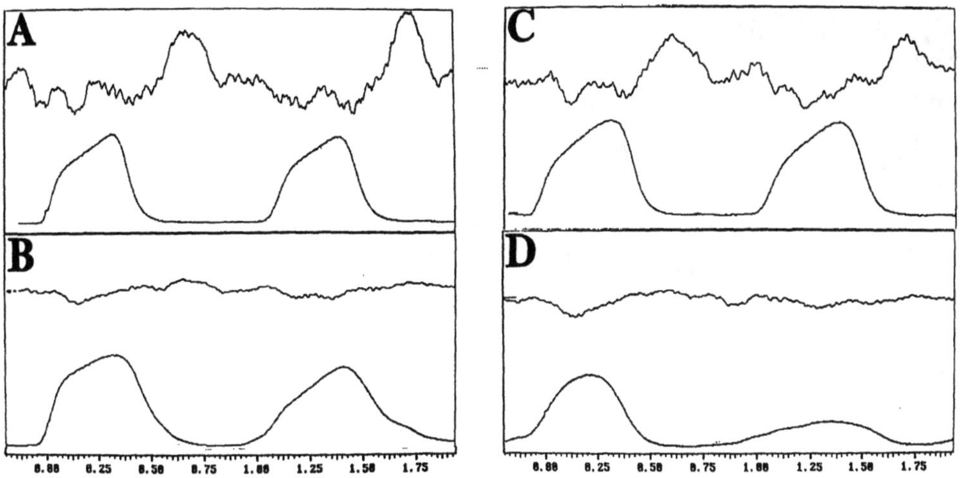

Figure 1. Averaged lumbar trunk (top) and phrenic (bottom) activity. A - control, B - after $MgCl_2$ microinjection into the RVLM, C - return to control, D - after kynurenate microinjection into the RVLM.

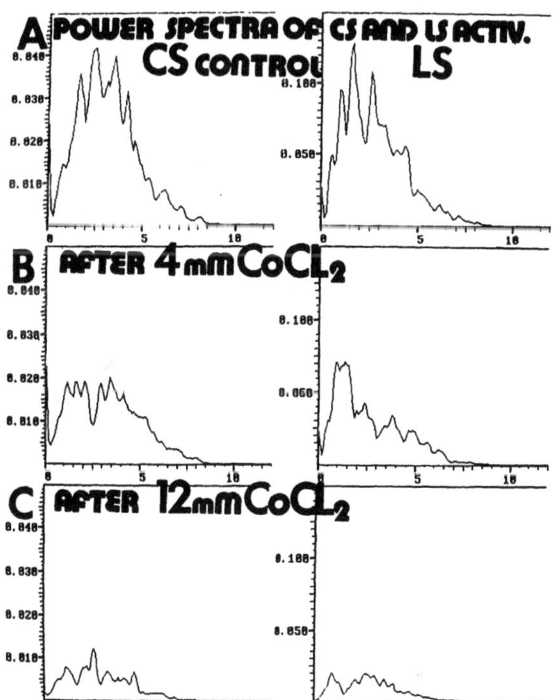

Figure 2A. Simultaneously recorded and computed averaged power density spectra of the sympathetic preganglionic cervical trunk integrated activity, CS (left), and of the postganglionic sympathetic lumbar nerve activity, LS (right). A - control, B - desynchronizing effect of low concentration of $CoCl_2$ microinjected into the ipsilateral RVLM, C - desynchronizing effect of the repeated higher concentration of $CoCl_2$ microinjected into the ipsilateral RVLM.

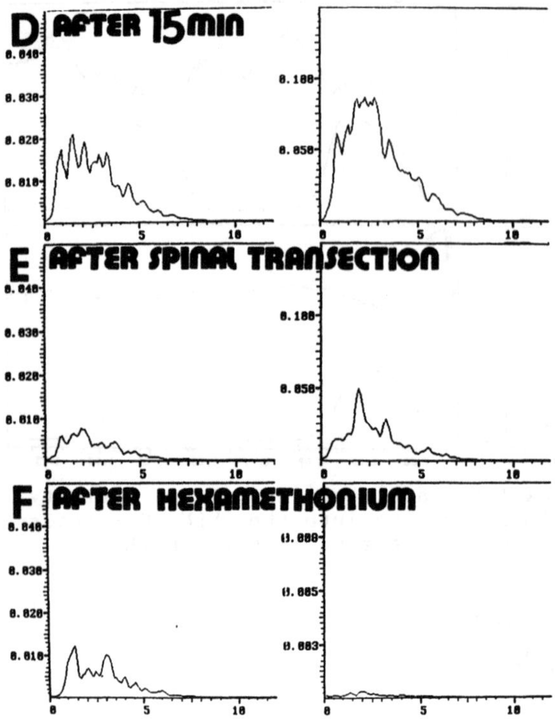

Figure 2B. Continuation of the recordings in Figure 2A. D - incomplete return to control PDS after 15 min., E - PDS after spinal cord section. Some rhythmicities of reduced power similar to that following Ca-channel blockade within the RVLM (Figure 2C), F - blockade of ganglionic transmission by i.v. hexamethonium administration entirely abolished rhythmicities in the postganglionic lumbar nerve (right) while the PDS of the preganglionic cervical nerve activity remained unchanged (left).

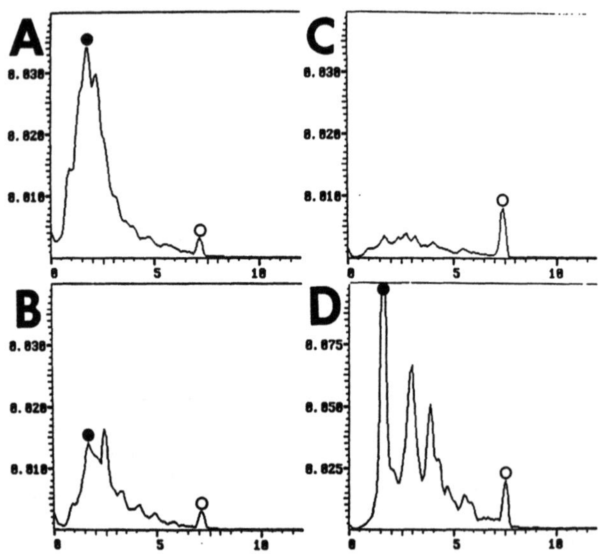

4. Discussion

Present findings indicate that RVLM neurons are critical for the expression of afferent sympathetic periodicities of supraspinal origin. Some SA periodicities still remain after spinal cord section. However, no respiratory rhythm or cardiac rhythm related peaks remained in the PDS in spinal rats. Neuronal organization of the apparent spinal cord sympathetic rhythm generator is to be elucidated. As unilateral, either ipsi- or contralateral, synaptic blockade within the RVLM significantly reduced the power of synchronized SA analyzed by spectral analysis, both symmetrical populations of the RVLM neurons have to contribute to the generation of the efferent sympathetic rhythmicities. Microinjections of drugs into the circumscribed para-ambigual area selectively abolished respiratory modulation of SA activity, other frequencies remaining intact in the PDS (Figure 3C). Simirarly, small volume microinjections into other areas of the RVLM eliminated reversibly only the cardiac rhythm related peak in the PDS. Taken together, these findings suggest that there is a functional subdivision of the RVLM neuronal pool into subpopulations involved more specifically in generation of distinct SA frequencies.

Pacemaker neurons within the RVLM, which project to preganglionic sympathetic neurons in the spinal cord, have been recently identified both "in vitro", on slices, and "in vivo" after intracisternal administration of kynurenatic acid in rats /8,10/. The intrinsic rhythm of the sympathetic pacemaker neurons in the RVLM was 22 Hz /8,10/, much higher than the frequency of firing of a single sympathetic fiber in resting conditions /5/. Our data suggest synaptic, glutaminergic connections between RVLM neurons. Such local circuits could shape out and synchronize the output of sympathetic discharge in a lower frequency range as compared to intrinsic activity of a single pacemaker cell. Blockade of the synaptic transmission within the RVLM by a glutamate receptor antagonist, kynurenate, or by calcium channel blockers, Mg^{2+} and Co^{2+} ions, disrupts this postulated local neuronal network. This local nouronal system could possibly be functioning as a sympathetic rhythm generator. Furthermore, an apparent disinhibition of the pacemaker cells' fast, yet unsynchronized intrinsic activity by local blockade of synaptic input could account for both the overall increase and desynchronization of SA and, consequently, for a blood pressure rise as observed in our experiments.

Evidently, central sympathetic rhythm generation appears to be critically dependent on the RVLM neuronal circuitry. Local neuronal subpopulation of the RVLM within the para-ambigual area is involved in a more specific way in generation and/or transmission of respiratory-related sympathetic rhythmicity, a finding in agreement with our previous results /11/.

Figure 4 presents a simplified block model of the sympathetic rhythm generator within the RVLM neuronal circuitry. The para-ambigual area appears to consist of a local subpopulation of the RVLM

Figure 3. Averaged PDS of the sympathetic lumbar trunk activity. A black dot marks the peak in PDS which corresponded to respiratory frequency. A white circle marks the peak in PDS related to cardiac frequency. A - control, B - after ipsilateral $MgCl_2$ microinjection into the para-ambigual area of the RVLM, C - after bilateral microinjection of $MgCl_2$ into the RVLM, D - reappearance of the respiratory frequency-related peak after 20 min. Carotid baroreceptors left intact. In C augmented cardiac frequency-related peak in PDS paralleled blood pressure rise and baroreceptor activation.

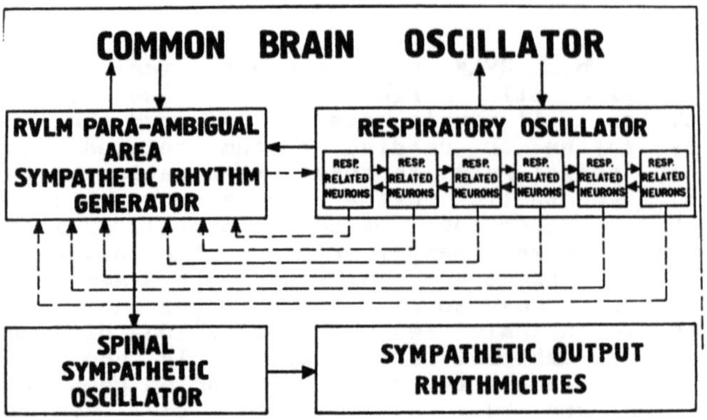

Figure 4. Schematic block diagram of a hypothetical sympathetic rhythm generator within the RVLM. The para-ambigual area represents a neuronal subpopulation of the RVLM connected synaptically with different categories of respiratory neurons. Separate spinal sympathetic oscillator. All rhythm generating neuronal pools are coordinated with a common brain system.

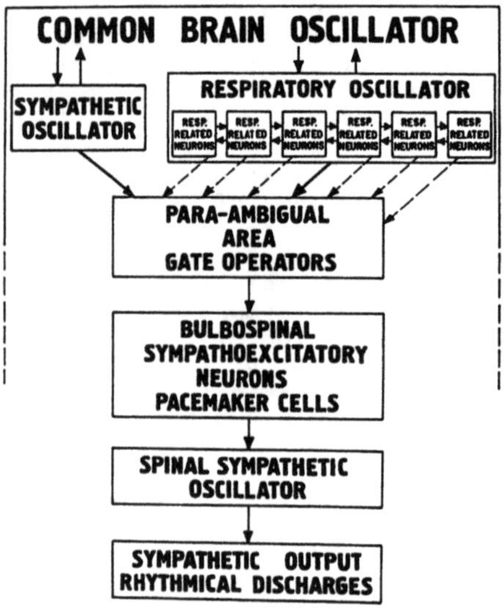

Figure 5. Alternative hypothetical model of neuronal populations within the RVLM functioning as gate-operators for the inputs coming from an antecedent neuronal circuitry responsible for generation of sympathetic oscillations. RVLM neurons represent a final common pathway in the medulla mediating the transmission of brain rhythmicities to the spinal neurons. Spinal sympathetic preganglionic neurons and/or possibly some interneurons are responsible for generation of spinal sympathetic rhythmicities.

neurons coupled with different categories of respiratory-related cells identified by Richter et al. /12/. A spinal sympathetic oscillator is also postulated in the model, because consistent, yet significantly reduced, rhythmicities are still preserved in the PDS of sympathetic discharge, after acute spinal cord section in the rat.

The central sympathetic rhythm generator appears to be integrated with mutual interactions between diffuse neuronal systems, called here, for simplicity, a common brain oscillator. Numerous data presented by Koepchen /6 and this Symposium/ and by others /7/ support the view that a sliding coordination of different rhythms is accomplished in the brain, probably according to the concept of synergetics introduced and developed by Haken /13 and this Symposium/.

Another, alternative model is proposed in Figure 5. A proposed sympathetic oscillator is represented here by a neuronal network antecedent to the RVLM neurons according to the concept of the probabilistic oscillator advocated by Gebber and Barman /4/. Both the RVLM neuronal circuitry and para-ambigual neurons, related specifically to respiratory neurons, would be merely a kind of gate-operator /14/, shaping out and processing synaptic input from other antecedent neuronal circuits responsible for generation of the sympathetic rhythm. Present data appear rather more consistent with the former hypothetical scheme, which emphasizes pacemaker properties of the RVLM neurons. In particular, the central desynchronizing effect of calcium channel blockers, applied to the RVLM supports the pacemaker hypothesis, as calcium channels are known to be responsible for pacemaker potentials /15/. Also postsynaptic glutamate neuronal receptors and glutaminergic excitatory synaptic transmission operate via modulation of multiple types of neuronal calcium channels facilitating intracellular calcium influx (for references see /16/).

Further research is still needed to clarify which, if any, of the proposed schematic models approximates better the real central synaptic mechanisms responsible for sympathetic rhythm generation.

References

1. J.M. Lundberg, T. Hökfelt: Trends Neurosci. **62**, 325 (1983)
2. M.F. Czyzyk-Krzeska, A. Trzebski: J. Physiol. (Lond.) **426**, 355 (1989)
3. E.D. Adrian, D.W. Bronk, G.P. Phillips: J. Physiol. (Lond.) **74**, 115 (1932)
4. G.L. Gebber, S.M. Barman: In: J. Ciriello, M.M. Caverson, C. Polosa (eds.) The Central Neural Organization of Cardiovascular Control, Progress in Brain Research, vol. 81, 131, (1989)
5. G. Preiss, F. Kirschner, C. Polosa: Brain Res. **87**, 363 (1975)
6. H.P. Koepchen, H.H. Abel, D. Klussendorf: In: J. Ciriello, F.R. Calaresu, L.P. Renaud, C. Polosa (eds.) Organization of the Autonomic Nervous System. Central and Peripheral Mechanisms, (Alan R. Liss, New York 1987, p. 179)
7. M. Bachoo, C. Polosa: J. Physiol. (Lond.) **385**, 545 (1987)
8. P.G. Guyenet, J.R. Hasselton, M.K. Sun: In: J. Ciriello, M.M. Caverson, C. Polosa (eds.) The Central Neural Organization of Cardiovascular Control, Progress in Brain Research, vol. 81, 105, (1989)
9. J. Ciriello, M.M. Caverson, C. Polosa: Brain Res. Reviews **11**, 359 (1986)
10. M.K. Sun, P.G. Guyenet: Brain Research **438**, 23 (1988)
11. S. Baradziej, A. Trzebski: In: G. Kunos, J. Ciriello (eds.) Central Neural Mechanisms of Cardiovascular Regulation (Springer, New York 1991) (in press)

12. D.W. Richter, D. Ballantyne, J.E. Remmers: News Physiol. Sci. 1, 109 (1986)
13. H. Haken: Advanced Synergetics (Springer, Berlin, Heidelberg 1987)
14. I. Ninomiya, T. Akiyama, N. Nishiura: Am. J. Physiol. 259, R499 (1990)
15. N. Hagiwara, H. Irisawa, M. Kameyama: J. Physiol. (Lond.) 359, 233 (1988)
16. R.W. Tsien, D. Lipscombe, D.V. Madison, K.R. Bley, A.P. Fox: Trends Neurosci. 11, 431 (1988)

Noninvasive Methods for Studying Rhythmic Nervous Control of Human Circulation

A. Lindqvist

Cardiorespiratory Research Unit, University of Turku,
SF-20520 Turku, Finland, and Research Institute of Military Medicine,
Central Military Hospital, SF-00300 Helsinki, Finland

Abstract. Oscillations of heart rate and skin blood flow were studied during standardized deep breathing, intermittent tilting and thermal stimulations. Deep breathing and intermittent tilting synchronized periodic heart rate variability. Rhythmic warm and cool stimulation of the skin synchronized skin blood flow. Thermal stimulation also affected the periodic heart rate variability. The synchronization may be an integrated central and peripheral function. Normal response at deep breathing frequency suggests normal efferent vagal fibres to the heart. Low-frequency (<0.15 Hz) periodic heart rate variability is affected by sympatho-vagal interactions during rhythmic cardiovascular stimulations and may provide noninvasive means to measure integrated sympathetic and parasympathetic autonomic reactivity.

1. Introduction

Basic observations of cardiovascular nervous control are mainly obtained from experimental animal studies. The role subnormal or supranormal nervous control of circulation plays in human physiology and pathophysiology is poorly understood.

Centrogenic rhythms in brainstem networks with facultative coupling with the respiratory oscillator, negative feedback from baroreceptors and chemoreceptors and the autorhythmicity of the vascular smooth muscle are a multicausal theory of the generation of vasomotor waves [1]. Cardiovascular perturbations initiate responses to control and regulate blood pressure or blood flow via arterial and cardiopulmonary baroreceptors and chemoreceptors. Peripheral and central thermoreceptors trigger vasomotor thermoregulatory reflexes. Information is carried to the central nervous system through afferent nerve fibres. Blood pressure and local blood flow are adjusted by cardiac, vascular and renal effectors (Fig.1). Most viscera receive a nerve supply from each of the generally antagonistic sympathetic and parasympathetic efferent autonomic systems. The oscillations of heart rate are affected by sympatho-vagal interactions in the cardiovascular control system [2-7]. Efferent vagal activity cannot be measured directly [8]. Although efferent sympathetic nervous traffic can be measured directly in conscious human volunteers, this technique has not been used clinically [8,9].

We have investigated oscillations of heart rate and skin blood flow during standardized respiratory, gravitational and thermal stimulations [10-14]. The aim of the studies has been to develop methodology for measuring peripheral and central autonomic reactivity.

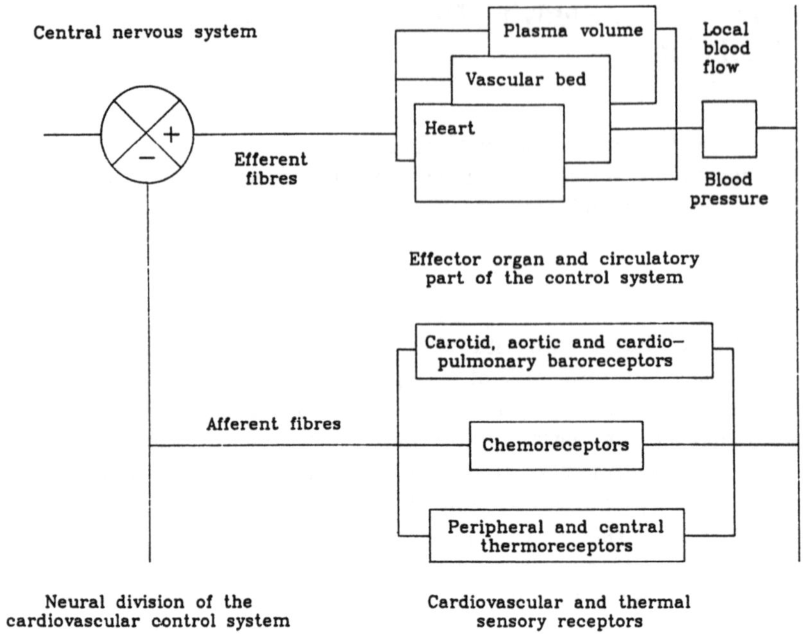

Figure 1. The organization of the cardiovascular control system.

2. Respiratory Sinus Arrhythmia during Deep Breathing

An irradiation of a central rhythm through cardiovascular effectors, cardio-cardiac reflexes, baroreceptor reflex or reflex activity from pulmonary receptors may initiate respiratory sinus arrhythmia [15]. The respiratory sinus arrhythmia requires intact vagal function and central inspiratory activity [15,16]. Average respiratory variation of the cardiac interbeat intervals correlates highly significantly to the decrease of the average cardiac interbeat interval with atropine [17]. Propranolol does not affect this correlation in dogs over a wide range of parasympathetic activity. Maximal doses of atropine abolish the respiratory sinus arrhythmia in humans [17]. Slowing of the heart rate by propranolol does not restore the respiratory sinus arrhythmia after its abolition by atropine [see 17]. The respiratory sinus arrhythmia is a clinically valuable index of parasympathetic activity of the heart.

The increasing tidal volume increases the respiratory sinus arrhythmia [18]. The peak corresponding to the breathing frequency becomes the dominant component of the heart rate power spectral density function when the breathing becomes deeper [12]. The periodic heart rate variability diminishes at frequencies below the respiratory rate. Deep breathing at frequencies between 0.10 - 0.34 Hz synchronizes the oscillations of heart rate. The mean heart rate remains generally unchanged [12]. The maximal response takes place at the respiratory rate of 6 cycles/min or 0.10 Hz (Fig. 2). Increased respiratory rate diminishes respiratory sinus arrhythmia [12,18,19]. The deep breathing test has been used to standardize the respiratory sinus arrhythmia and to detect parasympathetic failure [10].

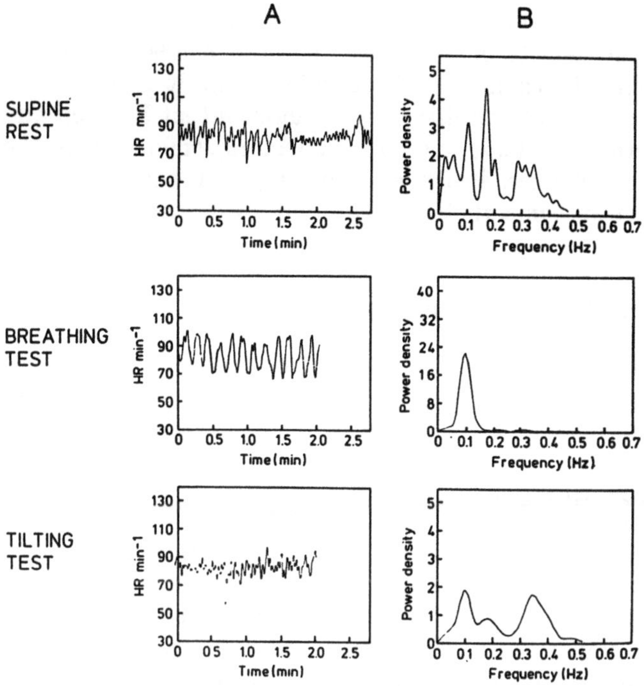

Figure 2. Instantaneous heart rate signal (A) and power spectral density function of heart rate (B) of one subject during spontaneous activity at supine rest and deep breathing and intermittent tilting stimulation at 0.10 Hz.

3. Periodic Heart Rate Variability during Intermittent Tilting

The arterial systolic blood pressure increases up to 30 mmHg 1-3 s after beginning an active standing up by means of arterial compression with postural muscles. An exercise reflex accelerates the heart rate within 3 s the same way as after commencement of isometric contraction [20]. Within 7 s after active standing up the arterial systolic blood pressure falls up to 30 mmHg below the control value [21]. The neural control mechanisms aim to maintain the normal arterial blood pressure. The arterial systolic blood pressure recovers within about 20 s to 20 mmHg above the control value [21]. Reflex alterations in heart rate include tachycardia which is maximal about 10 s after standing up. It is followed by a relative bradycardia which is maximal 20-30 s after standing up. This tachycardia-bradycardia oscillation is mainly under vagal control because the normal pattern of response is altered by atropine to small and gradual tachycardia over 30 beats with no rebound bradycardia [22].

The reactions of blood pressure and heart rate attenuate during passive head-up tilting compared to active standing up [23]. In response to the intermittent head-up tilting from 0° to 45° at 0.10 Hz (6 cycles/min) the mean heart rate is lower than during the spontaneous activity at supine rest [12]. The periodicity of the sinusoidally intermittent orthostatic stimulus can be clearly seen in the individual and average power spectral density functions of heart rate during head-up tilting (Fig. 2). The power density of the nearby spectral bands around the stimulus frequency diminishes indicating synchronization [12]. It may be that during a dynamic cardiovascular state produced by an intermittent head-up and head-down tilts, the fast parasympathetic autonomic effectors dominate over the slowly reacting sympathetic effectors. Prolonged standing and head-up tilting provoke a prominent increase of sympathetic activity and low-frequency periodic heart rate variability [5,6].

4. Cardiovascular Oscillations Related to Thermoregulation

Any thermosensitive structures in the brain, spinal cord and skin can cause vasoconstriction or vasodilatation in the skin vessels [24]. Both peripheral and central thermoreceptor reflex activity terminates at the cutaneous blood vessels [25].

Afferent nervous impulses arising in the heated skin of the trunk or the legs vasodilate the hand. After cessation of heating the hand blood flow soon returns to its original level. Vasoconstriction of the hand is caused by afferent impulses from a cooled skin elsewhere [26]. Radiant heating of the trunk elicits reflex skin vasodilatation in 15 to 30 s as a result of the reduction of sympathetic vasoconstrictor activity. The impairment of this response in tetraplegics shows that the involved neural pathways pass above the cervical spinal cord [16]. The reflex activity initiated in the peripheral cutaneous thermoreceptors functions by regulating the cutaneous blood flow through the same efferent sympathetic vasoconstrictor pathway as the blood pressure control [25-29]. This reflex activity has been used to measure sympathetic efferent nervous function.

In body temperature control the thermal gradient of the skin determines the heat exchange between the body and the environment. It is necessary to control the thermal gradient of the skin, not the skin temperature. The oscillations of skin blood flow continuously change the thermal gradient of the skin. The sympathetic vasoconstrictor activity initiated from the cutaneous thermoreceptors may be a thermoregulatory control process which adjusts the thermal gradient of the skin [19,27-30].

The oscillations of skin blood flow may be mediated to the periodic heart rate variability via the cardiovascular reflex activity [19,31].

5. Cardiovascular Responses to Periodic Thermal Skin Stimulation

We have designed an automatic water bath stimulator to induce standardized rhythmic warming and cooling of peripheral skin [10,13,32] (Fig. 3).

Rhythmic cooling and warming of the lower or upper extremity in the water baths at frequencies of 0.01-0.10 Hz induces reproducible oscillations in the forearm skin blood flow of young adults. Spontaneous oscillations of skin blood flow close to the stimulus frequency simultaneously attenuate indicating synchronization to the thermal stimulation [10,13,32].

The 0.01-0.12 Hz periodic heart rate variability of supine subjects increases when the thermal stimulus amplitude on the lower extremities is increased from

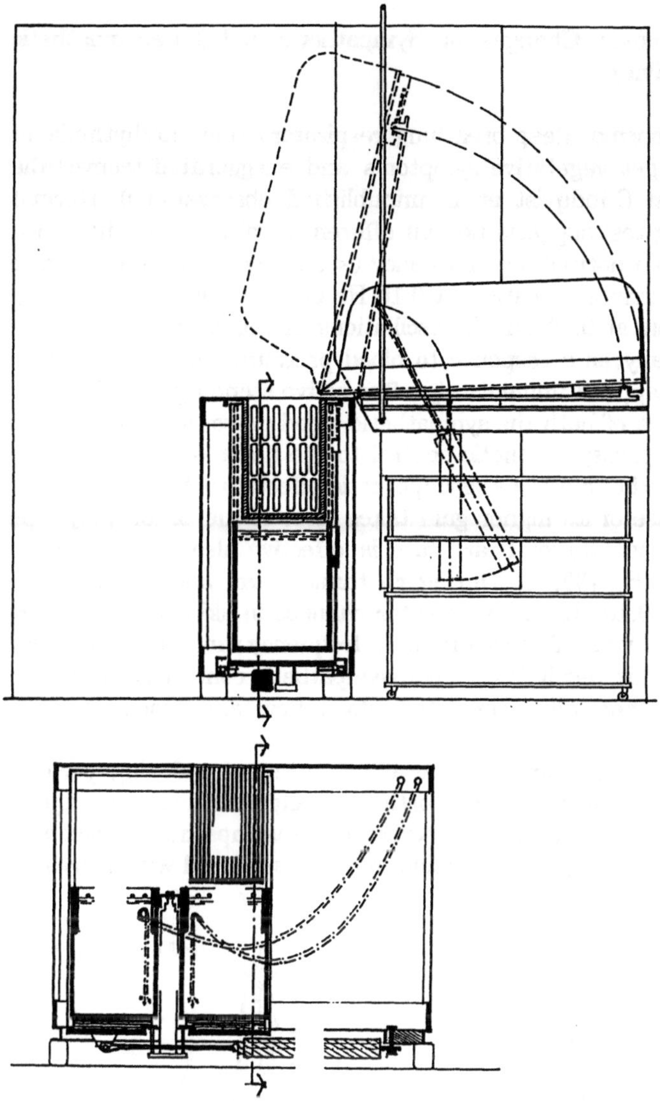

Figure 3. Water bath and intermittent tilting stimulator. Two water tanks containing 18°C and 38°C water are moved automatically upwards and horizontally by a pneumatic system. The water tanks induce rhythmic warm and cool stimulus on the skin of the feet and calves or hands and arms. The subject is lying on a tilt table which can be tilted automatically between 0° and 70°.

0°C to 20°C in the water bath stimulator [10]. The supine adults' periodic heart rate variability is entrained to the 0.10 Hz warm and cool water shower stimulation of the hand and arm. The overall heart rate variability decreases [14]. Thermal stimulation has a frequency-selective effect on the periodic heart rate variability of neonates [11].

6. Measurement of Dynamic Changes in Sympathetic and Parasympathetic Autonomic Nervous Reactivity

We have generally found normal deep breathing respiratory sinus arrhythmia in supine test subjects who get vegetative symptoms and exaggerated tachycardia during active standing test (Lindqvist et al, unpublished observations). Normal response in deep breathing test suggests normal efferent vagal fibres to the heart. The test subjects have also a normal low-frequency (<0.15 Hz) periodic heart rate variability during thermal skin stimulation at 0.10 Hz in the supine position [14]. During thermal stimulation at 0.10 Hz the oscillations of the heart rate may be mediated parasympathetically as a response to blood pressure perturbations in a similar way as normally happens immediately after active standing up (Fig. 4).

The open loop dynamics of both the sympathetic blood pressure control to the resistance vessels and the parasympathetic control to the heart is effective below 0.05 Hz. When both sympathetic and parasympathetic effectors act in parallel the sympathetic effector because of its higher gain takes most of the action [33]. The vagal baroreceptor reflex activity of heart rate is effective also at frequencies between 0.05 Hz and 0.15 Hz [33,34]. During rhythmic warm and cool stimulations at frequencies below 0.10 Hz, the vasomotor changes in skin blood flow are normally buffered by oscillations of heart rate on a frequency range up to 0.15 Hz [14]. An interpretation of this result is that parasympathetic control of heart rate is increased during slow thermally stimulated blood flow and blood pressure perturbations.

Low-frequency (<0.15 Hz) heart rate variability attenuates during periodic thermal skin stimulations at 0.01-0.03 Hz in orthostatically sensitive test subjects despite normal basal heart rates (Fig. 4). Decreased parasympathetic heart rate variability can explain this finding. This reaction may be associated with increased

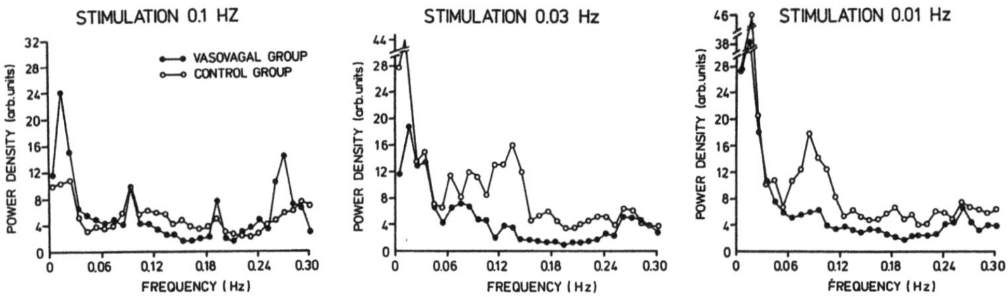

Figure 4. Power spectral density function of heart rate during rhythmic warm and cool stimulation of the hands and arms. The vasovagal group consisting of subjects who have vegetative symptoms and initially exaggerated tachycardia and later a tendency to vasovagal reaction in prolonged active standing (●—●) and control group (o—o). Both groups show similar entrained oscillations of heart rate during thermal stimulation at 0.10 Hz. The vasovagal group has on average a smaller periodic heart rate variability at the frequency band of 0.01-0.15 Hz during thermal stimulation at and below 0.03 Hz than the control group except at the stimulus frequencies.

low-frequency dynamics of the oscillations of skin blood flow and heart rate [14]. We have also found a heart rate pattern suggesting attenuated parasympathetic and/or increased sympathetic heart rate variability immediately after active standing up in similar subjects with increased sympathetic reactivity to orthostatic stress (Lindqvist et al., unpublished observations). In hyperdynamic circulation, reaction to sympathomimetic agents is exaggerated and an adrenergic blockade may attenuate the symptoms [35,36]. However, this exaggerated sympathetic function does not always explain the autonomic dysfunction of the hyperdynamic heart [37] which may also be caused by an impaired vagal function [38].

7. Conclusions

Deep breathing and intermittent tilting synchronize periodic heart rate variability. Rhythmic warm and cool stimulation of the skin synchronizes skin blood flow. Thermal stimulation also affects the periodic heart rate variability. The synchronization may be an integrated central and peripheral function. Mainly vagally mediated oscillations of heart rate can be synchronized by 0.10 Hz deep breathing and intermittent tilting. Periodic warm and cool skin stimulation at 0.01-0.10 Hz can synchronize sympathetically mediated oscillations of skin blood flow and affect periodic heart rate variability both via sympathetic and parasympathetic fibres. Cardiovascular oscillations may provide noninvasive clinical methods for studying integrated sympathetic and parasympathetic autonomic reactivity.

8. References

1. H.P. Koepchen, Mechanisms of blood pressure waves, K. Miyakawa, H.P. Koepchen & C. Polosa (eds.) Japan Scientific Societies Press, Tokyo & Springer-Verlag, Berlin. 3 (1984).
2. K. Antila, Scand J Clin Lab Invest 39, Suppl 153 (1979).
3. S. Akselrod, D. Gordon, F.A. Ubel, D.C. Shannon, A.C. Barger, & R.J. Cohen, Science 213, 220 (1981).
4. R.W. De Boer, J.M. Karemaker, & J. Strackee, Am J Physiol 253, 680 (1987).
5. B. Pomeranz, R.J.B. Macaulay, M.A. Caudill, I. Kutz, D. Adam, D. Gordon, K.M. Kilborn, A.C. Barger, D.C. Shannon, R.J. Cohen, & H. Benson, Am J Physiol 248, H151 (1985).
6. M. Pagani, F. Lombardi, S. Guzzetti, O. Rimoldi, R. Furlan, P. Pizzinelli, G. Sandrone, G. Malfatto, S. Dell'orto, E. Piccaluga, M. Turiel, G. Baselli, S. Cerutti, & A. Malliani, Circ Res 59, 178 (1986).
7. J.U. Grönlund, K.J. Antila, A.S.I Siimes, T. Metsälä, R. Oja, J. Tuominen, & I.A.T. Välimäki, Med Biol Eng Comput 27, 163 (1989).
8. D.L. Eckberg, Am J Physiol 239, H581 (1980).
9. G. Wallin, Autonomic failure. A textbook of clinical disorders of the autonomic nervous system, R. Bannister (ed.) Oxford University Press, Oxford. 36 (1983).
10. A. Lindqvist, Acta Physiol Scand 138, Suppl 588 (1990).

11. A. Lindqvist, R. Oja, O. Hellman, & I. Välimäki, Early Human Development 8, 37 (1983).
12. A. Lindqvist, R. Erkolahti, E. Heinonen, & I. Välimäki, Scand J Clin Lab Invest 46, 771 (1986).
13. A. Lindqvist, J. Jalonen, P. Parviainen, L. Halkola, K. Antila & L.A. Laitinen, Med Biol Eng Comput 27, 429 (1989).
14. A. Lindqvist, P. Parviainen, P. Kolari, J. Tuominen, I. Välimäki, K. Antila, & L.A. Laitinen, Cardiovascular Research 23, 262 (1989).
15. A. Melcher, Acta Physiol Scand Suppl 435 (1976).
16. R.H. Johnson, D.G. Lambie, & J.M.K. Spalding, Neurocardiology. The interrelationship between dysfunction in the nervous and cardiovascular systems, W.B. Saunders Company, London (1984).
17. F.M. Fouad, R.C. Tarazi, C.M. Ferrario, S. Fighaly, & C. Alicandri, Am J Physiol 246, H838 (1984).
18. A. Angelone, & N.A. Coulter Jr., J Appl Physiol 19, 479 (1964).
19. B.W. Hyndman, R.I. Kitney, & B.McA. Sayers, Nature 233, 339 (1971).
20. C. Borst, W. Wieling, J.F.M. Van Brederode, A. Hond, L.G. De Rijk, & A.J. Dunning, Am J Physiol 243, H676 (1982).
21. C. Borst, J.F.M. Van Brederode, W. Wieling, G.A. Van Montfrans, & A.J. Dunning, Clinical Science 67, 321 (1984).
22. D.J. Ewing, L. Hume, I.W. Campbell, A. Murray, J.M.M. Neilson, & B.F. Clarke, Am J Physiol 243, H676 (1980).
23. W. Wieling, C. Borst, J.M. Karemaker, & A.J. Dunning, Clin Physiol 5, Suppl 5, 23 (1985).
24. R. Hellon, Handbook of Physiology, section 2: The cardiovascular system, Vol III, J.T. Shepherd, F.M. Abboud & S.R. Geiger (eds.), American Physiological Society, Bethesda, Maryland 659, (1983).
25. L.B. Rowell, J Invest Derm 69, 154 (1977).
26. D.McK. Kerslake, & K.E. Cooper, Clinical Science 9, 31 (1950).
27. A.C. Burton, Am J Physiol 127, 437 (1939).
28. A.C. Burton, & R.M. Taylor, Am J Physiol 129, 565 (1940).
29. C.J. Gordon, & J.E. Heath, Comp Biochem Physiol 74, 479 (1983).
30. R.I. Kitney, Med Biol Eng 12, 57 (1974).
31. R.I. Kitney The study of heart-rate variability, R.I. Kitney & O. Rompelman (eds.), Oxford Medical Engineering Series, Clarendon Press, Oxford 81, (1980).
32. A. Lindqvist, Med Biol Eng Comput, 27, 435 (1989).
33. K.H. Wesseling, & J.J. Settlers, Psychophysiology and cardiovascular control, J.B. Orlebeke, G. Mulder & L.J.P. van Doornen (eds.), Plenum Press, New York and London 69, (1983).
34. C. Polosa, Mechanisms of blood pressure waves, K. Miyakawa, H.P. Koepchen & C. Polosa (eds.) , Japan Scientific Societies Press, Tokyo & Springer-Verlag, Berlin 27, (1984).
35. A. Bollinger, M. Gander, P.O. Pylkkänen, & G. Forster, Cardiologia 49, Suppl II, 68 (1966).
36. E.D. Frohlich, R.C. Tarazi, & H.P. Dustan, Arch Intern Med 123, 1 (1969).
37. H. Hörtnagl, D. Magometschnigg, & J. Prager, Cardiology 69, 74 (1982).
38. S. Julius, A.V. Pascual, & R. London, Circulation 44, 413 (1971).

The High Pressure System of the Mammalian Circulation as a Dynamic Self-Organizing System

H. Schmid-Schönbein and S. Ziege

Institut für Physiologie, Klinikum der RWTH Aachen, Pauwelsstr. 30, W-5100 Aachen, Fed. Rep. of Germany

Abstract. Highly stable fluctuations of arterial pressure are seen when rhythmic changes of heart rate and arterial blood pressure (measured indirectly by the non-invasive technique of photoplethysmography in the earlobe made hyperemic) are continuously measured during progressive relaxation of healthy subjects. Associated with physical and mental relaxation, the aperiodic rhythmic fluctuations of arterial pressure become entrained into a characteristic mode (frequency ca. 0.15Hz) in 27 out of 28 subjects lying comfortably. This rhythm could be clearly distinguished from respiratory sinus arrhythmia and respiratory blood pressure waves (ca. 0.25Hz) as well as from slower waves reflecting sympathetic activity (below 0.1Hz). This rhythm, further stabilized during early sleep, was regularly abolished by physical activity, standing, and exposure to various stimuli. The rhythm, also known to occur in postganglionic sympathetic muscle nerve activity, is discussed as a macroscopic projection of a synergetic process taking place in a cooperative mechanical system (the Windkessel vessels) exchanging information via afferent, central and efferent neuronal systems within the CNS. In generalizing the principle of the coherence of mesoscopic subsystems, we postulate that many systems (heart, arteries and various types of arterioles, afferent neurons, variable neuronal circuitry, interneurons and efferent neurons) get into progressive coherence due to mutual enslaving producing synergetic cooperativity. The energetic and informational aspects of progressive coherence during the reduction of central and cardiovascular drive (as control parameters) and of progressively active enslaving mechanisms (information compression) are discussed. A phenomenological model for a large synergetic system ("WKS-CNS ensemble") is presented as the site of generation of stable synergetic order under the conditions of decreasing physical and mental activity.

1. Introduction: A Highly Stable Mode of Arterial Blood Pressure Waves in Human Subjects

In the attempt to record non-invasively the time course of higher order arterial blood pressure waves for comparative purposes in studies of the cutaneous microcirculation, the blood content of the earlobe made hyperemic by a vasodilator ointment was recorded by transmission or reflection photoplethysmography as described in [1]. In subjects sitting uncomfortably or standing upright, aperiodic fluctuations were regularly found, as well as occasional oscillations synchronized to quiet respiration. Upon forced

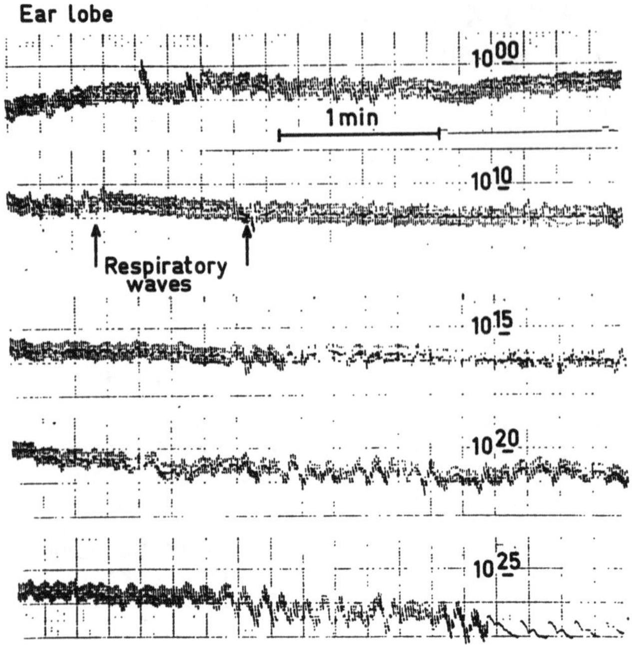

Fig. 1
Original optical plethysmogram trace (arbitrary units, reflecting fluctuations in the blood content) of the right ear (made hyperemic pharmacologically) of a relaxed awake human subject. Fluctuating changes in arterial pressure produce coherent changes in blood content, these are initially aperiodic and become progressively entrained into a highly stable rhythm. Note transient occurrence of waves equivalent to respiratory blood pressure.

ventilation, irregular respiration-associated waves were found, and also fluctuations accompanying the Valsalva (expirational apnoe) or the Müller maneuver (inspirational apnoe). About 30% of the subjects showed transient fluctuations below 0.1 Hz when sitting comfortably.

If, however, we continued photoplethysmographic recording, characteristic fluctuations in earlobe blood were regularly seen in 25 out of 26 healthy test persons when they were recumbant, accustomed to the experimental setting, feeling relaxed, and comfortably warm. The typical transients between aperiodic fluctations and the slow appearance of these highly regular sinusoidal waves (9/min), which reflect arterial blood pressure waves (see Section 6), are shown in Fig.1. The spectral analysis revealed a clearly distinguishable band between 0.12 and 0.18 Hz; they were well above the well-known 0.1 Hz and well below the ca. 0.25 Hz rhythm (respiratory blood pressure waves). They are strongly reminiscent of the blood pressure waves reported by SUNDLÖF and WALLIN [2] and related to sympathetic activity in muscle nerves. These waves will henceforth be referred to as the ca0.15Hz-rhythm. The latency between arrival of the subjects at the laboratory and the appearance of the ca. 0.15Hz rhythm varied substantially; in healthy subjects it appeared to reflect their

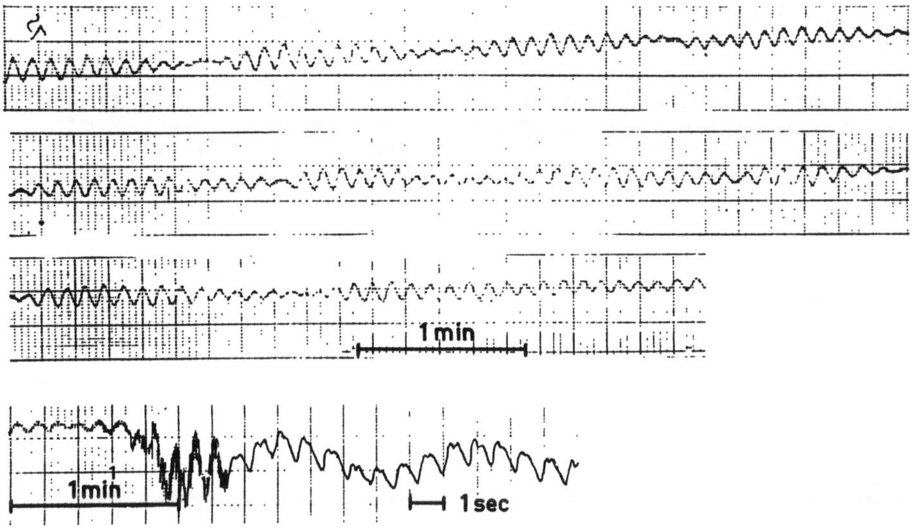

Fig. 2
Upper three traces: Typical recording of earlobe blood content in a fully relaxed human subject at the transition to sleep: note highly stabilized frequency (9/min = 0.15Hz) with increasing and decreasing amplitude resulting in "beat" phenomen due to interference of two rhythms of similar frequency (noticeable at the minima of the amplitudes). Bottom trace: Indirect photoplethysmographic record of arterial ca. 0.15Hz rhythm shown at higher resolution (arbitrary units).

ability to relax. This interpretation is substantiated by the fact that, when the subject felt asleep, this ca. 0.15Hz rhythm became much more stable, and often showed a characteristic beat phenomenon as shown in Fig.2, upper three panels. Details of a well-established ca. 0.15Hz photoplethysmographic rhythm are shown in the bottom panel of Fig.2., where the rhythm is displayed in higher amplification and at high recording speed.

As will be shown below (see Section 6, Fig. 5), once established, the ca. 0.15Hz rhythm became more and more pronounced and proved to be very stable against certain perturbations (voluntary change in respiration) but highly labile against others (especially the assumption of physical activity and pronounced cooling) [3].

The ca. 0.15Hz rhythm is clearly distinguishible from respiratory changes in heart rate (which mostly have twice their frequency) and is closely related in the majority of subjects to lower order fluctuations in heart rate [3] (see Section 6).

These highly regular ca. 0.15Hz rhythms are interpreted subsequently as a prototype of a synergetic rhythm. When the rules for the generation of synergetic order and of synergetic information compression as elaborated by HAKEN [4] are followed, the stable waves are taken to reflect the fully coherent state of the cardiovascular system when residuals of physical activity and of the equivalents of the so-called "defense reaction" have disappeared and when skeletomuscular perfusion is markedly reduced.

2. Mechanical Cooperativity of the Windkessel Vessels, the Heart and the Resistance Vessels

The arterial "Windkessel System" (WKS), an ensemble consisting of a pulsatile pump (the left ventricle) feeding into a series of elastic high conductance conduits and variable outflow resistors (the systemic arterioles), can be taken as the simplest case of a "self-organizing" mechanical dynamic system: during pumping, it is "pressurized" and therefore "displaced" from fluid-dynamic equilibrium. Obviously, the instantaneous pressure in the elastic arterial conduits is a direct function of the instantaneous volume of blood stored in them (Hooke's law), which in turn reflects the steady state volume of the "Windkessel Vessels" (WKVs) plus a fluctuating excess volume increment proportional to the instantaneous difference between the pulsatile inflow (restricted to the systolic phase of the cardiac cycle, see below) and the outflow (taking place throughout the whole cardiac cycle). Thence, fluctuating amounts of kinetic energy are stored (equivalent to the product of pressure and volume), keeping the WKS far from fluid-dynamic equilibrium and allowing it to act as a dynamic system in a "boosted steady state" (see SCHMID-SCHÖNBEIN [5])

In the semantics of HAKEN [4] the pulsatile variations of the intravascular pressure in the WKVs can be taken not just as a fluctuation in the energy content of a dynamic system, but as a source of synergetic information, convectible both to the outside world and to the central nervous system. We interpret the informational behavior of this central segment of the circulation as paradigmatic for that of a dynamic synergetic system. The WKS is obviously always operating far from fluid-dynamic equilibrium and it can be easily shown that it is exhibiting "spontaneous order" (reflecting cooperativity and functional coherence of control parameters and order parameters) and the accompanying "information compression" (see below). Following HAKEN [4] one can conclude that: 1) such a system is able to take up and transfer informational order in the form of "preferred modes"; 2) the macroscopic behavior of the entire system reflects the "mesoscopic order" of various "enslaved" subsystems; and 3) order parameters assume a dualistic informational role of enslaving subsystems and conveying information to adjacent dynamic systems (the baroreceptor afferents as well as dynamic man-made measuring devices).

In the present communication, it is postulated that **exclusively** under conditions of **physical rest**, such "preferred modes" (attributable to a variety of characteristic fluctuating order parameters) are found in the WKVs, can be monitored externally and are identifiable by spectral analysis of the time course of these fluctuations. We go on to assume that the very same fluctuations of the elastic energy stored in the arterial walls are monitored by the stretch receptors (in aorta and carotid sinus) and they are relayed with high fidelity to the central nervous system (see textbooks on physiology). As the order parameter for baroreceptor afferent excitation they transmit rhythmic inhibitory information to the central nervous system, which is highly correlated to the control parameter for blood volume in the hyperemic earlobe and hence can be "read" from the information obtained by our measuring system. It is in turn logical to assume that the information contained in arterial pressure fluctuations is not just a control parameter for the transfer of energy in the mechanical WKS, but for the flow of information in the central nervous system. This information is transduced into time series of electrical impulses which - by virtue of the specific circuitry in the bulbar centers - transmit the preferred modes to the brain stem via ascending and descending tracts, see below.

The cooperativity of mechanical and informational dynamic systems converts order parameters into control parameters for the flow of both information and mechanical energy. For example, we can expect that direct baroreceptor reflex activation of the vagal efferents to the heart reduces the import of energy into the WKS, whereas indirect reflex inhibition of the sympathetic nervous constrictor system enhances the export of energy. In this system, we can assume full cooperativity and highly non-linear response provided that spinal cardioaccelerator sympathetic neurons are coactivated with vasoconstrictor neurons. In the latter case, the contractile activity of the myocardium and smooth muscle in the arteriolar resistance vessels of the resting skeletal muscle, the intestine and the skin are coherently modulated: this mechanical cooperativity is ideally suited to **charge** the WKVs. As depicted in Fig. 3B, this kind of self-organization is restricted to phases of physical rest. It is essential to appreciate that there is also a different type of supraspinal "activation" of skeletomuscular vasoconstrictor neurons leading to **export** of energy: in the course of a supraspinal program for physical activity ("defense reaction"), the blood vessels in the skeletal muscle are relaxed and are highly perfused (see JÄNIG [6]). Since this is associated with cardioaccelerator neuron activation (enhanced cardiac output), energy **transfer** is augmented as a basis for so-called "ergotropic reactions" [7]. This situation is schematically depicted in Fig. 3A.

3. Energetics of the "Windkessel" Activity and Supplementary Non-linear Action of "Myogenic" Reactions

Upon initiation of cardiac pump activity the very construction of the WKS, in which a pulsatile pump, low resistance elastic container and high resistance outflow are put in series, automatically leads to the storage of a certain amount of the energy in it [9]. This kind of "self-organization" thus produces an arteriovenous pressure gradient, causing the circulatory motion of blood to the right atrium (and from there through the pulmonary circulation back to the left atrium) by energy import from the right ventricle whose contribution is henceforth neglected. As long as the import rate of energy and matter in the WKS (product of stroke volume, cardiac frequency and pressure, m^3 x 1/s x N/m^2, Nm/s) is higher than its export rate (Nm/s), volume and pressure are steadily increasing as is their product, potential energy, m^3 x Nm^{-2}, Nm. At equilibrium between import and export, i.e. when the power of cardiac pumping (flow rate x pressure) is identical to the power associated with export of kinetic energy and of frictional energy dissipation in the peripheral microvessels, the WKS is "boosted" by an energy content (0.5 x 10^{-3} m^3 x 133 x 10^3 N/m^2 = 66.7 N/m) equivalent to the work done in about 10 cardiac cycles (work of one stroke volume taken as 0.05 x 133 x 10^3 N/m^2. The pulsatile action of the left side of the heart thus leads to fluctuations in energy content of roughly 10%. Even without external control any analogue mechanical system will automatically "charge itself": it will be self-stabilizing only if the time averaged import and export rates of energy and matter become

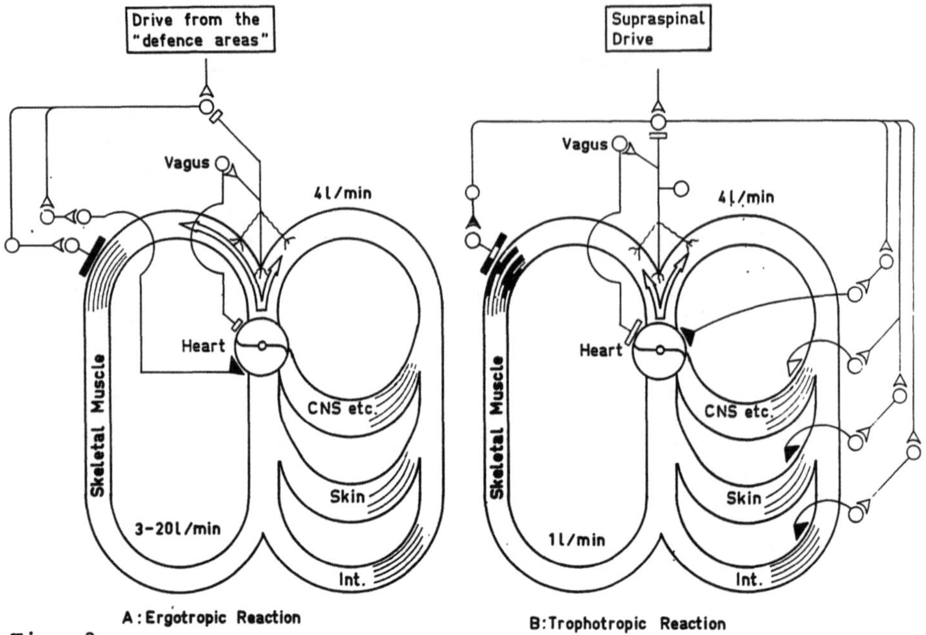

Fig. 3
Highly schematized representation of the mammalian Windkessel system consisting of a pump (heart), elastic high conductance conduits (arteries) and parallel outflow resistors (arterioles) representing the skeletomuscular bed (left side), the cutaneous, intestinal and other vascular beds (right side). Afferent baroreceptors directly activating the cardiac vagus and making inhibitory synaptic contact with arterioles are depicted. A) Representation of ergotropic reactions in the sense of [6], with a control program from the "defense area" for maximum energy transfer (activation of cardioaccelerator neurons, inhibitory innervation for skeletomuscular arterioles) leading to a strong increase in cardiac output and flow through the relaxed skeletomuscular arterioles. B) Representation of the trophotropic reaction with abolition of the defense reaction program and strongly reduced cardiac output. The conditions are ideally suited for energy storage in the WKS, with cooperative increments in energy import (cardioacceleration) and energy export (myogenic vasoconstriction in skin and intestine, neurally modulated constriction of arterioles in resting skeletal muscle). The synergetic WKS-CNS cooperativity with maximum coherence and information compression is assumed exclusively under these conditions, which prevail in physical rest.

equal. When, however, the import rate changes inversely to the export rate, the system is bound to be highly unstable. A crude model for this system, in which the pulmonary circulation is omitted for simplification, is depicted on the right side of Fig. 3.

Interestingly, the myocardium energy import and the arteriolar smooth muscle cells are endogenously capable of amplifying the WKS energy content of the "boosted WKS". The arterioles, as the energy export limiting structures are known to be "myogenically" activated, i.e. they automatically amplify their efficacy as export controllers as soon as they are extended by increases in transmural pressure (BASAR and WEISS [8]). Stretch-sensitive mechanisms enhance the contractile activity of arterioles, in the extreme to critical closure.

The so-called Frank-Starling mechanism, another "myogenic" reaction, acts cooperatively. In the myocardium, not only increased ventricular filling during the diastolic phase but also a residual volume after incomplete ejection automatically increases both the stroke volume and the tension developed during the subsequent systole. Speaking in cardiological terms, any increase in "afterload" (increase in arterial pressure and/or peripheral resistance) elicits this response, in essence acting as a built-in molecular amplifier for chemo-mechanical energy transfer.

There are built-in upper limits to this kind of cooperativity, namely the passive extension of myogenetically activated arterioles and the myocardium. At WKS pressures above the so-called "autoregulatory range" (roughly 180×10^3 N x m^{-2}) arterioles no longer contracted and the export rate rises in proportion to the energy content (flow rate in proportion to arterial pressure). Similarly, since there is an optimum of afterload (aortic pressure), under which the ventricle can do work (see textbooks on physiology), the mechanisms responsible for chemo-mechanical energy conversion are displaced from their functional optimum. Contraction against a resistance (and thence the performance of external work) requires a certain amount of overlap between the actin and the myosin components of the sarcomere. The two limits of the myogenic response combine in establishing a certain range of stability: energy import falls and export becomes constant as the WKS is strongly charged.

In the absence of external perturbations, the two myogenic control systems automatically establish a wide "window" of ordered behavior ("synergetic cosmos"), that is to say, the pressure dependence of both of these mechanisms widens the range of blood flow rates in which the system operates with high stability. Below this range, i.e. in all pathological "low flow states", the system is not capable of self-organization and can be said to be in "subsynergetic chaos" (see SCHMID-SCHÖNBEIN [5]). Above this range, i.e. in states of pathological cardio-vascular hyperactivity, there is evidence that it might go into suprasynergetic chaos (see below and [4, 10]).

4. External Control of the WKS: Synergetic Interpretation of Cardiovascular Control Effectors

Obviously, through the operation of the antagonistic cardiac nerves (vagus, sympathetic nerves) and through the antagonism between peripheral myogenic and metabolic reactions (e.g. the relaxation of the arteriolar smooth muscle by nitric oxide produced by the vascular endothelium (see BURNSTOCK [11]), it is possible to modulate either the transfer of energy (during physical activity) or its storage (during physical rest). We focus here on the latter situation: due to their action via inhibitory interneurons, the negatively fed back arterial blood pressure reflexes effectively "buffer" the "boosting action" of two myogenic systems alluded to above. This interpretation is in line with their primary role as "depressor reflexes". It is well known that the "pressor response" requires a substantial central "drive", which is easily abolished by brain lesions and/or drug action and is obviously strongly reduced in non-REM sleep (see below). Synergetically speaking, depressor reflexes attenuate the control parameter for the operation of the WKS (transfer rate of energy and matter) by their combined action in retarding heart rate (vagus effect) and reducing the tone of arteriolar constrictors in the resting skeletal muscle, and to a lesser degree in the skin and the intestine.

Again, we wish to stress that under all trophotropic reactions (i.e. when the nervous control apparatus for the so-called "defense reaction" is silenced and when the blood vessels in the skeletomuscular bed are in a state of low perfusion due to marked vasoconstriction), the sympathetic control of both boosting effectors acts in a strictly cooperative way, thence producing highly non linear mechanical effects. In terms of cybernetics, we take the resulting baroreceptor reflex as "down-regulation". In synergetic terms, we take it as information transfer establishing - under appropriate conditions - full cooperativity between the Windkessel system and the entire central nervous system. In the present first order approximation to a future mechanistic synergetic interpretation of the "enslaving mechanisms" of central nervous and cardiovascular control, we consider both the WKS and the CNS (here especially the sympathetic autonomous system) as a large boosted cooperative synergetic ensemble and as being continuously "driven".

The cooperativity between the WKS and the CNS can in fact be described in the unified semantic scheme of synergetics. The individual neurons as well as neuronal chains (or networks) operate as boosted self organizing cooperative electro-chemical systems [12]. We see them as generators of fluctuating order parameters for their own flow of energy and information and for the energy transfer rates in the mechanical system of the WKS.

Hemodynamic parameters and those attributable to nervous activity (potentials, ionic and transmitter flows) can all be discussed in terms of mechanical, electrical and chemical work or potential energy "stored" in the "boosted steady state" (Nm, mol, Coulomb) and/or power (Nm/s, mol/s, Coulomb/s). The very existence of a "resting" membrane potential" reflecting the cooperativity between ionic pumps, DONNAN equilibria and passive ionic leaks driven by metabolic expenditure, produces electro-chemical non-equilibrium by the appropriate combinations of fluctuating control parameters (e.g. metabolic rate, conversion rate of chemical energy to electrical and/or mechanical energy; see SIEGEL, these proceedings) they also become "boosted". The potential-dependent and oscillating ionic channels (i.e. the rapid Na^+ system, see textbooks on physiology) allow them to take up and amplify external information, thence they can be enslaved by fluctuating order parameters (acting by inducing phasic changes in the membrane conductance for small ions).

In polarized neurons, a "flow" of information occurs from the highly charged dendritic pole (continuously taking up mechanical and/or chemical information from its environment). The integrated information is transduced into a generator potential at the axon hill. By a process of amplitude-to-frequency modulation this triggers the generation of sequences of action potentials moving "down" the axons as a wave of "excitation". The travel without decrement along the axon is entropy driven, as is the activation of the neurosecretory pole and the transformation of electrical into "preformed" chemical information stored in vesicles and "released" to the synaptic cleft by yet another entropy-driven process (vesicle fusion after Ca^{++} influx) [12].

The qualitative effects of the different types of chemical free energy stored in synaptic vesicles on the postsynaptic membrane can vary substantially and this has pivotal effects on the flow of energy and information. The transmitters in the autonomic nervous system have dual effects (see BURNSTOCK [11], since they are 1) capable of transmitting information to the postsynaptic neuron (and/or effectors) in proportion to the product of the power (mol/sec) of transmitter flow and the time of action. 2) Transmitters also influence the susceptability of the postsynaptic membrane to chemical information reaching them, i.e. they modulate the transfer characteristics for conversion of chemical into electrical energy (and thence coherence between pairs of neurons). There are highly complex interactions of the multitude of transmitters acting in the central nervous system, which are based on the qualitative plurality of transmitter actions [10] as well as on complex anatomical connectivity.

From the standpoint of information transfer between synergetic subsystems, the following qualitative actions must be taken into account (see Fig. 4):

a) By isolated action of **excitatory** transmitters, the energy stored in the charged membrane of any presynaptic neuron is used to create steep chemical gradients in the synaptic cleft. Using preformed information channels, this allows strictly localized import of energy across the postsynaptic membrane. Kept far from electrochemical equilibrium, connections ("potential dependent gates") to deep attractor valleys for these ions in the postsynaptic membranes are "opened". As the threshold is reached, typical self-amplifying ("autoregenerative") processes result in explosive import of Na^{+} and Ca^{++} ions down a steep electrochemical gradient: the resulting oscillatory opening and closing of repetitively active channels allows highly efficient information transfer in a frequency modulated channel and initiation of a "flow" of trains of action potentials down the axon.

b) The isolated action of **inhibitory** transmitters on the postsynaptic membrane has the effect of increasing the electrical stability of the postsynaptic neuron. When the membrane potential is shifted to the electrochemical equilibrium for these ions, the potential is displaced - as it were - very close to the valley of a weak attractor for these ions. The resulting hyperpolarization, in turn, "displaces" the rapid Na^{+}-channels from their instability point (at the threshold) and thereby stabilizes the membrane. In the hyperpolarized state, the flow of information is interrupted, since the rapid Na^{+}-system is pushed far away from its threshold where enslaving sets in. Such a mechanism is ideally suited to control, as it were, the direction and the flow rate of electrochemically transported information in neuronal networks. It is interesting to speculate that the well-known "contrasting" mechanisms exerted by inhibitory interneurons upon excited ones might create specific "dissipative structures" in neuronal networks that act as transient "attractor networks" for "unidirectional flow", as well as for "circulating flows of information", as found in autonomic centers (see below).

c) **Simultaneous** action of **excitatory and inhibitory** transmitters on an individual postsynaptic neuron enhances the outflow of K^{+} and the simultaneous inflow of Na^{+} and Cl^{-}, reducing the order maintained by the ionic pumps at the resting state. Energetically

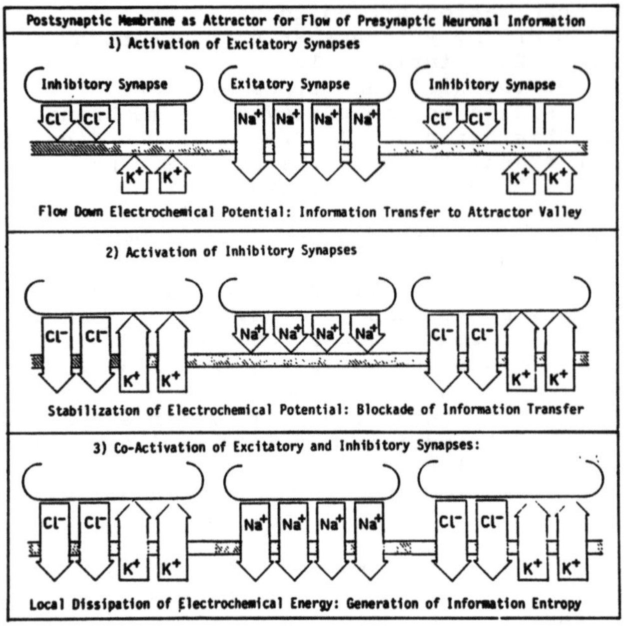

Fig. 4
Schematic representation of the synaptic transfer of energy and information by cooperation of excitatory and inhibitory synapses. 1) Exclusive activation of excitatory synapses: influx of Na^+ and nformation transfer by connection to postsynaptic attractor valley for Na displaced far from its electrochemical equilibrium. Effect: information transfer to postsynaptic membrane. 2) Exclusive activation of inhibitory synpase: stabilization of postsynaptic membrane by allowing K^+ and Cl^- to move to bottom of attractor valley (due to displacement of membrane potential to the respective electrochemical equilibrium). Effect: postsynaptic blockade of information transfer. 3) Simultaneous activation of excitatory and inhibitory synapses: simultaneous movement of Na^+, Cl^- (into postsynaptic neuron) and K^+ into synaptic cleft following the electrochemical gradients for the respective ions. Effect: dissipation of chemical energy in local entropy generation. (For details see text.)

speaking, the simultaneous generation of excitatory and inhibitory postsynaptic potentials leads to enhanced local electrochemical entropy generation due to strictly localized dissipation of chemical energy in a bidirectional flow of ions between any inhibitory (K^+ efflux, Cl^- influx) and any neighboring excitatory synapse (Na^+ influx) (see Fig. 4). Attractor basins are flattened, causing destruction of preexisting synergetic order simply by entropy generation due to local dissipation of preformed electrochemical gradients by antiparallel motion of K^+ and Na^+ and Cl^-. By this mechanism, the structural or functional connectivity between two synergetic systems forming an information transfer chain is temporarily destroyed [13].

For the interpretation of our results in awake humans, we can apply only with utmost caution (but not without justification) the increasingly clearer picture of the neurophysiology of the autonomic system as recently reviewed in a comprehensive videograph [14]. The autonomic nervous "drive", i.e. the flow rate of information down the sympathetic pathways from the central and medullar to the bulbar centers are generally assumed to be modulated by excitatory central psychic influences, by excitatory central respiratory influences, excitatory peripheral afferent influences and by inhibitory peripheral vascular influences ("depressor reflex"). From a synergetic point of view, three distinct types of autonomous neurons are predestined for limiting centrifugal flow of information by way of tonic inhibition of the sympathetic thoraco-lumbar neurons (energy flow through vagal limitation of cardiac pump activity).

1) The neurons in the rostral ventrolateral medulla, which combine rich multidirectional synaptic contacts with the capability of generating rhythmic activity (either as genuine pacemakers or as a partner in oscillatory networks), seem to be the bulbar relay station responsible for the characteristic 2-6Hz rhythm found in thoracolumbar intermediolateral preganglionic sympathetic neurons and peripheral postganglionic sympathetic axons (see GEBBER [15]). Being under tonic excitatory influence from supraspinal input and under phasic inhibitory influence from baroreceptor (and respiratory afferent) input, they are obviously not just capable of generating stable rhythms, but they act as a special band pass filter limiting the supraspinal as well as the propriospinal excitation of the spinal preganglionic neurons. It is now believed that these neurons are partners in an ensemble generating a "network rhythm" [15], where obviously excitatory and inhibitory influx can be readily integrated into circulating flows of information, the resultant of which is passed down as the 2-6Hz rhythm to spinal sympathetic neurons. From a synergetic point of view, GUYENET's concept of true individual brainstem pacemaker [16] cells is less plausible (see below).

2) Driven by supraspinal excitatory input at this mode (but also under depolarizing influence from other sources), the spinal preganglionic neurons display a characteristic limitation in discharge frequency (2-6 Hz). It appears that this is not caused by any specific biophysical peculiarity of the membranes, but must be taken as a consequence of tonic-descending inhibitory synaptic input. The rate limiting behavior appears to be significantly modulated by the duration of various forms of so-called "after-hyperpolarization". This phenomenon can be reduced in membranological terms to the presynaptic activation of multiple postsynaptic K+ channels (see CABOT [17]) (and therefore in synergetic terms to a synaptic entropy generation due to the postulated effect of dendritic leakage described above).

3) The neurons of the nucleus tractus solitarii, the relay station for afferent baroreceptor axons, for ascending and descending information from and to viscera and higher forebrain centers and, of course, to the vagal and sympathetic neurons by direct and indirect synaptic contact (see LOEWY [18]), possess a unique membranological machinery. As shown by MIFFLIN [19], sequential excitatory-inhibitory postsynaptic fluctuations can be generated, for which a specific chloride dependent hyperpolarization could be made responsible: conversion of coherent information into a random noise or informational entropy generation in major autonomic relay centers.

5. Synergetic Interpretation of Different Cardiovascular States: Maintenance of Coherent Order and Its Breakdown

Taking both the WKS and the descending sympathetic nervous control apparatus as independent, boosted and "self-organizing systems", we will henceforth refer to them as the synergetic WKS-CNS ensemble. We ascribe their operation to the action of a set of order parameters and control parameters driving the ensemble in a coherent state, placing them in their highly ordered working range (window of "synergetic cosmos") removed from either microscopic (infrasynergetic) or macroscopic (suprasynergetic or determinstic) chaos (see [5, 10] for details). Interestingly, in the cardiovascular system, this window establishes itself under conditions of physical rest. This may not be a coincidence, since the transition to physical rest is associated with the transition from the ergotropic state (and thence a situation with high transfer through the WKS) to the trophotropic state (high energy storage in the WKS). This, however, has not just quantitative, but pivotal qualitative consequences: in "re-recruiting" vascular tone in the skeletomuscular arterioles, there is not just less (and quantitatively highly variable) export of energy and matter from the WKS, but a highly modulated effector for energy export and thence energy content is coming into play. More specifically, with the restoration of skeletomuscular vascular tone, the one effector is brought into play that is most clearly governed by baroreceptor activity (via the influence of the latter on the phasic activity of the preganglionic skeleto-muscular vasoconstrictor neurons) and begins to establish cooperative action in boosting the WKS. As is well known from the work of the Uppsala group [2, 20, 21, 22], there is 1) a strong dependence of the activity of muscular vasoconstrictor neurons on inhibitory baroreceptor input, 2) a phasic oscillation of the neuronal activity (conspicuously with about 0.15Hz basic rhythm in man) and 3) a strict correlation with the slow changes in arterial pressure (again with the ca. 0.15Hz frequency in fully relaxed human subjects (see Fig. 5 in [2]). We are postulating that the resting state of the WKS-CNS ensemble is therefore governed by the equilibrium between phasic cardiac import of energy, and peripheral export through the cutaneous, the visceral and the tonically active skeletomuscular beds, of which the latter is modulated by a ca. 0.15Hz activity as a resultant of the baroreceptor-dependent sympathetic inhibition [20,21]. The "re-recruitment" of the skeletomuscular arteriolar vasomotion during the transition to physical rest is therefore seen by us as an event which completes synergetic cooperativity of the WKS-CNS ensemble, the components of which respond more and more coherently and thereby "enslave" each other. These, beginning to be subject to common order parameters and becoming mutually entrained, produce the phenomenon of "information compression", which in turn facilitates the establishment of stable synergetic order.

Our observation of this progressive "entrainment" into the very state with most stable correlation between sympathetic outflow and arterial pressure prompted the present work, and we accept the relaxed state as one of maximum functional coherence and minimum entropy generation ("synergetic cosmos"). The highly variable behavior seen under various other states of activity should not, however, be termed "chaotic" or attributed to "deterministic chaos" in the conventional sense (see POOL [23]) but represents a kind of desynchronization with more degrees of freedom and thence a higher requirement of informational expenditure for its adequate description. The system then utilizes or "recruits" more of its functional potentials. In the highly variable "ergotropic" states, there may also be a kind of destabilization, simply because the ratio of energy expenditure (power multiplied by time) and energy content (pressure multiplied by volume) is decreased; the system is a state of being boosted less stably.

Our experimental setting, which allows us to record the transition to a state with fewer and fewer degrees of freedom, made it possible to monitor very easily the state of optimal synergetic order. This, of course, is not the only stable state, since the system might be stable under the action of other order parameters as well. The robustness of the WKS-CNS ensemble proposed allows a wide range of synergetic liberty: in keeping with arguments put forward by BASAR [24], we refute the very idea of "chaos" in healthy physiological control systems; in keeping with HAKEN [4] we find it "counterintuitive" to speak about "informational entropy generation" in these states. The system has a wide spectrum of synergetic liberty, its potentials of the system should perhaps in the future rather be tested under a wide variety of test conditions uncovered by "spectral shift" as revealed in the transition from activity to rest.

In the same logic, we propose to accept as "spectral narrowing" the apparently regular order found in pathological states: for example that found in patients with surgically denervated hearts (heart transplant recipients) and with autonomic neuropathy (diabetics) or dysfunction (hypertensives) can then be quantified and taken as a sign of curtailed synergetic liberty (PAGANI et al. [25]) as reflected in "spectral narrowing" of the order parameter periodicities.

Accordingly, we can identify as "infrasynergetic or microscopic chaos" (see SCHMID-SCHÖNBEIN [10]) all quantitative deficits in energy transfer (various forms of cardiovascular shock due to pump failure, blood volume deficits, and loss of coherent control of peripheral resistance by spectral shifts (unpublished observation)). Conversely, true deterministic (or "suprasynergetic") chaos in the cardiovascular systems can be found under extreme natural or experimental conditions which are all highlighted by a loss of afferent and cerebral feedback control associated with persistance of efferent drive by the sympathetic neurons to the cardiovascular effectors. Since the classical experiments of COWLEY et al. [26], the effects of baroreceptor deafferentiation (and thence of irreversible loss of inhibitory control) are well known: the animals so treated can live orderly at rest but exhibit critical fluctuations of heart rate and arterial blood pressure when exposed to situations likely to enhance their central sympathetic drive. A strikingly similar effect is seen in patients and experimental animals suffering from high transsection of the spinal cord. In tetraplegic patients that have survived the so-called spinal shock phase, grossly exaggerated cardiovascular reactions to normally innoxious cutaneous peripheral stimuli ("autonomic dysreflexia", MATHIAS and FRANKEL [27]) are found, not just leading to blood pressure increase, but to extreme fluctuations in heart rate and blood pressure. These potentially life-threatening "chaotic reactions" are caused by autonomous and somatic spinal reflexes in the uninjured thoracolumbar sympathetic nervous system deprived of its physiological inhibition from supraspinal centers (severed from inhibitory reticulo-spinal tracts). When driven by either central input (in the deafferentiated dogs) or by peripheral afferent stimulation (in the tetraplegic individuals), uncontrolled drive to the fully functional heart and the resistance vessels produces a "dysreflectic" state similar to the clonic somatic motor reflexes found in these patients.

6. Interpretation of Experimental Results in the Steady State and in the Transient State

The measuring devices (photoplethysmograph, laser Doppler anemometer), data storage and computer assisted data processing are described in [1] and ZIEGE [28]: we obtained compressed time series about the fluctuations of cutaneous blood content and microvascular flow, of earlobe blood content and of heart rate. Log-frequency histograms for 15 min time windows allowed us to study either steady states or - in the majority of cases - the slow transitions between different states. This was important because the majority of records were taken when the subjects were "drifting" from physical activity to physical rest and sleep. As indicated above, these transients were faster when the subjects were also mentally relaxed (i.e. were assured about the non-invasive nature of our tests).

The prominence of the ca. 0.15Hz waves seen after full relaxation seems to have been overlooked in the past by investigators of the higher order of arterial blood pressure waves in awake human subjects. We offer as the cause of our regular success in virtually every subject simple psychological reasons (non-invasive technology and the progressive diminution of physical and mental activity of our subjects). The important influence of the time of relaxation can be seen in Fig. 5, where we continued to record when the ca. 0.15 Hz rhythm had transiently appeared and found a progressive increase in its power while the subject was still awake. The highest power of the ca. 0.15Hz rhythm was associated with the occurrence of the beat phenomenon, which in turn was very often associated with falling asleep. A typical example is depicted in Fig. 6, where events during a stable state of 15 min are recorded. These allow clearcut comparison to other physiological rhythms: close correlation of the ca. 0.15Hz rhythm to similar frequencies in the heart rate fluctuation can be seen. As an interesting observation, made quite often under these conditions, there are two bands in the 0.1Hz to 0.2 range of cardiac frequencies, in other words two incommensurate rhythms of very similar frequency which may well lead to wave interference and thus cause the conspicuous beat phenomenon found in the indirectly recorded arterial blood pressure waves.

The volatile nature of the ca. 0.15Hz rhythms has been mentioned repeatedly: once established, they are persistent and stable against voluntary respiration, during which they are maintained (although modified by rhythmic superposition in hypopnoe, hyperpnoe and apnoe (Fig. 7)). A spontaneous bout of laughing led to their immediate disappearance, but they were immediately restored once laughing stopped.

The rhythm can be abolished by all kinds of physical activity (e.g. upon passive tilt, extreme skin cooling test and inspiration of CO_2 (unpublished observations) that can be taken as inducers of "drive" in the CNS. The interesting transients found during an experimental setting with spontaneously occurring progressive reduction of central drive will be detailed elsewhere (SCHMID-

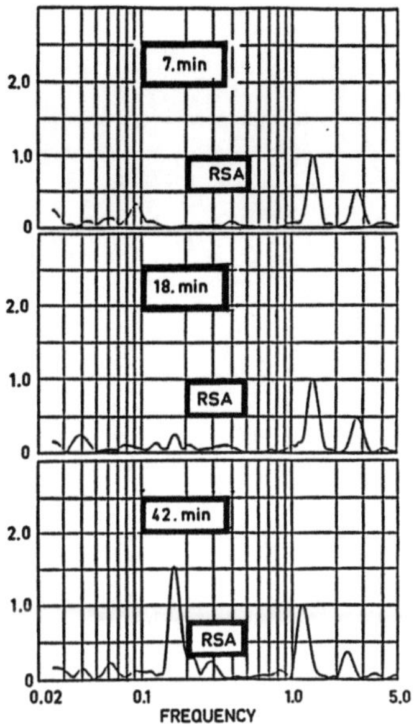

Fig.5

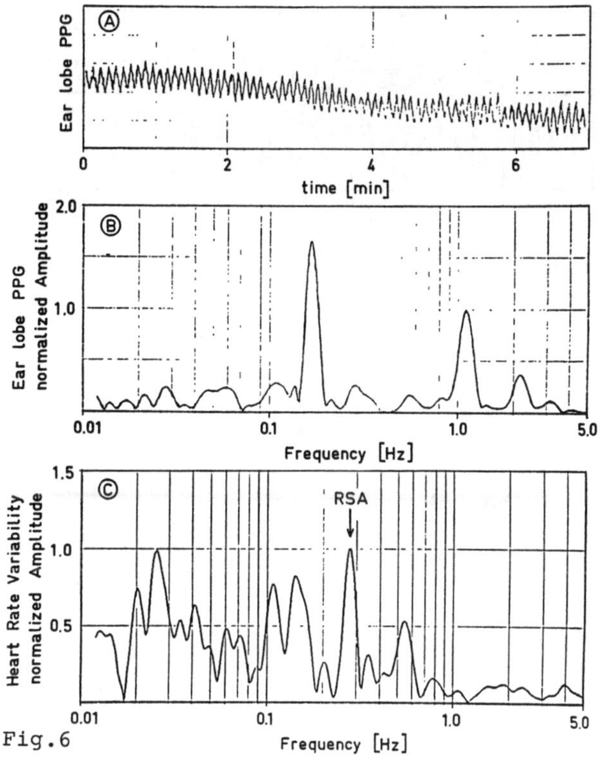

Fig.6

Fig. 5
Frequency distribution (amplitudes normalized for cardiac frequency) of earlobe blood content taken sequentially from the 1st to the 14th (7th min), from the 11th to the 25th (18th min) and from the 35th to the 49th minute (42th min) after first occurrence of ca. 0.15Hz waves in a relaxed human subject. Note the strong increase in the amplitude of the ca. 0.15Hz rhythm with time, associated with a shift in cardiac frequency to lower values.

Fig. 6
Frequency distribution of blood pressure waves and of heart rate variability during ca. 0.15Hz activity in full relaxation. A) Earlobe photoplethysmogram (indirect record of arterial blood pressure). B) Frequency distribution of arterial pressure fluctuations peaking at 0.16Hz. C) Frequency distribution of heart rate variability. Note two peaks between 0.1 and 0.2 Hz.

SCHÖNBEIN et al. [29]). Here, we just mention 1) the respiration-associated rhythm in blood pressure and heart rate first tends to appear during relaxation, whereas later the heart rate fluctuations at low frequencies (0.01 and 0.15Hz) become progressively more prominent; 2) at later stages there is a tendency in some - but not in all subjects - for the power of the ca. 0.25Hz rhythm to be reduced as the ca. 0.15Hz rhythms appear and grow (respiratory arrhythmia, see KOEPCHEN and PAGANI, these proceedings.

The stable rhythm found well below respiratory arrythmia might reflect some basal activity of the sympathetic nervous system as

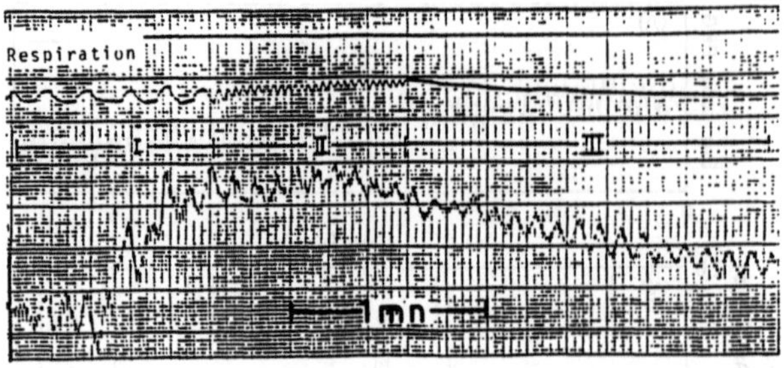

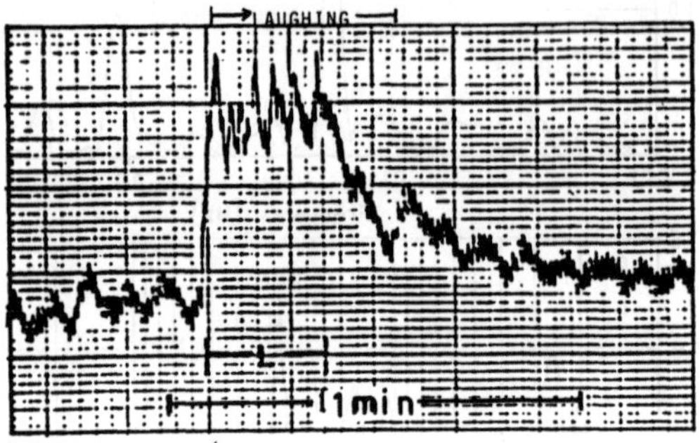

Fig. 7
Experimental alteration of established ca. 0.15Hz rhythm: upper panel: stability (but superposition) despite voluntary hypopmoe, hyperpnoe and apnoe.
Lower panel: Laughing: rapid disturbance and rapid restitution of ca. 0.15Hz rhythm.

proposed by PAGANI et al. [25]. However, the "injections of activity" that we find in the cardiac acceleration-deceleration and in the vasoconstriction (as seen by recordings of cutaneous perfusion) are not truly periodic but, instead, aperiodic (irregular intervals and asymmetric time course with rapid onset and slow release). The ca. 0.15Hz waves, however, are much more sinusoidal. We will address the question about the "nature" of such rhythms in terms of synergetics in the concluding section.

7. Summarizing Speculation: Blood Pressure Control as Synergetic Control; Information Compression Versus Feedback Control ?

The question of "cause" and "effect" in cardiorespiratory coordination has long been the topic of physiologists studying control of circulatory, cardiac and respiratory activity (see RICHTER and SPYER [30] and KOEPCHEN, these proceedings. There is increasing awareness that one can "explain" the observed phenomena only if one divorces oneself from strict positivistic linkages of the cardiorespiratory control system. Acceptance of "fluctuating" causalities in biology has its counterpart in the application of stringent synergetic principles as evidenced by this symposium. In synergetics, it has been derived as positive knowledge from first physical principles that even "circular causality" has a firm actual basis. We wish to apply HAKEN's theory [4] about self-organizing synergetic systems, which are not just uniquely capable of inducing coherent cooperativity of subsystems at the mesoscopic level, but also of inducing it to any number of other mesoscopic systems. The concepts outlined in this treatise start by accepting the validity of another fundamental rule of synergetics, namely that order is transmitted to the macroscopic system, and we derive from it the justification of monitoring arterial pressure fluctuations indirectly from the blood content of an organ made hyperemic. The present interpretation of our own experimental results and of the available literature shows that two apparently independent synergetic systems (WKS, sympathetic nervous control system) when driven to an intermediate non-equilibrium state of activity begin to "enslave" each other functionally. As they go into synergetic cooperativity, they can be shown to assume a stable state with maximum information exchange in only two modes. This is achieved by mutual entrainment into a preferred mode, a process that "compresses" information and thereby produces a very robust synergetic ensemble. However, it also has a more general effect on the whole organism, namely that of mimicking the well-known "sleep inducing effect", known for many centuries to be elicitable by massaging the carotid arteries (see MANCIA et al. [31]). We leave open the question about "cause or effect" and simply accept self-amplifying mechanisms. The synergetic interpretation of our findings allows us to state that we observed the subjective feeling of "relaxation" on the parts of the test persons and the objective fact of their falling asleep appears to be associated with an obvious state of information compression, allowing the transfer of a maximum of information into the one "channel" (the baroreceptor afferents) well-known to exert a maximum inhibitory influence all activities of the CNS.

Three facets of these observations deserve **physiological** emphasis: we monitor the ca.0.15 Hz rhythm in early non-REM sleep when 1) the cardiovascular system is known to be in a very stable steady state [31, 32] and when 2) the baroreceptor reflexes have been shown to have the highest gain [32]. 3) We can easily explain the coexistance of low arterial pressure and low cardiac frequency during non-REM sleep. This finding is diametrically contrary to the classical concept attributing control of arterial pressure and heart rate exclusively to negative feedback baroreflex regulation. We offer as an alternative explanation the synergic concept that the transition to the relaxed state studied by us has uncovered a range of "decrescendi" in control parameters (for informational flow and of energy flow into the WKS) which obviously allows optimal self-organization of the WKS-CNS ensemble by means of the mechanisms described in Sections 2,3 and 4 [33].

In the organism kept in complete physical rest, one can assume a minimum of sympathetic drive just sufficient to keep the energy content of the WKS sufficiently high to guarantee oxygen supply to the relaxing brain. Knowing that the descending influences upon sympathetic drive are also modulated by non-oxygen dependent mechanisms, e.g. the cerebral pCO_2-level, the drive from low pressure receptors and from thermoregulatory reflexes, we can explain that "stimulation" of any of these systems (e.g. by strong cooling and by CO_2-inhalation) shifts the system into a higher state of activity which abolishes the ca. 0.15 Hz rhythm.

At rest, however, high feedback "activity" due to information compression in afferent channels appears to activate bulbar and spinal **inhibitory** channels despite the fact that the arterial pressure is lowered. Here, a general rule established for dynamic systems in full synergetic coherence might be operational.

Based on the stringent theoretical concepts about "circular causalities" of general synergetics, we would like to address the "origin" of the low frequency fluctuations in autonomic nervous and/or effector "activity", here especially in the freqency range between 0.01Hz ("minute waves") and 0.25Hz (respiration-associated rhythms). PAGANI et al. ([25] and these proceedings) have presented convincing evidence that the cardiac rhythms in the above range are of "sympathetic origin. Their conclusions, in the last analysis based on pharmacological evidence, cannot be challenged. However, we feel it is of more than just academic interest to divorce oneself from such an unidimensional interpretation. Synergetics provide us with appropriate tools to interpret the experimental data in the literature and our own results as reflections of a dynamic system free to find its optimal "dissipative temporal order" in a preferred mode. We take the rhythms not as representing "activity states" but rather as representing the energetics of "processes". We are safe in assuming that the cooperativity of this "dissipative structuring of time" is governed by the same principle; of minimum energy expenditure as all other self-organizing dissipative systems studied so far. Coherent structuring of the temporal sequentiality of processes in highly different systems follows from the enslaving principle, but can also be taken as an indication that in each subsystem a principle of minimum energy expenditure is functional. Questions about "cause" and "effect" obviously lose their significance and are replaced by the question about the possibility of various optima at states of activity other than the resting one studied in the present experiments. Experiments to this end are in progress.

References

1. K. Dittmers: Diplomarbeit, Fachhochschule Hamburg, 1988.

2. G. Sundlöf, B.G. Wallin: J.Physiol. 274, 621-637 (1978)

3. In animal experiments with non-anaesthetized sheep (not shown) a similar arterial blood pressure rhythm can be seen when the animals are at rest and is abolished completely by the procedure of extracellular CO_2-removal, which renders the animals apnoic.

4. H. Haken: Information and Self-Organization. A Macroscopic Approach to Complex Systems (Springer, Berlin, Heidelberg, New York, London, Paris, Tokyo 1988)

5. H. Schmid-Schönbein: In: W. Bleifeld et al., Eds. Unstable Angina (Springer, Berlin, Heidelberg 1990), pp. 16-51

6. W. Jänig: Ann. Rev. Physiol. 50, 525-539 (1988)

7. W.R. Hess: Acta, Suppl. IV (1947)

8. E. Basar, Ch. Weiss: Vasculature and Circulation. The Role of Myogenic Reactivity in the Regulation of Blood Flow (Elsevier, North-Holland Biomedical Press Amsterdam, New York, Oxford 1981)

9. For a future synergetic description of the cardiovascular system and its subsystems, many more cellular, nervous, metabolic and energetic details must be filled in. For reasons of spatial constraint, the synergistic role of only a few more dynamic links can be mentioned. While capillaries can be taken as a purely passive system, the venous return is probably a pivotal dynamic link. Its behavior limits - via the diastolic filling of the right and left atrium and ventricle - the continuous re-import of energy and matter into the WKVs. We reduce the contribution of the low pressure system (and that of the control systems for it) to a principle of energy conservation: it is assumed that the kinetic energy of the systemic and pulmonary venous blood is converted into the mechanical work required to do frictional work in extending the relaxed contractile apparatus of the right and left ventricle (a topic beyond the scope of the present treatise).

10. H. Schmid-Schönbein: In: J.L. Vincent, Ed. Update in Intensive Care and Emergency Medicine (Springer, Berlin, Heidelberg, New York, London, Paris, Tokyo, Hong Kong 1990), pp. 3-21

11. G. Burnstock: Acta Physiol. Scand. 126, 67-91 (1986)

12. The quantal release of neurotransmitters is taken as a "transport" of chemical signal down a preformed chemical gradient. The postsynaptic sequelae of transmitter action are assumed to modulate the import of energy (and informational order) from one "exited" into any subsequent neuron linked by a synaptic "funnel of information" to the former by an exchange of free energy. This follows from the fact that in any presynaptic apparatus, electrical energy pulses arriving in the form of action potentials are converted to chemical energy and back to electrical energy. Transmitter quanta are considered to result from chemical work done in the resting periods and as a reservoir of free energy (which is more appropriately called "convertible" or "convectible" energy).

13. The extremely complex architecture of the reticular formation, in addition to the complexity of transmitter actions opens up the possibility that in the autonomous nervous system, due to the variable activation of excitatory and inhibitory neurons, information channels are far less "solidified" than in the somatic nervous system and that their ever changing plasticity in fact reflects their operation as a specific kind of "dissipative structure" (see contribution KOEPCHEN).

14. A.D. Loewy, K.M. Spyer: Central Regulation of Autonomic Functions (Oxford University Press 1990)

15. G.L. Gebber: In: K.M. Spyer Central Regulation of Autonomic Functions (Oxford University Press 1990), pp. 126-146

16. P.G. Gyyenet: In: K.M. Spyer Central Regulation of Autonomic Functions (Oxford University Press 1990), pp. 147-167

17. J.B. Cabot: In: K.M. Spyer Central Regulation of Autonomic Functions (Oxford University Press 1990), pp. 52-87

18. A.D. Loewy: In: A.D. Loewy, K.M. Spyer, Eds. Central Regulation of Autonomic Functions (Oxford University Press 1990), pp. 67-89

19. S.W. Mifflin: J. Physiol. 399, 349-367 (1988)

20. B.G. Wallin, D.L. Eckberg: Am. J. Physiol. 242, H185-H190 (1982)

21. E. Bath, L.E. Lindblad, B.G. Wallin: J. Physiol. 311, 551-564 (1981)

22. J. Fagius, B.G. Wallin, G. Sundlöf, C. Nerhed, S. Englesson: Brain 108, 423-438 (1985)

23. Pool, R.: Science 243, 604-608 (1989)

24. E. Basar, Ed.: In: Chaos in Brain Function (Springer Berlin, Heidelberg, New York, London, Paris, Tokyo, Hong Kong 1990), pp. 1 - 30

25. M. Pagani, F. Lombardi, S. Guzzetti, O. Rimoldi, R. Furlan, P. Pizzinelli, G. Sandrone, G. Malfatto, S. Dell'Orto, E. Piccaluga, M. Turiel, G. Baselli, S. Cerutti, A. Malliani: Circ. Res. 59, 178-193 (1986)

26. W. Cowley, J.F. Liard, A.C. Guyton: Circ. Res. 32, 564-565 (1973)

27. C.J. Mathias, H.L. Frankel: Ann. Rev. Physiol. 50, 577-92 (1988)

28. St. Ziege: MD Dissertationsschrift Rheinisch-Westfälische Technische Hochschule Aachen 1991

29. H. Schmid-Schönbein, W. Rütten, H. Heidtmann (manuscript in preparation)

30. D.W. Richter, K.M. Spyer: In: A.D. Loewy, K.M. Spyer Central Regulation of Autonomic Functions (Oxford University Press 1990), pp. 189-207

31. G. Mancia, A.L. Mark: In:.J.T. Shepherd, F.M. Abboud, Eds. Handbook of Physiology, Sect. 2 The Cardiovascular System (American Physiological Society, Bethesda, Maryland 1983)

32. H.S. Smyth, P. Sleight, G.W. Pickering: Circ. Res. 24, 109-121 (1969)

33. We see a striking analogy with the synergetics of motor systems as studied by KELSO [34]. In order to go into functional coherence of bilateral spinal motor neuron pools, KELSO comes from below and has to increase the supraspinal "drive". We reach the synergetic "window" of the slowly firing thoraco-lumbar preganglionic neurons by allowing the supraspinal drive to reduce itself (which occurs best once information compression into the inhibitory afferents sets in).

34. J.A.S. Kelso: Bull. Psychon. Soc. 18, 63 (1981)

Spectral Analysis of Circulatory Rhythms and Baroreflex Sensitivity in Man

N. Honzíková and B. Fišer
Masaryk University, Faculty of Medicine, Department of Physiology,
Komenského nám. 2, 662 43 Brno, Czechoslovakia

Abstract. The spectral analysis technique was applied for noninvasive assessment of heart-rate baroreflex sensitivity. The modulus (or gain) of the transfer function between variations in blood pressure and heart rate was calculated. During voluntarily controlled breathing intervals (3, 4, 5, 6, 8, 10 and 17 s), the amplitude of 0.1 Hz and respiratory peaks in spectra of heart rate and blood pressure changed markedly. Nevertheless, the average sensitivity of the baroreflex (modulus) changed insignificantly. The difference between the modulus at 0.1 Hz and at breathing rate indicates that baroreflex is only one of the factors causing respiratory arrhythmia.

1. Introduction

Spectral analysis was introduced into cardiovascular research in the sixties [1, 2, 3, 4]. Already in these studies the power spectra of several parameters (heart rate, plethysmogram and respiration) were compared, and later they were correlated together with blood pressure by means of cross-spectral analysis [5, 6], after Peňáz invented his noninvasive method of blood-pressure measurement in 1969 [7]. Usually, spectral peaks are divided into frequency ranges which are given by a different origin of corresponding fluctuations of circulatory parameters: 0.03-0.06 Hz (2-3.5 cpm) - low frequency range, 0.07-0.12 Hz (4-7.5 cpm) - 10-s rhythm and 0.13-0.50 Hz (8-30 cpm) - respiratory waves. Previously, the role of respiration in the origin of fluctuation in circulation was limited to this last range. But in 1980 [8], we found a new, previously undescribed relationship between the frequency and depth of respiration on the one hand, and the low-frequency circulatory waves on the other (frequency range 0.03-0.06 Hz, i.e. 2-3.5 cpm). Later we confirmed this finding by means of a coherence function. Thus it was demonstrated that the relationship between respiration and the regulation of circulation is more important than originally supposed. Additional evidence supported our idea. We found that the changes in circulatory rhythms during mental load [9] or during isometric exercise [10] correlate with the variations in respiration better than with any tested load.

In 1988, Mulder [11] applied the spectral analysis technique to the noninvasive assessment of heart-rate baroreflex sensitivity. The modulus (or gain) of the transfer function between variations in blood pressure and heart rate was calculated in the range of the 10-s rhythm. The modulus expresses the ratio between changes in RR intervals and changes in systolic pressure (ms/mmHg) in a specified frequency band. Therefore the modulus function in the frequency domain is comparable to the regression coefficient in the time domain. The modulus can be calculated only in the range where the coherence is high. It becomes unreliable if the coherence is low. The high coherence between blood pressure and heart-rate fluctuation at the 0.1-s rhythm and at respiratory rhythm was first observed in our laboratory [5, 6]. The values of baroreflex sensitivity assessed with the phenylephrin method and by means of spectral analysis correlate highly [11]. Thus Mulder invented a new method for the determination of baroreflex sensitivity on the basis of spectral analysis.

For man, two methods for testing baroreflex sensitivity we can take today as classical: neck suction as used by Eckberg's group [12, 13, 14] and the administration of phenylephrin [15, 16]. Both these methods measure only one part of the baroreflex - the change in the cardiac interval after arterial baroreceptor stimulation. This change is measured quantitatively. Phenylephrin causes a rise in blood pressure and the baroreflex mediates a slowing of the heart rate. The successive systolic pressure values are correlated with the length of the following RR interval. The regression coefficient of the correlation diagram gives the baroreflex sensitivity in milliseconds of interval prolongation per millimeter of mercury of pressure rise. Although these methods brought much information, they have some disadvantages. We cannot get continuous data from them.

An important question is the relationship between baroreflex sensitivity and respiration. Earlier studies conducted on animals have shown interactions between respiration and baroreceptor reflex responses [17]. It is also known that an important mechanism of the respiratory arrhythmia is a change in sensitivity of the sinus baroreceptor which occurs synchronously with the respiratory cycle. Sinus-node responses to baroreceptor stimulation are smaller when stimuli are delivered during inspiration than when delivered during expiration. They are simultaneously dependent upon the depth of respiration, which is evidence for a central respiratory-baroreflex interaction [14].

In this paper, we have used spectral analysis to answer three questions: Does baroreflex sensitivity depend on frequency of breathing? Is there a correlation between baroreflex sensitivity and the amplitude of the 0.1 Hz peak? Is it possible to differentiate between baroreflex-mediated response of the heart rate to the blood-pressure waves associated with respiratory movements and between other factors causing respiratory arrhythmia?

2. Methods

By means of modulus, we assessed the baroreflex sensitivity in 8 healthy adult subjects at various breathing frequencies. The blood pressure (by means of the Peňáz volume-clamp method) and cardiac intervals were recorded with the subject in a sitting position at spontaneous quiet breathing and during voluntarily-controlled breathing intervals. Duration of breathing intervals in experimental sessions was: 3, 4, 5, 6, 8, 10 and 17 s. The sequence of experimental periods was randomly chosen for each subject. The tidal volume was not controlled -- the subjects breathed according to their need. Examples of the analysis are seen in Fig. 1.

2.1 Baroreflex Sensitivity at Various Respiratory Rates

We observed various amplitudes of the 0.1 Hz peak in spectra of cardiac intervals at various breathing rates and despite this the modulus remained relatively constant (an example in Fig. 2). Also despite the diminution of the sinus-arrhythmia amplitude with an increase of the breathing rate, the modulus at the respiratory frequency was practically independent of the breathing rate. A small decrease of the modulus at a breathing rate of 15 and 20 cpm (0.25 and 0.33 Hz) is insignificant. Figs. 3 and 4 demonstrate the average values and standard deviations of power in the heart-rate spectrum and of the modulus at both frequencies (res-

Fig. 1

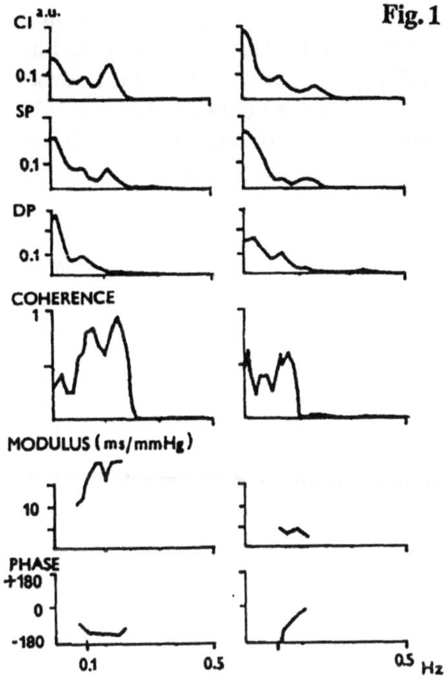

Fig. 2

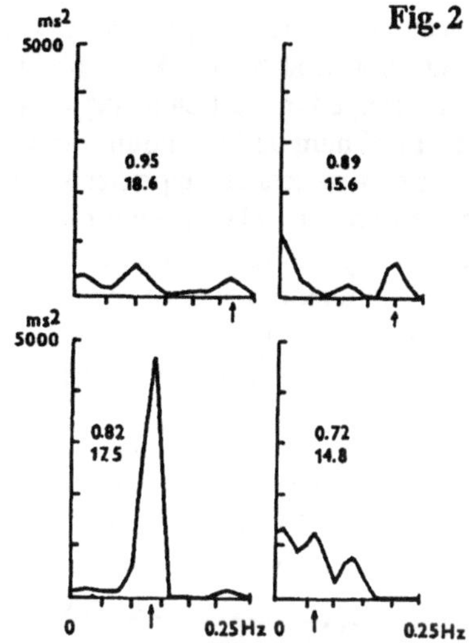

Fig. 1. Power spectra of spontaneous fluctuations in cardiac intervals (CI), systolic (SP) and diastolic pressure (DP), a.u. - arbitrary units. Coherence, modulus and phase angle determined for spectra and cross- spectra of CI and SP. Left and right - two different subjects.

Fig. 2. Spectra of cardiac intervals of one subject at 4 different respiratory rates (indicated by arrows). Coherence and modulus in ms/mmHg between cardiac intervals and systolic blood pressure at 0.1 Hz in each period of experiment are presented.

Fig. 3

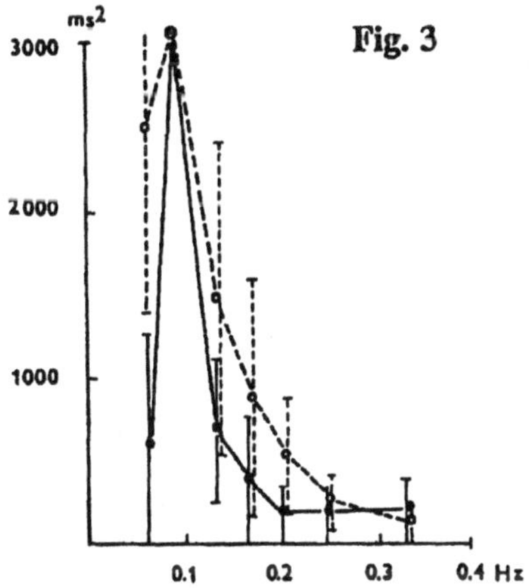

Fig. 4

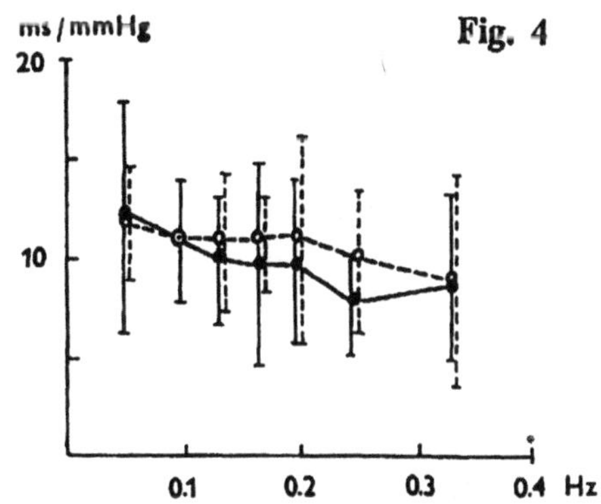

Fig. 3. The amplitude of the 0.1 Hz peak (dots) and of the respiratory peak (circles) in spectra of cardiac intervals at various respiratory rates (abscissa). Mean values and standard deviations of 8 subjects.

Fig. 4. Modulus between cardiac intervals and systolic blood pressure expressed in ms/mmHg at 0.1 Hz (dots) and at respiratory frequency (circles) at various respiratory rates (abscissa). Mean values and standard deviations of 8 subjects.

piratory and 0.1 Hz) in each experimental period. The ratio between maximum and minimum power at the 0.1 Hz peak in the spectrum of cardiac intervals at various breathing rates ranged in particular subjects between 2.31 and 57.24. Alternatively, the ratio between the maximum and minimum value of the modulus at 0.1 Hz ranged between 1.16 and 3.15. This contradicts the hypothesis that the frequency and amplitude of heart-rate oscillations depend on the relative gains of the elements that modulate cardiac vagal, cardiac sympathetic, and peripheral sympathetic efferent activity [18].

The phenomenon of resonance at the breathing frequency near the 10-s rhythm, described earlier [19, 20], manifests itself in a great increase in amplitude of the 10-s rhythm. It was present in half of the subjects. On the other hand, modulus was not influenced by resonance. The explanation could be in a primary increase in blood pressure oscillations and unchanged sensitivity of baroreflex control of the heart-rate frequency.

There is a question whether the modulus at respiratory frequency corresponds to the modulus at 0.1 Hz. The correlation between these two values of modulus is positive, but insignificant ($R = 0.58$). A significant difference between the average value of modulus at 0.1 Hz and at the breathing rate, calculated for each subject for all the experimental periods, was found in 5 subjects (in 3, $P < 0.05$; in 2, $P < 0.01$; t-test). It is necessary to note that, in 4 subjects, the modulus at 0.1 Hz was higher than at respiratory frequency and in 4 subjects an opposite relationship was observed. These facts indicate that baroreflex is only one of the factors causing respiratory arrhythmia at rest and that additional factors play a role.

2.2 Discussion

Several theories explain the role of the baroreflex in low- frequency waves in blood pressure. The most popular theory was introduced by Sayers [21]: the spontaneous feedback rhythm is centered at the frequency of 0.1 Hz due to time constants and delay in feedback circuits. This theory is accepted by most of the authors which is documented also by a recent overview [18]. The changes in amplitude of the 10- s rhythm accompanying changes in the respiratory rate are undoubtedly associated with resonance or synchronization [19, 20] or with "entrainment" [22]. The resonance phenomenon at frequencies of 0.03-0.09 Hz was also described in blood flow in a. mesenterica which reflects the resistance of this vascular area [23].

In our opinion, the most plausible theory was suggested by Wesseling et al. [24], which is based on two facts. The smooth muscle of blood vessels is not able to follow rapid changes of vasomotor nerve activity. Therefore the oscillations of peripheral resistance at frequencies higher than 0.066 Hz are attenuated [25]. The low-frequency oscillations of peripheral resistance in some regions are compensated by changes in other regions by means of the baroreflex. Thus slow and fast oscillations are eliminated and the 0.1 Hz peak remains. In this sense, the origin of this peak is in the baroreflex but the 0.1 Hz peak itself is not caused by lags in the feedback mechanism. This is important because according to this theory the magnitude of this peak does not depend on the sensitivity of baroreflex. On the other hand, the oscillation of blood pressure elicits changes of cardiac intervals and the modulus between systolic pressure and cardiac intervals at 0.1 Hz corresponds to the baroreflex sensitivity expressed in ms/mmHg. Baroreflex sensitivity at various respiratory rates remains constant.

An additional practical conclusion resulted from this analysis: In studies in which sympathovagal activity is determined from the amplitude of the 10-s and respiratory spectral peak, the respiratory frequency should be taken into account. It was proved in this study that the amplitude of the 10-s rhythm is dependent on respiratory frequency not only at a respiratory frequency of 6 cpm, where the respiratory and the 10-s rhythms merge into one peak. It is also dependent on respiration at breathing frequency lower than 0.2 Hz (12 cpm).

Acknowledgement. The authors thank Ir. K. H. Wesseling, who made possible the gathering of the data at BMI-TNO in Amsterdam, and B. de Wit and J. J. Settels for technical cooperation during experiments.

References

1. Peňáz J, Fišer B (1966) Amplitudové spektrum spontánních vln plethysmogramu a tepové frekvence u člověka (Amplitude spectrum of spontaneous fluctuations in the plethysmogram and heart rate in man). Čs fysiol 15:517
2. Taylor MG (1966) Use of random excitation and spectral analysis in the study of frequency-dependent parameters of the cardiovascular system. Circul Res 18:585-595
3. Peňáz J, Roukens J, vd Waal HJ (1967) Korelační a spektrální analysa spontánních rytmů v plethysmogramu a tepové frekvenci u člověka. Čs fysiol 16:259
4. Peňáz J, Roukens J, vd Waal HJ (1968) Spectral analysis of some spontaneous rhythms in the circulation. In: Biokybernetik I. Karl-Marx-Univ, Leipzig, pp 233-236
5. Peňáz J, Honzíková N, Fišer B (1978) Spectral analysis of spontaneous variability of some circulatory parameters in man. Physiol bohemoslov 27:349-357
6. Fišer B, Honzíková N, Peňáz J (1978) Power spectra of spontaneous variations of indirectly recorded blood pressure, heart rate and acral blood flow. Automedica 2:143-147
7. Peňáz J (1969) Czechoslovak patent no. 133205
8. Honzíková N, Fišer B, Peňáz J (1980) Relationship between power spectra of respiration and of some circulatory parameters in man. In: Proc. of the Internat. Union of Physiol. Science. XXVIII. Internat. Congress, Budapest, p 477
9. Honzíková N, Peňáz J, Fišer B (1988) Power spectra of blood pressure and heart rate fluctuations during mental load. J Interdiscipl Cycle Res 19:75-59
10. Honzíková N, Peňáz J, Fišer B (1987) Interpretation of differences in power spectra of blood pressure, heart rate and respiration in man. In: Hildebrandt G, Moog R, Raschke F (eds) Chronobiology and chronomedicine. Peter Lang, Frankfurt/Bern/New York/Paris, pp 172-176
11. Mulder LJM (1988) Assessment of cardiovascular reactivity by means of spectral analysis. Proefschrift, University Groningen
12. Eckberg DL, Cavanaugh MS, Mark AL, Abboud FM (1975) A simplified neck suction device for activation of carotid baroreceptors. J Lab Clin Med 85:167-173
13. Eckberg DL, Kifle YT, Roberts VL (1980) Phase relationship between normal human respiration and baroreflex responsiveness. J Physiol (Lond) 304:489-502
14. Eckberg DL, Orshan QR (1980) Central respiratory-baroreflex interaction in man. In: Koepchen HP, Hilton SM, Trzebski A (eds) Central interaction between respiratory and cardiovascular control systems. Springer, Berlin- Heidelberg-New York, pp 206-211
15. Smyth HS, Sleight P, Pickering GW (1969) Reflex regulation of arterial pressure during sleep in man: a quantitative method of assessing baroreflex sensitivity. Circ Res 24:109-121

16. Brooks D, Fox P, Lopez R, Sleight P (1978) The effect of mental arithmetic on blood pressure variability and baroreflex sensitivity in man. J Physiol (Lond) 280:75P-76P
17. Koepchen HP, Lux HD, Wagner PH (1961) Untersuchungen über Zeitbedarf und zentrale Verarbeitung des pressoreceptorischen Herzreflexes. Pfluegers Arch 273:413-430
18. Saul JP (1990) Beat-to-beat variations of heart rate reflect modulation of cardiac autonomic outflow. NIPS 5:32-37
19. Golenhofen K, Hildebrandt G (1962) Zur relativen Koordination von Atmung and Bludruckwellen dritter Ordnung. Z für Biologie 112:451-458
20. Peňáz J (1957) Oscillations de la fréquence cardiaque et du tonus vaso-moteur au cours de respiration ralenti et accélérée. J Physiol (Paris) 49:346-349
21. Sayers BMcA (1973) Analysis of heart rate variability. Ergonomics 16(1):17-32
22. Hyndman BW (1974) The role of rhythms in homeostasis. Kybernetik 15:227-236
23. Peňáz J, Buriánek P, Semrád B (1968) Dynamic aspects of vasomotor and autoregulatory control of blood flow. In: Hudlická O (ed) Circulation in Skeletal Muscle. Pergamon Press, Oxford
24. Wesseling KH, Settels JJ, Walstra van Esch HJ, Donders JJH (1983) Baromodulation as the cause of short term blood pressure variability? In: Alberi G, Bajzer Z, Baxa P (eds) Proc of the Internat Conf on Applications of Physics to Medicine and Biology. Singapore, pp 247-276
25. Peňáz J (1970) The blood pressure control system: A critical and methodological introduction. In: Koster M, Musaph H, Visser P (eds) Psychosomatics in Essential Hypertension. Karger, Basel-München-New York, pp 125-150

Spectral Analysis of Cardiovascular Variables as a Tool to Quantify Neural Cardiovascular Control in the Laboratory and Real Life Conditions

M. Pagani, O. Rimoldi, P. Pizzinelli, D. Lucini, and A. Malliani

Instituto Ricerche Cardiovascolari, CNR, Patologia Medica, Ospedale L. Sacco, Centro Fidia, Università di Milano, Italy

The neural control of the circulation depends upon a continuous interaction between opposite, i.e. negative and positive, feed-back mechanisms, mediated respectively by vagal/baroreceptive and sympathetic circuits.

The assessment, with a spectral analytical approach, of the short term high frequency, respiratory (HF), and 0.1 Hz, low frequency (LF) oscillations of heart period and arterial pressure variabilities provides indices of vagal and sympathetic modulatory activities, and of their balance.

This contention is based on human and experimental investigations which indicated that, under stationary conditions, LF of R-R interval variability is an indicator of sympathetic activity, HF of vagal activity, and LF/HF of sympathovagal balance.

This approach may furnish important insights into the role of neural mechanisms in cardiovascular pathophysiology.

1. Introduction

In recent years the view of the organization of the neural control of the circulation has undergone a radical change: passing from a relatively simple schema, to a more complex and dynamic picture, largely based on multifarious continuously interacting mechanisms.

The former view probably largely arose from experiments on anesthetized animals, such as those performed around 1930 by Heymans [1], who described the importance from baroreflex mechanisms in controlling the cardiovascular system. From the results of those experiments, the concept was derived from the overwhelming importance of inhibitory reflexes, mediated by supraspinal afferents, and characterized by negative feed-back properties [2].

Although this concept is still the leading one in the interpretations of many investigations on anesthetized animals, it does not take into account several new findings, which have been gathered more recently, particularly in studies performed on conscious animal preparations, in which the confounding effects of anesthesia and recent surgery, which are known to alter drastically cardiovascular neural control mechanisms [3], are avoided. These latter studies have highlighted the pathophysiological importance of a different set of cardiovascular control mechanisms that are based on sympatho-sympathetic circuits [4,5].

These circuits operationally appear to subserve positive feed-back mechanisms [6].

In brief, mechanical and chemical stimuli acting on sympathetic afferent fibers [7] have been shown to mediate excitatory cardiovascular reflexes. These responses, though rather blunted in anesthetized animals [8], where sympathetic tone is already elevated, are instead particularly evident and easy to obtain in conscious fully

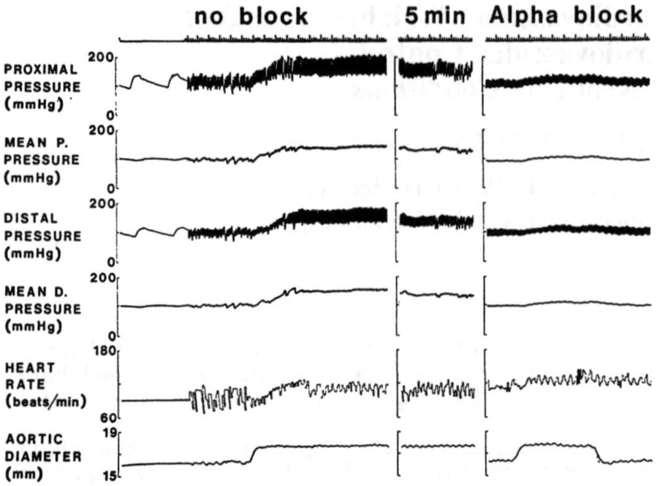

Fig.1 Example of the reflex sympathetic response to a mechanical distension of the descending thoracic aorta in a conscious dog. Note the marked increase in aortic pressure and in heart rate during the distension, which is indicated by the increase in aortic diameter. (From [9])

innervated animal preparations [9,10], where the basal sympathetic tone is low.

Just to provide an example, Fig.1 shows the reflex increase in arterial pressure and in heart rate produced in a conscious dog by a mechanical distension of a segment of the descending thoracic aorta in the absence of any pain reaction [9]. From the point of view of the neural control of the circulation, this stimulus, which mimicked on the aortic walls the effects of a physiological rise in pressure (which would mechanically distend its walls) [11], induced a further rise in the controlled variables (i.e. arterial pressure and heart rate), thus accounting for the positive sign in the operational schema of the feed-back circuit of Fig.2. This schema indeed emphasizes the continuous interaction between the positive (sympathetic) and negative (vagal/baroreflex) control mechanisms [6], and also points out the danger of singling out a given mechanism from the entire system in order to assess its contribution to the overall performance.

From a physiological and clinical point of view, however, the problem of inferring the properties of specific components of this complex control system is a very important one. In practice, the use of highly simplified experimental approaches, such as those offered by the model of the simple reflex, viewed as if they were operating independently of the rest of the system and following a simple linear causality, has been very fruitful. However, we should also consider, on the other hand, the practical consequences of making inferences about the overall performance of the system. In both cases, we might have to accept that, at least in biology, the myth of complete knowledge is unattainable, and take a more probabilistic approach [12]. In so doing, however, we will be able to make reasonable predictions of the system, based on all available knowledge: a typical synergetic approach [13].

In the following pages, we will briefly outline a spectral analytic approach that we have developed in the last decade, and which provides quantitative inferences about the dynamic balance between the sympathetic and vagal nerve activities controlling the

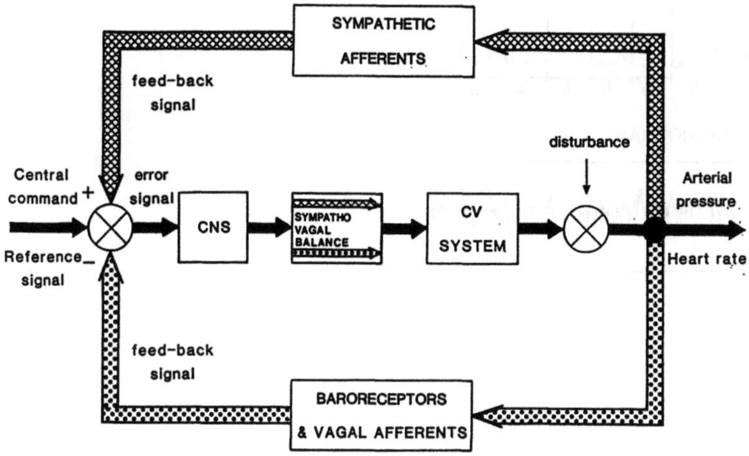

Fig.2 Schematic representation of the neural model of neural cardiovascular control, depicted as a continuous interaction between opposite, positive and negative, feed-back mechanisms. (From [6])

cardiovascular system, starting from simple non-invasive (or minimally invasive) measurements, such as the ECG, and systemic arterial pressure [14].

2.Methodology

Figure 3 schematically depicts the procedure of spectral analysis of R-R variability, as performed in our laboratory [14]. A similar procedure is employed in the case of other signals, such as arterial pressure or respiration.

From the original signal, and after appropriate acquisition, the computer program detects consecutive R peaks, and stores in the memory the series of successive R-R intervals as the tachogram. The length of the tachogram used for the subsequent spectral analysis is a compromise with the theoretical requirements of absolute stationarity, which is unknown to biology.

Various algorithms can be used to assess the possible rhythmicity hidden in the signal. We routinely use an autoregressive approach, which, we feel, has several potential advantages: it provides automatically the number, power and center frequency of spectral components, thus avoiding the need to preselect bands of interest. Furthermore, since autoregressive algorithms operate efficiently also on short segments of data, their use makes it possible to analyze shorter series, which are more likely to be stationary.

The spectrum of Fig.3 contains three components, with frequency centered, respectively, at 0.00 Hz (component 1), 0.09 Hz (component 2) and 0.24 Hz (component 3).

The study of very low frequency (0.00-0.03 Hz) components requires specific methodologies and algorithms [15] and is not addressed with the present approach.

Components 2 and 3, labelled respectively LF and HF, are evaluated in terms of frequency and power. In the case of R-R variability they are also assessed in normalized units (nu), which are obtained by dividing the power of a given component, expressed in $msec^2$, by the total variance (diminished by component 1) and multiplying the result by 100.

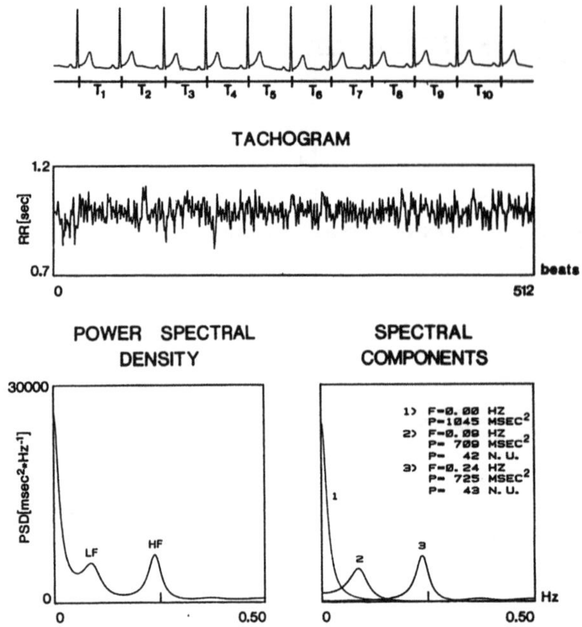

Fig.3 Schematic representation of the technique employed in our laboratory to compute with an autoregressive approach the spectral components of R-R interval variability. Note (bottom right panel) that for any given component the computer output provides both the center frequency (in Hz) and the power (in msec² and in normalized units - nu - in the case of components 2 and 3).

A recursive version of the program permits the analysis of long periods of recordings, up to 24 hours [16]. It should also be mentioned that with the simultaneous spectral analysis of R-R and systolic arterial pressure variabilities it is possible to obtain an assessment of the overall gain of baroreceptor reflex mechanisms, numerically represented by the index α, which corresponds to the square root of the ratio between the powers of spectral components of R-R and systolic arterial pressere variabilities [17]. Results obtained with the index α or the phenylephrine method are similar.

3.Results and Discussion

3.1.Animal Experiments

Initial experiments on decerebrate cats, using denervation procedures or pharmacological blockades [18], as well as studies on conscious dogs, using pharmacological blockades [19], led to the concept that the high frequency respiratory component of R-R interval variability could provide a quantitative marker of vagal control of heart rate. The possibility that, similarly, the analysis of the lower part of the spectrum of R-R variability could provide information about sympathetic activity was essentially discarded, as the spectral components with a frequency slower than respiration were thought to depend on other mechanisms, such as the frequency response of the baroreceptor reflex or the renin/angiotensin system [19].

In the following pages, we will summarize, briefly, experimental evidence in favor of the hypothesis that the LF component of R-R and

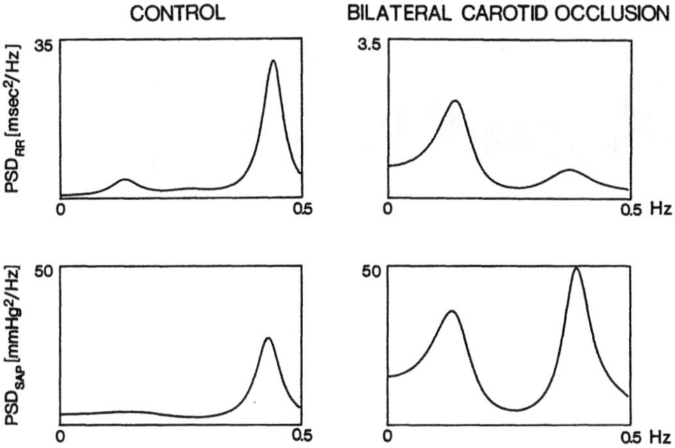

Fig.4 Example of the simultaneous analysis of R-R and systolic arterial pressure (SAP) variabilities in a conscious resting dog. Note the prominent HF component present at control and the marked LF component during the sympathetic excitation produced by carotid occlusion (From Rimoldi et al., unpublished)

SAP variabilities could be used as markers of sympathetic activity modulating heart rate and vasomotor tone, respectively [14, 20]. This hypothesis is based on the general concept that the neural control of the circulation is effected by a continuous interaction between different mechanisms, acting with opposite, positive and negative, operational signs Fig.1 [6].

As shown in Fig.4, in the conscious dog, under resting conditions (i.e. with a high vagal tone), the spectra of both R-R and systolic arterial pressure variabilities contain a major respiratory (HF) component. On the other hand, during the high sympathetic tone produced by the baroreceptor unloading accompanying an i.v. infusion of the vasodilator nitroglycerine [14, 20] or a transient bilateral carotid artery occlusion (Rimoldi et al., work in progress) Fig.4, there is a large increase of the low frequency (LF) component, in both R-R and systolic arterial pressure variabilities.

Similar increases in LF component were obtained whenever sympathetic tone was likely to be increased, by either physical activity with treadmill exercise (Fig.5) [21], transient coronary artery occlusion [20], or mental "stress" (Pagani et al., work in progress).

Denervation experiments provided further support for the concept of an instrumental role of sympathetic mechanisms in determining the LF spectral component. Indeed, after chronic bilateral stellectomy, a manoeuvre which drastically reduced the (afferent and efferent) sympathetic innervation of the heart, baroreceptor unloading no longer induced a rise in the LF component of R-R variability autospectra, while a marked increase in LF of systolic arterial pressure was maintained [14, 20]. Furthermore, during chronic α adrenergic blockade with prazosin, the increase in the LF component of systolic arterial pressure observed during control treadmill exercise was abolished, while that of R-R variability was not (Rimoldi et al., work in progress), thus pointing to the instrumental role of vasomotor sympathetic activity in determining the LF component of systolic arterial pressure variability.

Additionally, after atropine administration, a treatment which reduced total variance by three orders of magnitude (from 60000 to

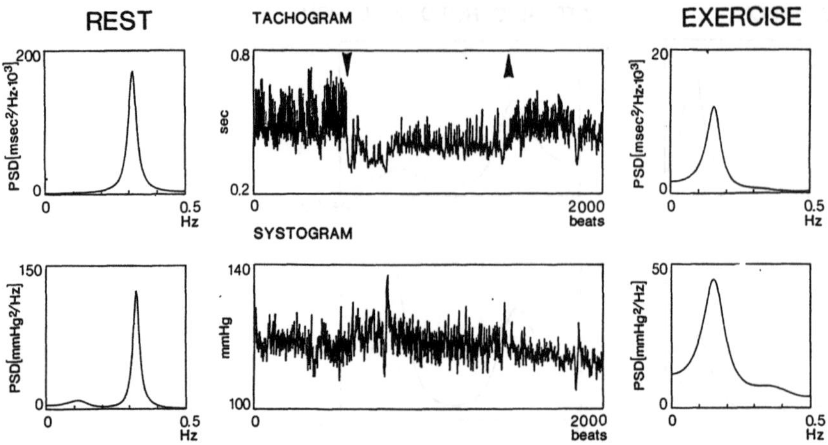

Fig.5 Example of the effects of treadmill exercise on R-R and systolic arterial pressure variabilities in a conscious dog, note the prevalence of HF at control, and the prevalence of LF during exercise (From [21]).

≈40 msec²), the major fraction of the remainder of the oscillations were found in the LF region [20].

Finally, after ganglionic transmission blockade, R-R variability was totally abolished, and systolic arterial pressure variability was markedly reduced. In this latter a small HF component was still present, as an expression of a direct mechanical influence of respiration [20].

When examining with a similar approach the nerve traffic directed to the heart via the vagi or the sympathetic nerves, both LF and HF components were found in both nerve systems ([22] and Lombardi et al., work in progress), thus indicating the potential risk of extrapolating from a marker of function to a neural circuit.

3.2.Human Studies

While the concept that vagal activity controlling the heart rate in man could be inferred from the power of the high frequency, respiratory component of R-R variability was readily accepted, on the basis of studies employing muscarinic receptor blockade [23], there was some disagreement about the possibility that sympathetic activity could be reflected, as well, by the power of the low frequency component.

In the following we will briefly summarize evidence in favor of the hypothesis that sympathetic modulatory activity to the heart and to the vasculature can be assessed in man by the power of the LF component of, respectively, R-R and systolic arterial pressure variabilities [14, 16].

As shown in Fig.6 for the case of R-R variability, the autospectrum of a resting normal man contained two components of similar amplitude (i.e. LF and HF), while during passive tilt (i.e. a manoeuvre which enhanced sympathetic drive to the SA node) LF was increased markedly (and HF conversely reduced) [14]. Such an increase was observed during moderate dynamic [16, 17] or static exercise [24], and during mental stress (Fig.7) [24, 25].

Acute β adrenergic receptor blockade blunted the tilt induced rise in the LF component of R-R variability, while chronic blockade reduced LF also at rest [14].

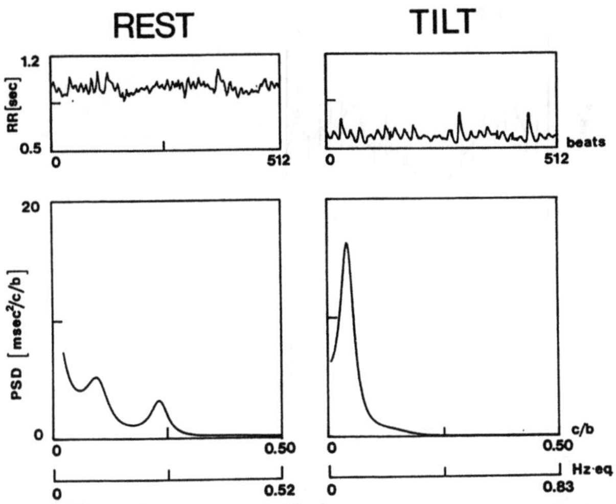

Fig.6 Example of the effects of tilt on R-R varibility in a healthy human subject. Note the prevalence of the LF component during tilt. (From [14])

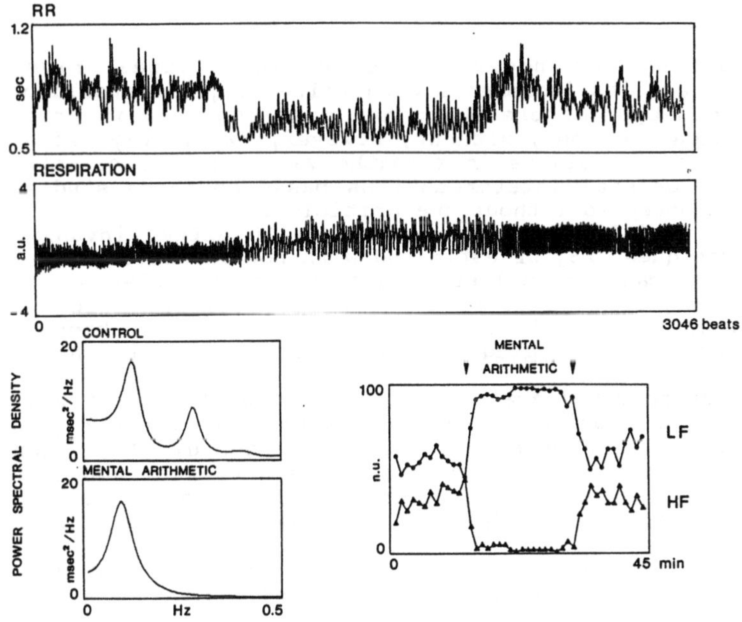

Fig.7 Example of the effects of mental stress on R-R variability. Note the marked increase of the LF component (and simultaneous decrease of the HF component) during mental arithmetic. (From [25])

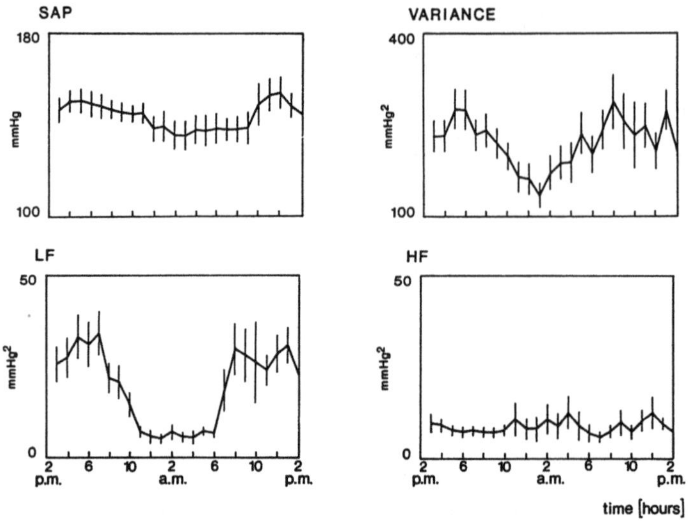

Fig.8 Circadian changes in systolic arterial pressure and its variability in man. Note the rapid rise of the LF component (an index of vasomotor sympathetic control) at the time of waking up. (From [16])

Metronome breathing, a manoeuvre which enhances vagal drive, reduced LF at rest and its increments with tilt [14]. It should be mentioned that this type of breathing had been employed in prior studies for technical reasons, without, however, explicitly recognizing the attendant alterations in sympatho-vagal control of heart rate: thus all the general interpretations based on those studies [23] should not be accepted without due criticism.

The study of arterial pressure, measured with either invasive high fidelity techniques [26] or estimated with non-invasive devices (Finapres) [27], indicated that likely increases in vascular sympathetic drive, during tilt [14], mental stress [25], and moderate dynamic physical activity with either treadmill [28] or bicycle exercise [17] were all associated, as well, with a rise in the LF component.

The simultaneous spectral analysis of R-R and systolic arterial pressure variabilities provided a tool to quantify the overall gain of baroreceptor arterial mechanism (using the index α) [17]. When analyzing high fidelity blood pressure measurements from ambulant subjects such a gain was found to undergo marked circadian changes, with minimal values during the daytime, and maximal values during the night [17].

Likewise, the LF component of R-R and SAP(Fig.8) underwent day-night changes, with high values during the day and low values during the night [16].

Opposite, parallel, changes were observed in the HF component of R-R variability.

3.3. Pathophysiological Studies

The possibility of obtaining quantitative information on the neural control of the circulation appears to offer a potentially important window into the pathophysiology of those cardiovascular diseases in

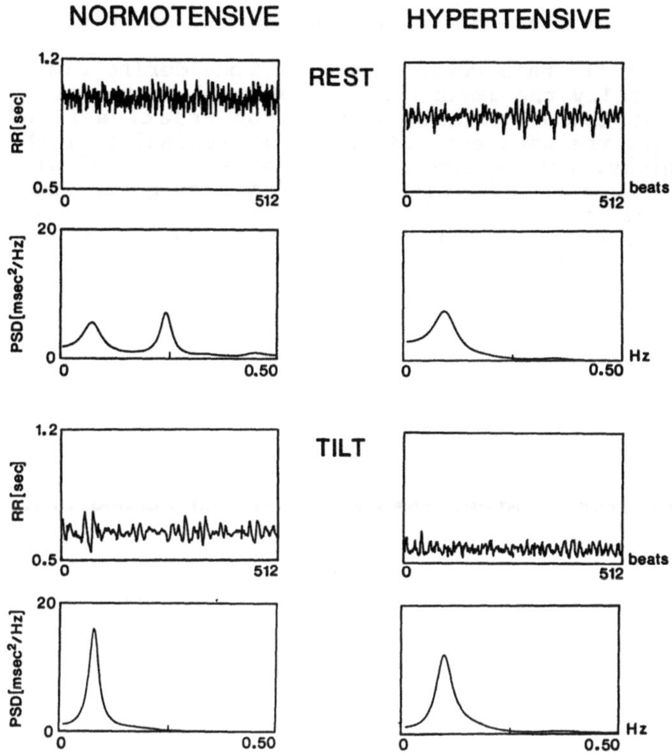

Fig.9 Example of the difference in R-R variability between normotensive and hypertensive subjects. Note the prevailing LF component at rest in the hypertensive subject. (From [29])

which a role of neural factors is suspected, such as arterial hypertension or ischemic heart disease.

Just to allude to the potentialities of the approach, evidence for an increase in sympathetic activity in early hypertension (Fig.9) [29] and in the first weeks of convalescence from myocardial infarction [30] was suggested by the findings of elevated values of the LF component of R-R variability. Furthermore, the effects of treatment could be assessed as well: for instance chronic ß adrenergic receptor blockade, in hypertensive patients, induced a reduction of the resting value of the LF component of R-R variability [29].

4.Conclusions: Neural Control of the Circulation and Synergetics

Neural control of the circulation depends upon a complex and dynamic interaction of several different components, which can be directly explored only in part. A synergetic approach, which is based on probabilistic concepts [13] about the behavior of complex systems, while not disregarding any available knowledge about them, appears to provide exciting suggestions about future directions.

For instance, in synergetic terminology, it is conceivable that "order parameters" could be found which not only consider the frequency domain (i.e. LF and HF relative powers), but also the amplitude modulation, as well as the non linear dynamic properties of the signal.

In terms of R-R interval, this would amount to considering simultaneously and synthetically the information contained in the absolute R-R value, in the total variance and in the relative LF/HF components together with the knowledge of the characteristics of the neural control mechanisms. Furthermore, in the context of a simplified dynamic balance of sympathetic and vagal activities governing a system whose performance is characterized by large and rapid swings, this would lead to the study of not only steady states, but also the changes in levels of performance, an aspect which seems of great relevance to the biological reality of the intact behaving organisms.

References

1. Heymans,C. (1929) Le sinus carotidien. London, H K Lewis.

2. Abboud, F.M. and Thames, M.D. (1983) Interaction of cardiovascular reflexes in circulatory control. In: Handbook of Physiology, The Cardiovascular System, Shepherd J.T. and Abboud F.M. (Eds). Am. Physiol. Soc., Bethesda, MD, pp 675-753.

3. Vatner,S.F. and Braunwald,E. (1975) Cardiovascular control mechanisms in the conscious state. N. Engl. J. Med. 293: 970-976.

4. Malliani,A., Schwartz,P.J. and Zanchetti,A. (1969) A sympathetic reflex elicited by experimental coronary occlusion. Am. J. Physiol. 217: 703-709.

5. Bishop,V.S., Malliani,A. and Thoren,P. (1983) Cardiac mechanoreceptors. In: Handbook of Physiology, The Cardiovascular System. Shepherd J.T. and Abboud F.M. (Eds). Am. Physiol. Soc., Bethesda, MD pp 497-555.

6. Malliani,A., Pagani,M. and Lombardi,F. (1986) Positive feedback reflexes. In : Handbook of Hypertension, vol 8: Pathophysiology of Hypertension. A Zanchetti and RC Tarazi (Eds), Elsevier, Amsterdam, pp 69-81.

7. Malliani,A. (1982) Cardiovascular sympathetic afferents. Rev. Physiol. Biochem. Pharmacol. 94: 11-74.

8. Lioy,F., Malliani,A., Pagani,M., Recordati,G. and Schwartz,P.J. (1974) Reflex hemodynamic responses initiated from the thoracic aorta. Circ. Res. 34: 78-84.

9. Pagani,M., Bergamaschi,M. and Malliani,A. (1982) A positive feed-back sympathetic pressure reflex during stretch of the thoracic aorta in conscious dogs. Circ. Res. 50: 125-132.

10. Pagani,M., Pizzinelli,P., Furlan,R., Guzzetti,S., Rimoldi,O., Sandrone,G. and Malliani,A. (1985) Analysis of the pressor sympathetic reflex produced by intracoronary injections of bradykinin in conscious dogs. Circ. Res. 56: 175-183.

11. Pagani, M., Mirsky, I., Baig, H., Manders, T., Kerkhof,P. and Vatner,S.F. (1979) Effects of age on aortic pressure-diameter and elastic stiffness-stress relationship in unanesthetized sheep. Circ. Res. 44: 420-429.

12. Blois,M.S. (1988) Medicine and the Nature of Vertical Reasoning. N. Engl. J. Med. 318: 847-851.

13. Haken,H. (1983) Synergetics: An introduction. 3rd edn Springer -Verlag, Berlin

14. Pagani, M., Lombardi, F., Guzzetti, S., Rimoldi, O., Furlan, R., Pizzinelli, P., Sandrone G, Malfatto G, dell'Orto S, Piccaluga E, Turiel M., Baselli G., Cerutti S., Malliani A. (1986) Power spectral analysis of heart rate and arterial pressure variabilities as a marker of sympathovagal interaction in man and conscious dog. Circ. Res. 59: 178-193.

15. Saul,J.P., Arai,Y., Berger,R.D., Lilly,L.S., Colucci,W.S. and Cohen,R.J. (1988) Assessment of autonomic regulation in chronic congestive heart failure by heart rate spectral analysis. Am. J. Cardiol. 61: 1292-1299.

16. Furlan,R., Guzzetti,S., Crivellaro,W., Dassi,S., Tinelli,M., Baselli,G., Cerutti,S., Lombardi,F., Pagani,M. and Malliani,A. (1990) Continuous 24 hour assessment of the neural regulation of systemic arterial pressure and RR variabilities in ambulant subjects. Circulation 81: 537-547.

17. Pagani,M., Somers,V., Furlan.R, dell'Orto,S., Conway,J., Baselli,G., Cerutti,S., Sleight,P. and Malliani,A. (1988) Changes in autonomic regulation induced by physical training in mild hypertension. Hypertension 12: 600-610.

18. Chess,G.F., Tam,R.M.K. and Calaresu,F.R. (1975) Influence of cardiac neural inputs on rhythmic variations of heart period in the cat. Am. J. Physiol. 228: 775-780.

19. Akselrod,S., Gordon,D., Ubel,F.A., Shannon,D.C., Barger,A.C. and Cohen,R.J. (1981) Power spectrum analysis of heart rate fluctuations: A quantitative probe of beat-to-beat cardiovascular control. Science 213: 220-222.

20. Rimoldi, O., Pierini, S., Ferrrari, A., Cerutti, S., Pagani, M. and Malliani, A. (1990) Analysis of short-term oscillations of R-R and arterial pressure in conscious dogs. Am. J. Physiol. 258: H967-H976.

21. Rimoldi,O., Pagani,M., Pagani,M.R., Baselli,G. and Malliani,A. (1990) Sympathetic activation during treadmill exercise in the conscious dog: assessment with spectral analysis of heart period and systolic pressure variabilities. J. Auton. N. Syst. 30: S129-S132.

22. Lombardi,F., Gnecchi-Rusconi,T., Montano,N., Finocchiaro,M.L., Contini,M. and Malliani,A. (1989) Restraining effect of Captopril on cardiovascular sympathetic efferent neural activity. J. Hypert. 7 (S. 7): S55-S56.

23. Pomerantz,B., Macauly,R.B.J., Caudill,M.A., Adam,D., Gordon,D., Kilborn,K.M., Barger,A.C., Shannon,D.C., Cohen,R.J. and Benson,H. (1985) Assessment of autonomic function in humans by heart rate spectral analysis. Am. J. Physiol. 241: H151-H153.

24. Pagani,M., Mazzuero,G, Ferrari,A., Liberati,D., Cerutti.S., Vaitl,D., Tavazzi,L. and Malliani,A. (1991) Sympathovagal interaction during mental stress. A study using spectral analysis of heart rate variability in healthy control subjects and patients with a prior myocardial infarction. Circulation. 83 (suppl 2: in press).

25. Pagani,M., Furlan,R., Pizzinelli,P., Crivellaro,W., Cerutti,S. and Malliani,A. (1989) Spectral analysis of R-R and arterial pressure variabilities to assess sympatho-vagal interaction during mental stress in humans. J. Hypert. 7(suppl 7): S14-S15.

26. Pagani,M., Furlan,R., dell'Orto,S., Pizzinelli,P., Lanzi,G., Baselli,G., Santoli,C., Cerutti,S., Lombardi,F. and Malliani,A. (1986) Continuous recording of direct high fidelity arterial pressure and electrocardiogram in unrestricted subjects. Cardiovasc. Res. 20: 384-388.

27. DeBoer,R.W. (1985) Beat-to-beat blood pressure fluctuations and heart rate variability in man: Physiological relationships, analysis techniques and a simple model (Ph. D. Thesis). Amsterdam: Univ. of Amsterdam.

28. Furlan,R., dell'Orto,S., Crivellaro,W., Pizzinelli,P., Cerutti,S., Lombardi,F., Pagani,M. and Malliani,A. (1987) Effects of tilt and treadmill exercise on short term variability in systolic arterial pressure in hypertensive men. J. Hypert. 5(suppl 5): S423-S425.

29. Guzzetti,S., Piccaluga,E., Casati,R., Cerutti,S., Lombardi,F., Pagani,M. and Malliani,A. (1988) Sympathetic predominance in essential hypertension: a study employing spectral analysis of heart rate variability. J. Hypert. 6: 711-717.

30. Lombardi,F., Sandrone,G., Pernpruner,S., Sala,R., Garimoldi,M., Cerutti,S., Baselli,G., Pagani,M. and Malliani,A. (1987) Heart rate variability as an index of sympathovagal interaction after acute myocardial infarction. Am. J. Cardiol. 60: 1239-1245.

Cardiorespiratory Variability: Fractals, White Noise, Nonlinear Oscillators, and Linear Modeling. What's to Be Learned?

J.P. Saul

Department of Cardiology, Children's Hospital, and Harvard-M.I.T. Division of Health Sciences and Technology, 300 Longwood Ave., Boston, MA 02115, USA

Abstract. Over the last 20 years, the study of heart rate (HR) variability has progressed from a phenomenological description of the spontaneous oscillations of HR to an understanding that relatively complex physiological control systems underlie the observed beat-to-beat fluctuations. Many of the current concepts have evolved from the use of either linear analysis techniques or linear systems modeling to evaluate classical physiological control systems such as respiratory sinus arrhythmia or the arterial baroreflex. Despite these efforts, linear techniques have only been successful at explaining a portion of the total variance in HR and arterial blood pressure (ABP). The remaining variance must be accounted for in one of two ways: 1) more complex linear models, or 2) nonlinear system elements which may contribute both locally and globally to the process of physiologic equilibrium known as homeostasis. This paper attempts to briefly examine both possibilities.

1. Introduction

Since the 1930's, the control of physiological systems has been understood in terms of **homeostasis**, a process by which system variables are maintained within a relatively narrow range by the system control elements. For the intact cardiovascular system, feedback and control mechanisms are particularly efficient at maintaining the mean values of arterial blood pressure (ABP) and central venous volume within a narrow range by constant regulation of heart rate (HR) and vascular tone. Thus, quantifying changes in the mean value of ABP or HR may not unveil substantial responses in the underlying control systems during periods of hemodynamic stress.

Alternatively, spontaneous fluctuations of HR and ABP reflect the interaction between ongoing perturbations to the cardiovascular system and the response of its regulatory mechanisms. In addition, analysis of beat-to-beat variability of HR and ABP can provide specific quantitative information about modulation of cardiac vagal, cardiac sympathetic and peripheral sympathetic nervous activity [1, 2, 3].

In the early 1970's power spectral analysis was first employed to quantify spontaneous HR variability in normal humans [4]. These investigators described three typical peaks in the HR power spectrum (Figure 1A): 1) a peak at the respiratory frequency corresponding to respiratory sinus arrhythmia (RSA), 2) a peak centered near 0.1 Hz considered important in ABP control, and 3) a peak at 0.04 Hz thought to be related to peripheral vasomotor regulation. Since then a number of points have been clarified. First, HR fluctuations occur preferentially in these frequency bands because of a combination of the spontaneous frequency content of various perturbing "noises", such as respiration, and

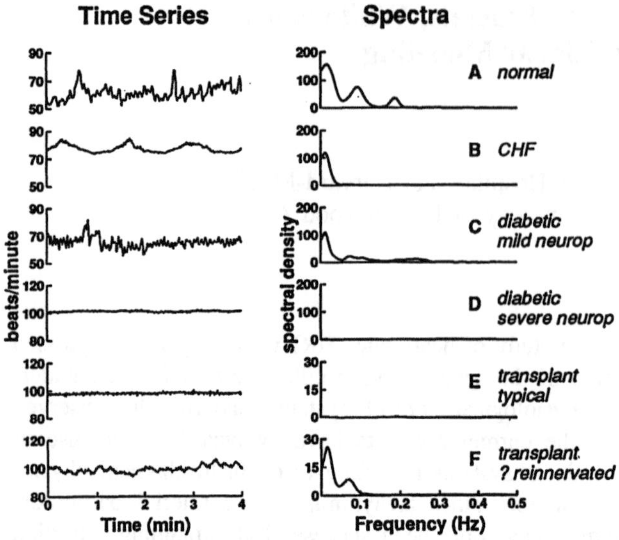

Figure 1. Examples of 4-minute segments of HR and corresponding spectra from a normal subject (A), and 5 patients with cardiovascular pathology. Although the range of heart rates in beats/minute on the y-axis is different for subjects A-C than for subjects D-F, the scale of 40 beats/minute is the same for all the tracings. Note changes in the spectral density scale for subjects E and F.

the frequency response of the cardiac and peripheral vascular end organs which control HR and ABP. Specifically, respiration typically occurs at frequencies above 0.15 Hz, and is the only physiological noise in that frequency range. From the standpoint of the end-organ, early investigations demonstrated that HR fluctuations at frequencies above 0.15 Hz are mediated solely by changing levels of vagal activity, while fluctuations below 0.15 Hz can be mediated by changing levels of both cardiac vagal and sympathetic activity [1, 2]. HR fluctuations below 0.03 Hz (period of 50 seconds) may be also mediated by changing plasma levels of neurohormones. Second, although there is a tendency for HR fluctuations to occur in the three described frequency ranges, the frequency as well as the amplitude of the spectral peaks is constantly changing in response to various hemodynamic perturbations, such that over the course of an entire day, few preferential frequencies stand out in normal individuals [5]. Finally, HR variability represents the response of the heart to **changing** levels of vagal and sympathetic activity so that quantification of HR variations in a particular frequency band may not correlate with the mean level of activity of either autonomic branch.

Considering the above information, analyses of spontaneous fluctuations of HR and ABP have proven useful in understanding cardiovascular regulation in a number of clinical settings. The HR power spectrum can be used to demonstrate increased modulation of cardiac sympathetic and decreased modulation of cardiac vagal activity during congestive heart failure [3], the degree of autonomic neuropathy in diabetes, and the cardiac denervation which accompanies a heart transplant (Figure 1) [6]. Although these applications of HR spectral analysis support the notion that this simple linear technique is useful, we and others have found that there is a great deal of both inter- and intra-individual variance in the spectra, leading to poor sensitivity and specificity when it is used as a test of autonomic function in individuals. In addition, interpretation of these data has been restricted both by the limited number of oscillatory frequencies which occur spontaneously, and by the inability to differentiate the effects of various underlying control mechanisms when only system outputs such as HR are measured. For these reasons, we chose to pursue two paths simultaneously. The first was to use linear models which relate HR as a system output to known modulators of HR such as respiration and ABP. The goal is to account for changes in HR variance by relating them to measured inputs. In combination with the formulation of these models, we have attempted to stimulate the system with an imposed noise source which produces input excitation over the full range of physiological frequencies of interest (0-0.5 Hz). The second was to accept that significant nonlinearities exist in cardiovascular control systems, as with any biological system, and try to evaluate the nonlinearities with appropriate techniques.

2. A Linear Model

HR is influenced by a number of physiological mechanisms which operate at all frequencies between 0 and ~ 1 Hz (Figure 2). Reflex mechanisms and central command (high-level brain stimuli) usually produce sinus arrhythmia by modulating sympathetic and vagal outflow from the brainstem, while other mechanisms may induce changes in sinus

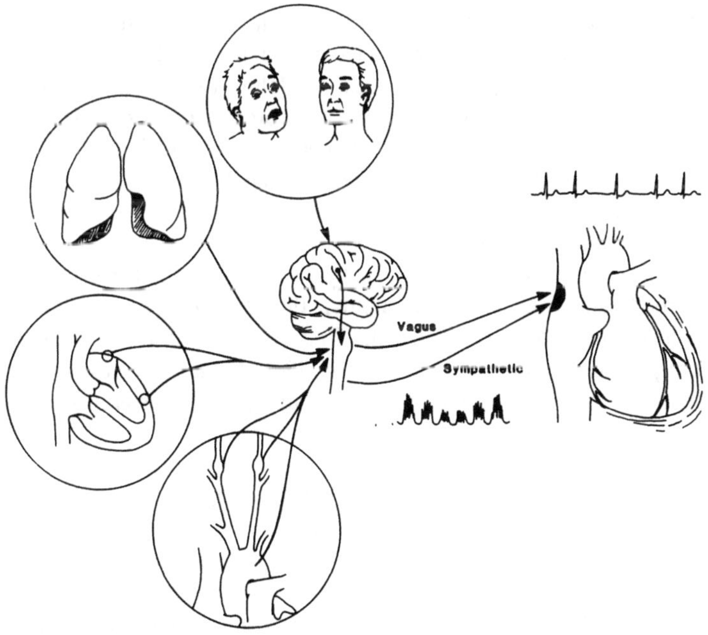

Figure 2. Physiological model of short-term HR control. Central command, respiration, and feedback from arterial and cardiopulmonary baroreceptors modulate cardiac vagal and sympathetic efferent activity. The HR response (i.e. HR variability) is then determined by the response characteristics of the SA node to modulation of the activity of each of the autonomic branches. See text for full details.

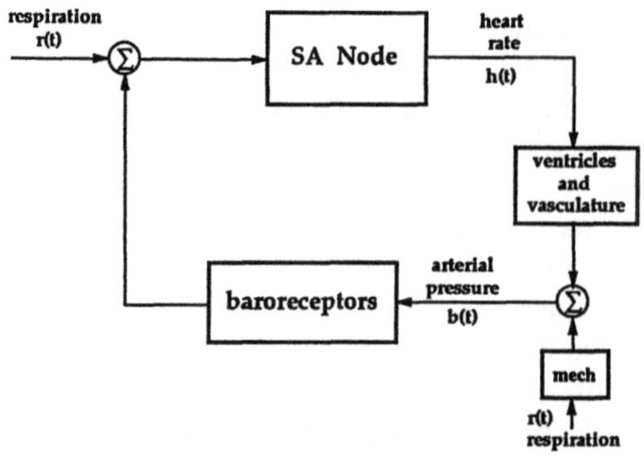

Figure 3. Simple feedback and control model of ABP and HR regulation. Respiration enters both sides of the loop as an external noise source.

rate by local stimulation of the sinus node (sinus node stretch or metabolic changes). Arterial baroreflexes, which buffers changes in ABP by causing HR to change in the opposite direction of the ABP change, and central command are generally considered the dominant influences on HR. Cardiopulmonary baroreflexes, which buffer changes in central blood volume primarily by increasing peripheral vascular resistance and decreasing venous capacity, are considered to be less important. RSA seems to result from a combination of influences which include a direct central interaction between respiratory drive and cardiac efferents in the brainstem, modulation of arterial baroreflex sensitivity with respiration, reflex responses to mechanically induced changes in both ABP and central volume, sinus node stretch, and chemoreceptor influences. Modeling these complex physiological mechanisms can be done with as little or as great a complexity as one desires. We chose to begin with the relatively simple closed loop model shown in Figure 3. The model's important features are that it allows for feedback and feedforward between HR and ABP, and it includes noise sources which affect both HR and ABP directly. Our first goal was to attempt to evaluate the open-loop characteristics of each block with linear transfer functions, using the cross-spectral technique [7]. In a series of animal and human experiments employing broad-band physiological inputs, the model has now been expanded (Figure 4). We have succesfully described the frequency response of the sinoatrial node to vagal and sympathetic stimuli, the response of HR to broad-band respiratory and ABP inputs, and the ABP response to broad-band fluctuations of respiration and HR [7, 8, 9, 10, 11].

Our data suggest that the accuracy and sensitivity of variability measurements in assessing autonomic modulation of HR can be enhanced by accounting for the amplitude of the inputs to HR control, ABP and respiration, and by simultaneously both broadening and standardizing the frequency content of those inputs between individuals. Despite these encouraging results, much remains unexplained. As demonstrated in Figure 5, the average coherence for some of our transfer functions is far from unity, indicating that a substantial portion of the output variance is not accounted for in our analysis scheme. Thus, the data further suggest that either: 1) there are additional unmeasured inputs which account for some of the output variance, 2) system nonlinearities are not adequately accounted for by the model, even for small fluctuations of system variables around their mean operating points, 3) physiological output "noise" (i.e. random, 1/f ...) contributes

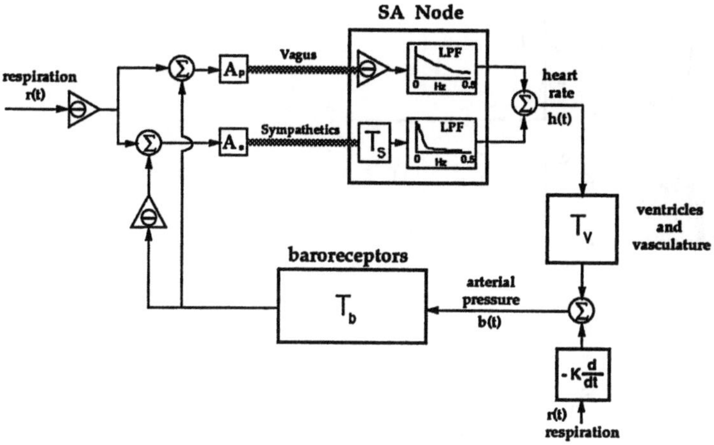

Figure 4. Model of Figure 3 which includes experimentally determined details of the open-loop transfer functions of the SA node, the mechanical feedforward from HR to ABP, the baroreflex feedback from ABP to HR, and the direct effects of respiration on both HR and ABP.

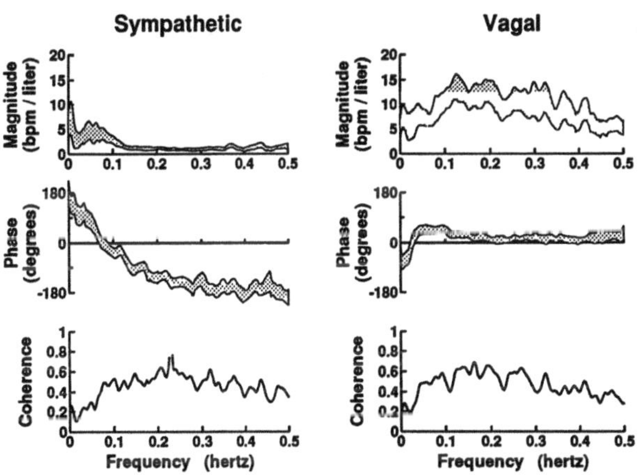

Figure 5. Group-average transfer functions between respiration and HR. Magnitude, phase and squared coherence are shown for two autonomic states: sympathetic (standing + parasympathetic blockade with atropine) and vagal (supine + sympathetic blockade with propranolol). Gray areas represent mean ± standard error.

significantly to the output variance, or 4) unmeasured inputs, nonlinearities, and physiological noises are important. Almost certainly the last explanation comes closest to the truth.

3. A Nonlinear Phenomenon

In contrast to the typical HR power spectrum in healthy adults, we have found that HR fluctuations in full term healthy neonates, adults with severe chronic congestive heart

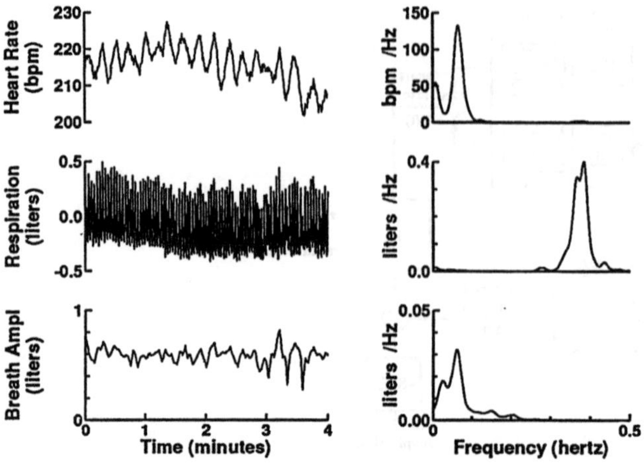

Figure 6. Heart rate, respiratory and breath amplitude time series and their corresponding spectra during periodic breathing in a dog after a 15cc/kg hemorrhage and complete pharmacologic parasympathetic blockade. The HR and breath amplitude spectra contain peaks at an identical low frequency, while the respiratory signal spectra has peaks at the primary respiratory frequency plus sidelobes at ± the modulation frequency of ~ 0.07 Hz. Amplitude modulation of the respiratory signal is most obvious towards the end of the time series.

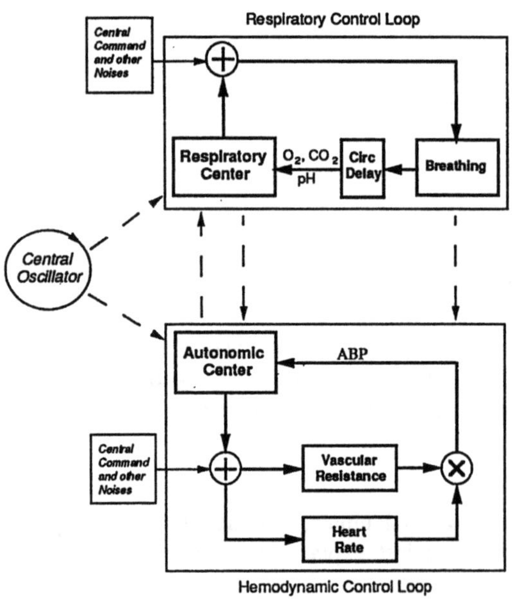

Figure 7. A model of cardio-respiratory control in which the respiratory and hemodynamic control loops operate somewhat independently, but have the potential for neural and mechanical "cross-talk". Central command and other physiologic noises such as body movement can enter both control loops. A hypothetical central oscillator is included which may influence the respiratory and autonomic control centers in the brainstem.

failure and animals after acute hemmorhage share different characteristics (Figure 6) [3, 12, 13]. The HR spectra of these groups often display very little activity above 0.1 Hz, and may be dominated by lower frequency peaks between 0.01 and 0.08 Hz. In all of these groups, we and others have also observed an association between these low frequency fluctuations of HR and fluctuations of respiratory tidal volume (and sometimes rate) at the same frequency. Because these respiratory fluctuations represent a type of amplitude (and/or frequency) modulation of the breathing pattern by a low frequency signal, the respiratory signal itself may have little spectral content at the modulation frequency. As Figure 6 demonstrates, however, the breath amplitude (tidal volume) signal formed by the subtraction of the expiratory minima from the inspiratory maxima has a clear spectral peak at the modulation frequency. Thus, for such cases, the relationship between the instantaneous lung volume signal and the HR time series has a nonlinear component. This phenomenon, common to a range of species in a variety of physiologic situations, is in contrast to the more recognized linear links established between the respiratory waveform and HR (i.e. - the typical RSA).

Regardless of the physiological mechanism, the mode of transmission between the respiratory and hemodynamic control loops seems to have a nonlinear component which either modulates or demodulates the respiratory signal to impose or extract the periodic component (Figure 7). There are a number of mechanisms which might explain this nonlinear interaction between the respiratory and HR control loops.

3.1 Classical Feedback and Control

An explanation for the demodulation might involve the mechanical coupling between the *breathing* element in Figure 7 and central venous or arterial pressure, or might depend on neural coupling between the two brain stem centers. A simple amplitude demodulation could take place, similar to an AM radio receiver, where a sensitivity threshold allows only positive deflections to be transmitted, returning the "correct" low frequency spectral content to the signal. Because intrathoracic pressure remains negative during all phases of normal respiration, such a mechanism might be present in the effects of respiration on venous return to the heart. Modulation of the primary respiratory drive signal by a low frequency signal also requires a nonlinear mechanism. A multiplicative process between the low frequency signal and the respiratory center in the brain stem might be employed, or the process could be viewed as the interaction of a forcing function with a nonlinear oscillator.

3.2 Forcing of a Van der Pol Oscillator

If we consider the basic respiratory drive signal x to be the output of a generalized Van der Pol oscillator,

$$\ddot{x} = \mu(1-x^2+\alpha x^4-\beta x^6)\dot{x} - x + f, \tag{1}$$

then in the absence of any sort of external forcing ($f=0$), the respiratory signal will oscillate at some constant frequency ω_c. However, if the external forcing function, f, which might arise from the various feedback loops, is a sine wave with amplitude ε,

$$f = \varepsilon \sin \omega t \tag{2}$$

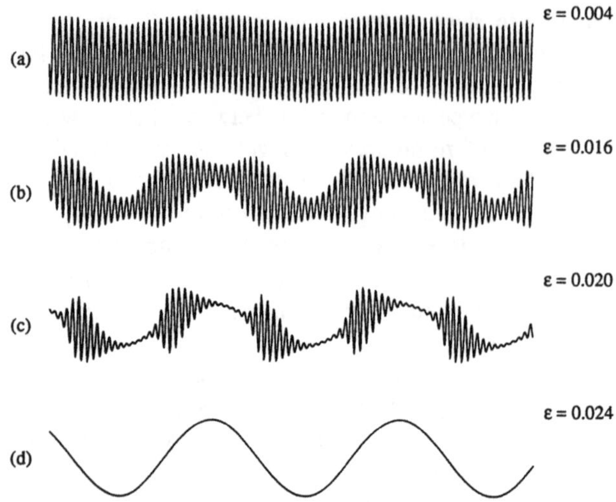

Figure 8. Output x of the forced Van der Pol oscillator described in Equations 1-3 for four values of ε, the amplitude of the sine wave forcing function. Note how the signal has increasing amounts of amplitude modulation as ε increases. Finally, in (d) only the modulating signal is present. The parameters chosen are $\mu=0.3$, $\alpha=0.1$ and $\beta=0.1$.

and

$$\omega \ll \omega_c, \tag{3}$$

then the behavior exhibited in Figure 8 will be seen as the effect of the external forcing increases with increasing ε [14]. Although the patterns in Figure 8 are derived from an entirely hypothetical set of equations, it is surprising how closely the rich behavior of a forced nonlinear oscillator mimics patterns of respiratory and HR variability seen clinically. If we consider the oscillator outputs to be respiratory signals, Figure 8d would represent death.

The above proposal is simply one hypothesis for how a low frequency fluctuation arising in one of the feedback loops or in the central oscillator of Figure 7 might modulate the respiratory frequency fluctuations of both the HR and respiratory signals. An entirely alternative explanation can be formed by considering the respiratory drive signal as emanating from a population of coupled nonlinear oscillators, rather than a single Van der Pol oscillator.

3.3 Coupled Nonlinear Oscillators

Multiple biological clocks, including the human circadian system and the cardiac pacemaker complex, have been hypothesized to be under the control of two or more coupled oscillators whose net output depends on the intrinsic frequency and phase of each of the component oscillators and the strength of the coupling between them [15]. We now consider the possible behaviors of a respiratory control system in which two populations of coupled oscillators (x and y) comprise the basic oscillatory drive for respiration (Figure 9). The two oscillators will be housed in the the respiratory center of Figure 7, which

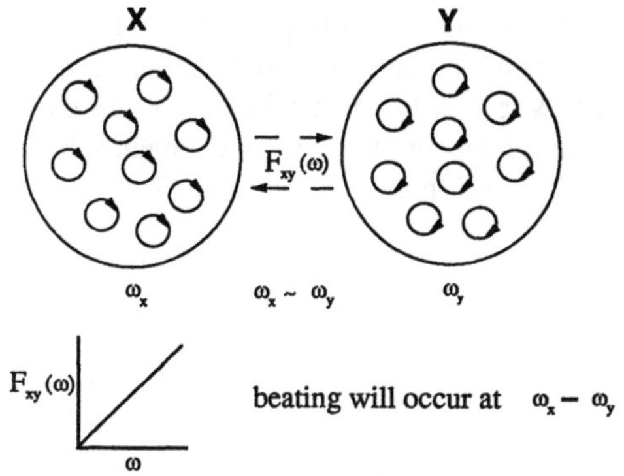

Figure 9. Two populations of hypothetical respiratory oscillators, X and Y, which together produce the respiratory drive signal. Their intrinsic oscillatory frequencies, ω_x and ω_y, are relatively close, and the coupling between them, $F_{xy}(\omega)$, is proportional to their average frequency.

is in the brainstem. Consequently, their intrinsic frequencies, which are close to each other and close to the basic respiratory drive frequency, are increased and decreased by the gain of the respiratory control loop. We propose that the degree of coupling between them F_{xy} is directly proportional to their frequency

$$F_{xy}(\omega) \propto \omega. \tag{4}$$

That is, the coupling force increases as the respiratory frequency increases, and thus, the frequencies of the two groups of oscillators will be closer to each other. We also assume that respiratory activity radiates to the hemodynamic control loop both through neural links in the brainstem and mechanical links which result from breathing, but the hemodynamic effects of the radiation may be limited by the autonomic response characteristics of the heart and vasculature (e.g. the frequency dependence of RSA).

The net output of the two oscillator groups will have two frequency components, one near their primary frequency and one at a beat frequency, $\sim (\omega_x - \omega_y)$. During resting conditions, the frequencies of the two oscillators will be low, allowing full transmission of the respiratory signal to HR (i.e. a prominent RSA). Also, because the coupling between them is weak, the basic respiratory rate and the beat frequencies may be quite variable, as is observed during relative quiet or mild activity. As respiratory rate is increased either in response to increased activity or cardiorespiratory stress, the coupling between the oscillators will increase, and the beat frequency will decrease. That is, periodic breathing will ensue at lower and lower frequencies as the gain of the respiratory control loop and the basic respiratory rate increase. Such a scenario is consisitent with the observation of periodic breathing in newborns, patients with congestive heart failure, and animals under hemorrhagic stress - all 3 are groups with increased respiratory rates and cardiorespiratory stress. In addition, the loss of HR fluctuations at the primary respiratory rate (RSA) can be explained by both the high respiratory rate and the withdrawal of vagal tone, which is also a physiologic component of these groups.

Whether or not this example of a clear nonlinear relation between two of the physiological variables in our linear model represents a true nonlinear interaction between respiration and HR, or simply a nonlinear association, the cardio-respiratory control network seems to have mechanisms for nonlinear modulation and demodulation of signals such as respiration. Because the feedback and cross-talk in this control network are complex, it is difficult to identify these mechanisms precisely, but we have qualitatively presented three theoretical explanations for these observations.

4. 1/f Noise and Fractals

As noted above, the HR power spectra of short (~ minutes) segments of data display a great deal of both inter- and intra-individual variability, which is reflected by changes in both the frequencies and amplitudes of the various spectral peaks. In fact, when the spectra of consecutive 5 minute segments of data are examined over the course of 24 hours, no single frequency is most prominent [5]. These observations suggest first that there are significant nonstationarities over time in the HR signal, and second that there may be a more global process underlying long term HR control. To examine both of these possibilities, we computed the low frequency (0.00003-0.1 Hz) slope of the HR spectrum of 24 hour segments of data in 5 healthy young adults. In agreement with a previous limited study in one subject [16], we found that the power of the entire 24 hour segment was inversely proportional to frequency, **1/f**, over 4 decades of frequency from 0.00003 to 0.1 Hz (period approximately 10 hours to 10 seconds)(Figure 10). For the group of subjects, the slope on a log-log scale of this low frequency linear portion of the spectra ranged from -0.93 to -1.07 [5]. Two additional observations were that no particular frequency stood out in the 24 hour spectra, and that the power in the usual frequency range of interest, above 0.1 Hz, was orders of magnitude smaller than that at lower frequencies.

The implications of these findings are multiple. First, frequencies well below the resolution of the spectra of shorter observation periods have an important contribution to HR variability. Second, the variance of a 1/f process is not well defined. Consequently, estimates of HR variance are a function of the observation time (i.e. data record length). Third, a 1/f waveform is scale invariant, that is, the waveform appears the same regardless of time scale, and its spectral power in any decade of frequency (e.g. 0.00001 to

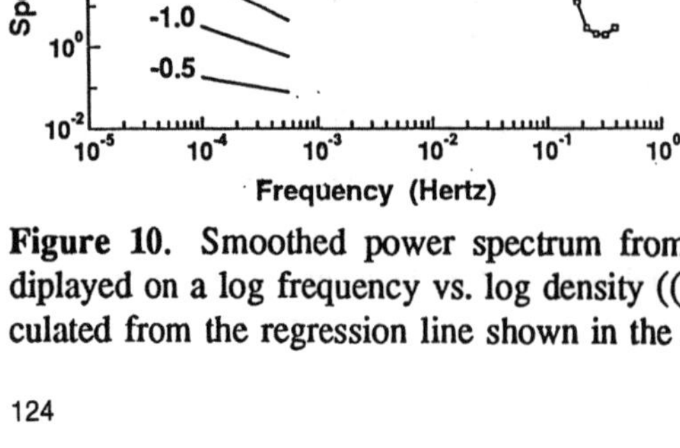

Figure 10. Smoothed power spectrum from 0 to 0.4 Hz for a 24.27 hour time series diplayed on a log frequency vs. log density ($(beats/min)^2/Hz$) scale. The slope was calculated from the regression line shown in the figure.

0.0001, 0.0001 to 0.01 and 0.01 to 0.1 Hz) is the same. Finally, one might speculate that this biologic 1/f process is in some way similar to the 1/f processes found in a host of physical systems, ranging from earthquakes to semiconductors [7]. Further work will be necessary to elucidate the origin and importance of 1/f noise in HR.

5. Conclusion

Cardiorespiratory variability is the result of a complex control network which includes multiple inputs, outputs, and feedback loops, as well as unidentified physiological noise sources. Because many of the control systems behave linearly over a small operating range, relatively sophisticated linear models seem to offer reasonable accuracy for interpreting experimental and clinical observations, particularly in a controlled and limited setting. However, when physiological conditions operate over their full range during both normal activities and periods of marked stress, nonstationarities and nonlinearities arise, which require different and perhaps more global analyses. Clearly, much remains to be learned.

REFERENCES

1. Akselrod, S., D. Gordon, F.A. Ubel, D.C. Shannon, A.C. Barger, and R.J. Cohen. Power spectrum analysis of heart rate fluctuations: a quantitative probe of beat-to-beat cardiovascular control. *Science* **213:** 220-222, 1981.
2. Pomeranz, B., R.J.B. MaCaulay, M.A. Caudill, I. Kutz, D. Adam, D. Gordon, K.M. Kilborn, A.C. Barger, D.C. Shannon, R.J. Cohen, and H. Benson. Assessment of autonomic function in humans by heart rate spectral analysis. *Am. J. Physiol.* **248:** H151-H153, 1985.
3. Saul, J.P., Y. Arai, R.D. Berger, L.S. Lilly, W.S. Colucci, and R.J. Cohen. Assessment of autonomic regulation in chronic congestive heart failure by heart rate spectral analysis. *Am J Cardiol* **61:** 1292-1299, 1988.
4. Kitney, R.I. and O. Rompelman. *The Study of Heart-rate Variability.* New York: Oxford Univ. Press, 1980.
5. Saul, J.P., P. Albrecht, R.D. Berger, and R.J. Cohen. Analysis of long term heart rate variability: methods, 1/f scaling and implications. *Comp. in Cardiol.* **14:** 419-422, 1987.
6. Sands, K.E.F., M.L. Appel, L.S. Lilly, F.J. Schoen, G.H. Mudge, and R.J. Cohen. Power spectrum analysis of heart rate variability in human cardiac transplant recipients. *Circulation* **79:** 76-82, 1989.
7. Berger, R.D., J.P. Saul, and R.J. Cohen. Transfer function analysis of autonomic regulation: I - the canine atrial rate response. *Am. J. Physiol.* **25:** H142-H152, 1989.
8. Saul, J.P., R.D. Berger, M.H. Chen, and R.J. Cohen. Transfer function analysis of autonomic regulation: II - respiratory sinus arrhythmia. *Am. J. Physiol.* **25:** H153-H161, 1989.

9. Berger, R.D., J.P. Saul, P. Albrecht, S.P. Stein, and R.J. Cohen. Respiratory effects on arterial pressure: a novel signal analysis approach. *Proc. IEEE EMBS* **10:** 533-534, 1988.

10. Saul, J.P., R.D. Berger, P. Albrecht, S.P. Stein, M.H. Chen, and R.J. Cohen. Transfer function analysis of the circulation: unique insights into cardiovascular regulation. *Submitted for publication*, 1990.

11. Berger, R.D., M.F. Fogaca, J.P. Saul, and R.J. Cohen. Transfer function analysis of cardiovascular regulation in an open-loop animal model. *Comp. in Cardiol.*, 1989.

12. Dykes, F.D., P.A. Ahmann, K. Baldzer, T.A. Carrigan, R. Kitney, and D.P. Giddens. Breath amplitude modulation of heart rate variabiliy in normal full term neonates. *Ped. Research* **20:** 301-308, 1986.

13. Madwed, J.B., K.E. Sands, J.P. Saul, and R.J. Cohen. Spectral analysis of beat-to-beat variability in heart rate and arterial blood pressure during hemorrhage and aortic constriction. In: *Neural Mechanisms and Cardiovascular Disease,* edited by B. Lown, A. Malliani, M. Porsdecimi. Padova: Liviana Press, 1986, pp. 291-302.

14. Kaiser, F. Entrainment-quasiperiodicity-chaos-collapse: bifurcation routes of externally driven self-sustained oscillating systems. In: *Nonlinear electrodynamics in biological systems,* edited by W. Ross Adey and A.F. Lawrence. New York: Plenum, 1983, pp. 393-411.

15. Winfree, A.T. In: *The geometry of biological time,* New York: Springer-Verlag, 1980.

16. Kobayashi, M. and T. Musha. 1/f fluctuation of heartbeat period. *IEEE Trans. Biomed. Eng.* **BME-29:** 456-457, 1982.

Spectral Analysis as an Assessment of the Neural Control of the Heart: A Methodological Comparison

R. Stark and D. Vaitl

University of Giessen, Department of Clinical Psychology,
Otto-Behaghel-Str. 10, W-6300 Giessen, Fed. Rep. of Germany

Abstract. It is commonly accepted that heart rate variability provides information on the neural control of the heart. Normally, three components are found in the spectrum of the cardiotachogram: A low frequency component (LF) with a frequency around 0.1 Hz, a high frequency component (HF) with a frequency around 0.25 Hz and a third component with a frequency around 0.03 Hz. Different experiments designed to relate the dominant frequency components to the amount of sympathetic and parasympathetic influence on the heart have resulted in contradictory findings. In order to examine the influence of the different methods of spectral analysis on these contradictory findings, we compared three calculatory procedures: FFT, AR1 and AR2. FFT is believed to represent the different methods based on the deterministic Fourier transform whereas AR1 and AR2 are two different versions of autoregressive spectral analysis. The results of our study indicate that the discrepancies of the findings, with respect to the different methods applied, are not caused by differences in the underlying theories (interpretation of the time series as a deterministic or random process), but by the different definitions of LF and HF.

1. Introduction

In the last twenty years, interest in relating the short-term changes of heart rate to the neural regulatory processes on the heart has increased. Although there have been some attempts to describe the heart rate variability in the time domain [1,2,3], only the transformation from the time domain into the frequency domain, via Fourier transform, results in additional information about hidden oscillations in the data. In the past, three components were normally found in the spectrum of the cardiotachogram [4,5,6]: a low frequency component (LF) with a frequency around 0.1 Hz, a high frequency component (HF) with a frequency around 0.25 Hz and a third component with a frequency around 0.03 Hz; the latter is mainly believed to be connected to the regulation of temperature [7], and to the renin-angiotensin-system [6]. LF corresponds to the frequency of the Mayer-Traube-Hering waves of blood pressure, and HF is closely connected to the breathing frequency.

In order to link the dominant frequency components of the spectrum to the neural control mechanisms, different research strategies have been employed. In order to examine this relationship, experiments using pharmacological and surgical blockades on humans and animals have been conducted [6,8,9,3,10,11,12,13,14]. Yet another research strategy has studied the spectra of people with diseases which are believed to be related

Eds.: H. Haken and H.P. Koepchen

to disturbances in neural control (heart diseases [15], hypertension [16,17], and diabetes [18,19]). Finally, different physical and psychological tasks were utilized, in order to vary the levels of sympathetic and parasympathetic activity [4,5,13].

The results of the different experiments suggest that HF is solely influenced by the parasympathetic nervous system. The findings relating to LF, however, are inconsistent. For example, Pagani et al. [13] found that LF reflects the activity of the sympathetic nervous system, whereas Akselrod et al. [6] found that the very low frequency component is mediated by both the parasympathetic and sympathetic nervous system, and LF is only influenced by the parasympathetic nervous system. Mulder [20], however, demonstrated that LF relates inversely to the effort of a mental load task.

It may be possible that the frequently observed differences between the various experimental results might be caused by the different methods used in the quantification of the frequency components. In fact, researchers have used many diverse methods ranging from derivations of the Fast Fourier Transform of either the raw data or the autocorrelation function [4,5,6] to autoregressive spectral analysis [13,14,15,16,17], to quantify the different frequency components.

To our knowledge, there have only been a few attempts to directly compare the results of the different quantification procedures [19]. Due to this lack of empirical data, we considered comparing three methods (FFT, AR1, AR2), one of which is a simple FFT approach, which is believed to represent all the different versions of the deterministic Fourier transform. In contrast, the calculation algorithm of the Italian working group [13,14,15,16], AR2, is the preferred version of autoregressive spectral analysis. These two strategies differ with respect to both their theoretical and methodological aspects. First they consider time series either as a deterministic or a random process. Second, they differ methodologically (in definition of frequency bands of interest and in the normalization procedure to determine relative values of LF and HF). For this reason, we introduced a third method, AR1, which is comparable to the FFT approach in its definition of the frequency components, and is an autoregressive method.

We applied all three of these methods of spectral analysis to the cardiotachograms of a recent experiment, the goal of which was two-fold: First, we wanted to replicate the findings, in which a tilting procedure resulted in a predominance of LF over HF, which was interpreted as an increase of the sympathetic influence on the heart [13]. Second, we wanted to examine whether LF decreases under mental load, as suggested by the results of Mulder [20], or increases, as suggested by the hypothesis that mental load is a sympathetic stimulus, and that LF represents the sympathetic influence on the heart.

2. Subjects and methods

2.1. Subjects

Twenty-four healthy male volunteers (age: 24.9 ± 3.9 years), most of whom were psychology students, took part in the experiment. Blood pressure measurement was used to ensure the exclusion of any hypertensives from the experimental group.

2.2. Experimental conditions

Physical load. Physical load was induced by a tilt table procedure. The subjects were placed on a pneumatic driven tilt table. A head-up tilt position (45^0) was used to evoke a strong sympathetic outflow [13].
Mental load. Mental load was induced by a concentration test called KVT [21]. This computer assisted test required subjects to decide if one, both or neither of two preselected numbers were hidden in a 6 x 6 matrix presented on a slide in front of them. The difficulty of the task depended on the presentation time of a slide; whenever subjects became more successful, the presentation time decreased. However, whenever subjects became less successful, the presentation time increased. Consequently, the level of difficulty could be kept constant for all subjects.

2.3. Protocol

The experiment consisted of four conditions: Rest, Tilt, KVT, Tilt/KVT. Under Rest condition, subjects were simply lying on the tilt table. Under Tilt condition, subjects were passively moved into a 45^0 head-up position. Under KVT condition, the concentration test was applied to subjects lying in a supine position on the tilt table. Finally, under Tilt/KVT condition, subjects were placed in the upright position, while mental load was induced by the KVT. In order to avoid position effects, the sequence of the four conditions was random for each subject. Each experimental condition consisted of a period of stabilisation (duration: 2 minutes), and of measurement (duration: 235 sec) during which respiration and ECG were continously measured; the respiration was assessed by means of an aneroid chest bellows, and ECG by means of Einthoven II electrode positions; respiration was digitized at 25 samples/sec, ECG at 1000 samples/sec.

2.4. Methods of spectral analysis

From the cardiotachograms of a total of twenty-four subjects sixteen were chosen which exhibited sufficiently stationary as well as artifact-free recordings. The spectra were calculated on the basis of the interval tachogram [4], in which R-R interval times are related to the interval number (Figure 1). The X-axis of the spectrum is expressed in cycles/beat. This frequency can be converted into Hertz equivalents by dividing by the mean R-R interval [4]. Three different methods were used to quantify the portion of total variance of inter-beat intervals (P_{LF} and P_{HF}), which were determinated by LF and HF: FFT, AR1 and AR2.
FFT. Neither trend removal, tapering nor smoothing were applied to the data; only zero-padding was used to supplement the available data-points with zero-values up to a full number of two, in order to permit the use of Fast Fourier Transform. P_{LF} and P_{HF} were defined as the fraction of the power, determined by frequencies between 0.07 and 0.15, and between 0.15 and 0.40. The boundaries of the frequency components were the same as those used by Mulder [22].

In addition to absolute units of LF and HF, relative units were calculated as

$${}_rP_{LF} = (P_{LF} * 100) / P_{TOT}$$

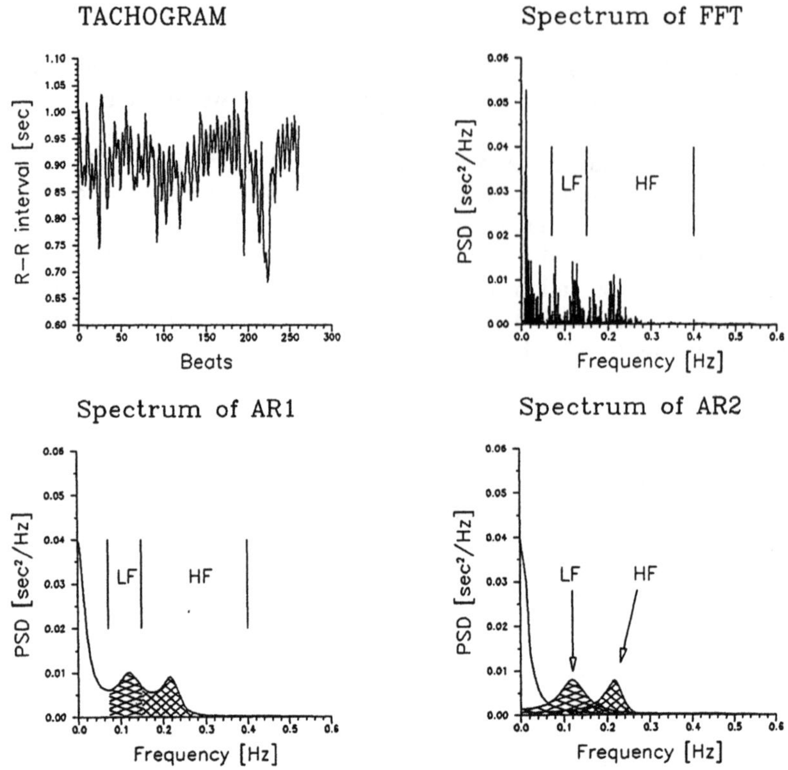

Figure 1. Cardiotachogram and three different methods of spectral analysis which were used for the comparison of the quantification procedures to obtain powers of the low (LF) and high (HF) frequency components.

$$_{r}P_{HF} = (P_{HF} * 100) / P_{TOT} \tag{1}$$

where $_{r}P_{LF}$ and $_{r}P_{HF}$ are the relative powers of LF and HF, P_{LF} and P_{HF} are the absolute powers of the components expressed in absolute units (sec^2), and P_{TOT} is the total variance.

In contrast to FFT, the other two methods of spectral analysis, AR1 and AR2, are based on autoregressive processes. An autoregressive process is a specific random process in which the current value of the process is expressed as a finite, linear aggregate of previous values of the process and of an unpredictable error e_t [23].

$$x_t = a_1 x_{t-1} + a_2 x_{t-2} + ... + a_p x_{t-p} + e_t \tag{2}$$

where e_t is a white noise (mean = 0, variance = s^2) and p is the order of the autoregressive process.

An autoregressive process can be thought of as a linear time invariant (LTI) system, with white noise as input and the observed time series as output. The transfer function of the system is given by

$$H(z) = (1 - a_1 z^{-1} - a_2 z^{-2} - \ldots - a_p z^{-p})^{-1} . \tag{3}$$

The power spectral density (PSD) can be calculated by

$$P(f) = \frac{2 s^2}{(1 - a_1 e^{-j2\pi f} - a_2 e^{-j4\pi f} - \ldots - a_p e^{-j2p\pi f})^2} \tag{4}$$

(Box & Jenkins [23]).

When using autoregressive models, it is necessary to estimate $a_1, a_2, a_3, \ldots, a_p$ and s^2 from the data. The Levinson-Durbin-Algorithm was used to estimate, stepwise, the unknown coefficients $a_1, a_2, \ldots, a_p$ by means of the empirical covariance for the different model orders p. In order to check the accuracy of the fitting model, the Anderson test was used to examine whether the prediction error e_t represented only white noise [23]. The model order p was determined by the Akaike Information Criterion [24]. Stationarity was examined by a pole diagram analysis.

AR1. By means of equation (4) the PSD was calculated. For the definition of LF and HF the same fixed boundaries were used, in order to increase the similarity between AR1 and FFT. In order to obtain relative units of P_{LF} and P_{HF}, the equation (1) was used.

AR2. The second autoregressive method was introduced by Bartoli et al. [25]. It is a modified version of the standard autoregressive spectral analysis. The poles of the transfer function of a LTI system are related to different frequency components which are specified through their center frequencies, and the portion of power of the time series, which is explained by these different components [26]. Therefore, a decomposition of PSD into different components is possible. HF was determined by the decomposed component with a center frequency equal to the main breathing frequency; LF was defined as the most dominant frequency component with a frequency lower than the breathing frequency (no fixed frequency range was used). The relative power units of LF and HF were calculated as

$$_rP_{LF} = (P_{LF} * 100) / (P_{TOT} - DC)$$

$$_rP_{HF} = (P_{HF} * 100) / (P_{TOT} - DC) \tag{5}$$

where P_{LF} and P_{HF} are power values of the components in absolute units, $_rP_{LF}$ and $_rP_{HF}$ are the relative power values and P_{TOT} is the total variance, and DC is the power of the frequency component with a center frequency of zero.

Figure 1 shows an example of a cardiotachogram, and the three methods of spectral analysis.

2.5. Statistics

To examine the similarity of the three methods, the Pearson correlation was calculated. By means of MANOVA, the effect of the different experimental conditions was tested. Data are presented as means $\pm$ standard deviations.

3. Results

3.1. Changes in mean heart period and variance under different experimental conditions

The mean values and the variance of R-R intervals decreased under both tilting and KVT conditions (Table 1). On the basis of mean R-R intervals, MANOVA revealed a statistically significant effect only for Tilt and for Tilt/KVT interaction, but not for KVT due to the pronounced interindividual variance.

Table 1. Mean values and variance of the R-R intervals and mean frequencies of the low (LF) and the high (HF) frequency components (± standard deviations).

Conditions	R-R Intervals		Frequency	
	Mean [msec]	Variance [$10^3 \cdot msec^2$]	LF [Hz]	HF [Hz]
Rest	925±108	6.45±8.46	0.13±0.02	0.28+0.06
Tilt	730±80 **	2.85±2.24	0.11±0.03**	0.27±0.06
KVT	887±100	5.08±7.91	0.15±0.02**	0.32±0.05**
KVT/Tilt	746±82 *	2.64±2.25	0.12±0.03	0.32±0.06

* Statistically significant effect (MANOVA; $p<0.05$)
** Statistically significant effect (MANOVA; $p<0.01$)

3.2. Comparison of the three methods of spectral analysis

In order to compare the three methods, Pearson correlation coefficients were calculated for the different absolute and relative units of P_{LF} and P_{HF}. The correlation matrix, shown in Table 2, demonstrates a very high correlation (for all conditions: $r > 0.95$) between FFT and AR1, both of which used fixed boundaries, regardless of the absolute or relative units for P_{LF} and P_{HF}. AR2 correlated significantly with the other two methods only for absolute, not relative, units. In particular, only under the Tilt/KVT condition did the correlation coefficient become less than 0.25.

In addition, the mean values of P_{LF} and P_{HF}, calculated for absolute and relative units, were compared under the different experimental conditions (Table 3 and Table 4).
Physical load. All three methods indicated that P_{LF}, using absolute units, decreased under the tilting condition, while the P_{LF}, using relative units, increased statistically significantly ($p<0.05$). The power of HF, however, diminished, regardless of whether absolute or relative units were used, under the tilting condition.
Mental load. Independent of the applied methods of spectral analysis, P_{LF} and P_{HF} tended to decrease when absolute units were used. A remarkable difference between the three methods was found for relative power units of HF. While AR1 and FFT indicated a statistically significant augmentation of the relative power units of HF under the KVT condition, AR2 did not result in a difference between Rest and KVT conditions.

Table 2. Pearson correlation coefficients between the relative ($_rP_{LF}$, $_rP_{HF}$) and absolute (P_{LF}, P_{HF}) power values of the low and the high frequency components calculated by the use of FFT, AR1 and AR2.

Conditions	P_{LF}	rP_{LF}	P_{HF}	rP_{HF}
	Correlation between FFT and AR1			
Rest	.99**	.97**	1.00**	1.00**
Tilt	.98**	.97**	1.00**	.99**
KVT	1.00**	.98**	1.00**	1.00**
Tilt/KVT	1.00**	.96**	1.00**	.96**
	Correlation between FFT and AR2			
	P_{LF}	rP_{LF}	P_{HF}	rP_{HF}
Rest	.99**	.74**	1.00**	.74**
Tilt	.83**	.62*	.92**	.48
KVT	.99**	.44	.99**	.35
Tilt/KVT	.80**	.54	.93**	.19
	Correlation between AR1 and AR2			
	P_{LF}	rP_{LF}	P_{HF}	rP_{HF}
Rest	.97**	.73**	1.00**	.73**
Tilt	.85**	.62*	.92**	.48
KVT	1.00**	.47	.99**	.36
Tilt/KVT	.81**	.56	.94**	.23

* Statistically significant correlation ($p < 0.01$)
** Statistically significant correlation ($p < 0.001$)

Overall, these results demonstrate that the differences between FFT and AR1 are only minimal. However, under certain conditions, there are significant differences between AR2 and the two other methods (FFT, AR1) especially with respect to the mean values of relative and absolute powers of LF and HF.

Table 3. Mean values ± standard deviations of the absolute and relative power values of the low frequency component using the three methods of spectral analysis (FFT, AR1, AR2).

Conditions	Low Frequency Component Absolute Power Values [10^{-3} sec^2] FFT	AR1	AR2
Rest	1.49±1.55	1.37±1.35	2.00±2.25
Tilt	0.78±0.50	0.71±0.45	1.06±0.67
KVT	1.10±1.69	1.01±1.49	1.96±3.08
KVT/Tilt	0.76±0.56	0.73±0.52	1.17±1.00
	Relative Power Values [%] FFT	AR1	AR2
Rest	25.3±6.5	23.9±6.7	59.3±18.8
Tilt	31.6±13.8*	29.4±13.5*	82.8±11.6**
KVT	23.4±9.3	22.0±8.4	60.8±18.1
KVT/Tilt	29.4±9.3	28.5±9.1	84.3±14.2

* Statistically significant effect (MANOVA; p<0.05)
** Statistically significant effect (MANOVA; p<0.01)

Table 4. Mean values ± standard deviations of the absolute and relative power values of the high frequency component using the three methods of spectral analysis (FFT, AR1, AR2).

Conditions	High Frequency Component Absolute Power Values [10^{-3} sec^2] FFT	AR1	AR2
Rest	2.65±5.95	2.66±6.11	1.86±4.53
Tilt	0.35±0.41	0.34±0.38	0.20±0.19
KVT	2.37±4.69	2.41±4.78	1.63±3.81
KVT/Tilt	0.46±0.47	0.46±0.46	0.18±0.24
	Relative Power Values [%] FFT	AR1	AR2
Rest	39.4±15.4	30.2±15.7	33.4±16.6
Tilt	11.2±4.26**	11.1±4.2 **	15.3±9.3 **
KVT	38.1±18.8*	38.0±18.2*	32.4±17.9
KVT/Tilt	17.1±5.9	17.0±6.3	13.9±11.1

* Statistically significant effect (MANOVA; p<0.05)
** Statistically significant effect (MANOVA; p<0.01)

4. Discussion

The goal of our study was to determine the influence of different methods of spectral analysis on the quantification of frequency components which are believed to represent different regulatory processes at the heart and to discover whether some of the contradictory results of previous research might be explained by the different methods used.

From the different methods of spectral analysis applied in the past, we have chosen to compare three methods: FFT, AR1, and AR2. We chose FFT to represent the different methods based on the deterministic Fourier transform, and AR2, an autoregressive method used by the Italian working group [13,14,15,16,17]. AR1, introduced as a mixture of the other two methods, was also an autoregressive method. However, AR1 was comparable to FFT with respect to fixed boundaries for the determination of the frequency components of interest.

Our results indicate:

1. There are no substantial differences between FFT and AR1. While FFT has the advantage of simple application to spectrum analysis, there are three important reasons for favouring autoregressive methods: First, an interpretation of an observed time series as a random process, rather than as a deterministic one, seems to be more satisfactory. Second, using autoregressive methods, a search for an appropriate filter design is unnecessary, since the filter is predefined by the data structure. Third, the FFT approach provides no possibility of separating different components as sophisticatedly as autoregressive methods.
2. There are considerable differences when absolute or relative power values are used. For example, under the tilting condition, there was a decrease of LF using absolute units, and an increase of LF using relative units. Since both types of units contain different information, it is advisable to calculate both absolute and relative units. Some contradictory findings of earlier research might be explained by the fact that researchers in this area rarely distinguished between absolute and relative units.
3. There are important differences between the method of spectral analysis used by the Italian working group (AR2) and the other two methods (FFT, AR1). These differences are mainly caused by the divergent definition of the frequency components (fixed vs adaptive boundaries). Since HF is adaptively determined by the main breathing frequency of subjects, using AR2, the absolute power values of HF are lower than when a wide range of frequencies defines HF (AR1 and FFT). Further, AR2 results in higher power values of the low frequency component. This is due to the adaptive definition of the low frequency component as the most dominant frequency component outside of the frequency component relating to the breathing frequency. In some subjects the peak of the low frequency component exceeded the fixed boundaries of 0.07 and 0.15. Consequently, the value of the low frequency component using AR2 is higher than using the other two methods.

The question arises whether fixed or adaptive boundaries are to be preferred.

Assuming that there is a close connection between parasympathetic activity and respiration, it seems logical to use the main breathing frequency as a reference point. In contrast, the definition of the low frequency component (using AR2) as the most dominant component with a frequency lower than the breathing frequency seems problematic for two reasons: First, some spectra demonstrated more than one dominant frequency component with a lower frequency than the breathing frequency. Second, the frequency of the low frequency component ranged from 0.05 to 0.18 Hz and the question arises: Does the low frequency component always reflect the same neural control process? Consequently, further research should focus on searching for an appropriate reference point for the low frequency component as is the case with HF.

Another problem caused by using the AR2 method is related to the procedure of obtaining relative power values of LF and HF. Our results indicate that although the correlation between AR2 and the other two methods is high using absolute power values, the normalisation procedure results in low correlation coefficients between AR2 and the other two methods under certain given experimental conditions. Consequently, a careful use of the normalisation procedure of AR2 (eq. 5) seems appropriate.

A final problem of AR2 concerns Zetterberg's method of spectral decomposition, which relates every pole of the characteristic equation of the system to a center frequency with a given power. Using AR2, all of the power of a component is related to the center frequency. However, especially in the case of a weak peak (which means that there is a great distance between the pole related to the weak peak and the unit circle), the distribution of power is spread over a higher range of frequencies. Consequently, it seems questionable whether it is then appropriate to concentrate all the power at the center frequency.

References

1. H. Luczak, W. Laurig: Ergonomics **16**, 85 (1973)
2. P. Grossman, S. Svebak: Psychophysiology **24**, 228 (1987)
3. F.M. Fouad, R.C. Tarazi, C.M. Ferrario, S. Fighaly, C. Alicandri: Am. J. Physiol. **246**, 838 (1984)
4. B.McA Sayers: Ergonomics **16**, 17 (1973)
5. L.J.M. Mulder. *In*: J.F. Orlebeke, G. Mulder, L.J.P. van Doornen (ed.): Psychophysiology of cardiovascular control - Models, Methods, and Data. Plenum Press, New York and London, 333 (1985)
6. S. Akselrod, D. Gordon, F.A. Ubel, D.C. Shannon, A.C. Barger, R.J. Cohen: Science **213**, 220 (1981)
7. R.I. Kitney, O. Rompelman (eds.): The study of heart-rate variability, Clarendon Press, Oxford (1980)
8. P.M. McCabe, B.G. Yongue, S.W. Porges, P.K. Ackles: Psychophysiology **21**, 149 (1984)
9. B.G. Yongue, P.M. McCabe, S.W. Porges, M. Rivera, S.L. Kelley, P.K. Ackles: Psychophysiology **19**, 426 (1982)
10. P.G. Katona, F. Jih: J. Appl. Physiol. **39**, 801 (1975)
11. B. Pomeranz, R.J.B. Macaulay, M.A. Caudill, I. Kutz, D. Adam, D. Gordon, K.M. Kilborn, A.C. Barger, D.C. Shannon, R.J. Cohen, H. Benson: Am. J. Physiol. **248**, 151 (1985)
12. G.F. Chess, R.M.K. Tam, F.R. Calaresu: Am. J. Physiol. **228**, 775 (1975)
13. M. Pagani, F. Lombardi, S. Guzzetti, O. Rimoldi, R. Furlan, P. Pizzinelli, G. Sandrone, G. Malfatto, S. Dell'Orto, E. Piccaluga, M. Turiel, G. Baselli, S. Cerutti, A. Malliani: Circulation Research **59**, 178 (1986)

14. O. Rimoldi, S. Pierini, A. Ferrari, S. Cerutti, M. Pagani, A. Malliani: Am. J. Physiol. **258**, 967 (1990)
15. F. Lombardi, G. Sandrone, S. Pernpruner, R. Sala, M. Garimoldi, S. Cerutti, G. Baselli, M. Pagani, A. Malliani: Am. J. Cardiol **60**, 1239 (1987)
16. S. Guzzetti, E. Piccaluga, R. Casati, S. Cerutti, F. Lombardi, M. Pargani, A. Malliani: J. Hypertension **6**, 711 (1988)
17. M. Pagani, F. Lombardi, S. Guzzetti, G. Sandrone, O. Rimoldi, G. Malfatto, S. Cerutti, A. Malliani: J. Hypertension **2**, 383 (1984)
18. R. I. Kitney, S. Byrne, M.E. Edmonds, P.J. Watkins, V.C. Roberts: Automedica **4**, 155 (1982)
19. T. J. Van Den Akker, A.S.M. Koeleman, L.A.H. Hogenhuis, O. Rompelman: Automedica **4**, 201 (1983)
20. G. Mulder. *In*: J.F. Orlebeke, G. Mulder, L.J.P. van Doornen (ed.): Psychophysiology of cardiovascular control - Models, Methods, and Data. Plenum Press, New York and London, 407 (1985)
21. W. Kuhmann, H. Lachnit, D. Vaitl. *In*: A. Steptoe, H. Rüddel, H. Neus (eds.): Clinical and methodological issues in cardiovascular psychophysiology, Springer, Heidelberg, 45 (1985)
22. L.J.M. Mulder, H.J. Van Dellen, P. Van der Meulen, B. Opheikens. *In*: F.J. Maarse, L.J.M. Mulder, W.S. Sjouw, A. Akkerman (eds.): Computers in Psychology: Methods, Instrumentation & Psychodiagnostics, 39, Lisse: Swets & Zeitlinger (1988)
23. G.E.P. Box, G.M. Jenkins: Time series analysis - forecasting and control, Holden-Day, Oakland, California (1976)
24. H. Akaike: The Statistician **27**, 217 (1978)
25. F. Bartoli, G. Baselli, S. Cerutti. *In*: Computers in Cardiology. Silver Spring, MD:IEEE Computer Soc., 485 (1982)
26. L.H. Zetterberg: Mathematical Biosciences **5**, 227 (1969)

Power Laws, Transients, Attractors, and Entropy: Possible Implications for Cardiovascular Dynamics

J.P. Zbilut

Departments of Physiology and OR/Surgical Nursing,
Rush-Presbyterian-St. Luke's Medical Center, 1653 W. Congress Parkway,
Chicago, IL 60612, and Section of Cardiology,
VA Edward Hines, Jr. Hospital, Hines, Illinois, USA

Joseph P. Zbilut, Departments of Physiology and OR/Surgical Nursing, Rush-Presbyterian-St. Luke's Medical Center, 1653 W. Congress Parkway, Chicago, Illinois 60612; and Section of Cardiology, VA Edward Hines, Jr. Hospital, Hines, Illinois, USA

Abstract. Considerable work has been done to demonstrate that the dynamics of the cardiovascular system is nonlinear. What has not been well recognized is that the system is essentially governed by transients. Conversely, then, if a system component's dynamics become dominated by basins of attraction, pathology may ensue. It is heuristically demonstrated that a critical feature of possible pathology is the relation of the system entropy to component entropy: if a component's (sub)basin of attraction is accompanied by a decrease in entropy, it may become informationally isolated. A method, originally suggested by Eckmann, et al., is proposed which may be able to detect such changes without the need to reconstruct a global attractor.

1 Introduction

The heart has long been the subject of investigation from a nonlinear dynamics perspective, and a considerable theory of its function has been developed [1]. Experimentally, much interest has been focused on the demonstration of pathologic dynamics with suitable change of critical control parameters.

An often unstated supposition in such experimental time series is that they are obtained from autonomous systems with characteristic times shorter than that of the series itself. Clearly, such assumptions cannot often be substantiated, and a given explanation for a system may yield variable results depending on which "modes" are captured for a given calculation. As a remedy to this situation, Ruelle has proposed the existence of adiabatically fluctuating parameters such that [2]

$$dx/dt = F(x, \lambda(t)) \quad (continuous\ \ time), \tag{1}$$

$$x_{n+1} = f(x, \lambda(n)) \quad (discrete\ \ time). \tag{2}$$

The suggestion is that instead of a fixed attractor A, there is a family of attractors, A_λ, depending on $\lambda = \lambda(t)$ or $\lambda(n)$. This, however, is dependent upon a faithful representation of the dynamics in $\Re^n$.

In the case of cardiovascular dynamics, it has been demonstrated for the case of inter-beat intervals that there are a number of degrees of freedom acting often in nonlinear ways on time scales much slower than the mean heart rate, and with differing feed-back delays. This is to

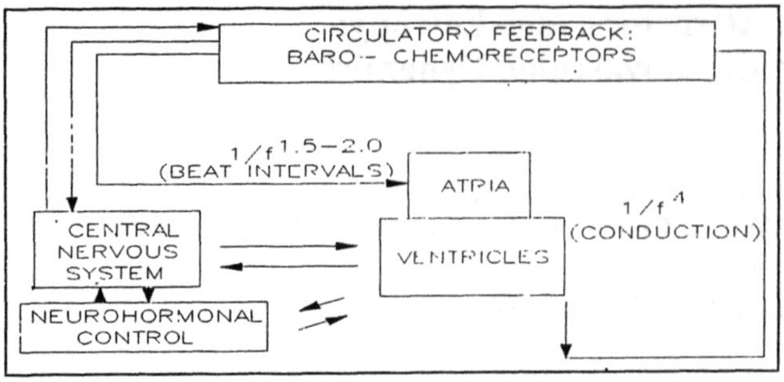

Fig. 1. Schematic representation of major cardiovascular controls.

say that the cardiovascular system is a spatio-temporal system of many degrees of freedom, whose inputs typically exhibit periodicities. With relatively long relaxations relative to the heart's beating, their effects are transients and noise, which proscribe any reasonable global modeling of its dynamics. And certainly this is not unremarkable in that the controls of the heart must constantly be readjusted in order to accommodate the requirements of the larger organism to which it is strongly coupled. (Fig.1)

Although such a description would suggest pessimism relative to an actual rigorous model of the cardiovascular system, the opposite may be true. By suggesting that the normal "stability" of the system is characterized by transients, pathology may be characterized by the absence of the ability to respond to such transients. It is argued that attractors and their (sub)basins of attraction are nonetheless relevant to the dynamics of the cardiovascular system as markers for pathologic states. This is to say that if the organism were allowed to fully relax, the heart dynamics would settle down to an attractor (chaotic or otherwise). But this is not the case, and a resultant approach to an attractor suggests a limitation in phase space in terms of being able to respond to a multidimensional dynamics by virtue of changes in its entropy.

In Section 2, possible models for transient generation in the context of power laws are presented. Section 3 briefly reviews some experimental evidence for the transient nature of cardiovascular controls. Section 4 proposes a heuristically derived theoretical dynamical framework; and Section 5 presents a method which might evaluate these dynamics.

2 Power Laws and Transients

In recent years, it has been emphasized that, at least on the phenomenological level, many cardiac processes exhibit a power law scaling so typical in many natural processes. Considerable attention has been generated by propositions that heart beat and electrocardiogram (ECG) spectra exhibit an inverse power law. The hypothesis is that the normal, healthy heart is characterized by scaling relationships, whereas pathology is characterized by a loss of "spectral reserve," i.e., by increased regularity [3]. Along with these observations, there has been the speculation that this represents deterministic chaos. This derives from the fact that 1/f scaling can be a signature for chaotic dynamics.

Although such a hypothesis is intriguing, attention to the actual values and form of the power spectrum questions the specific formulation of such a hypothesis. Specifically, it is noted that the exponent of scaling is typically > 1 in normal subjects, and depending upon the actual number of beats as well as realizations (scales of seconds to

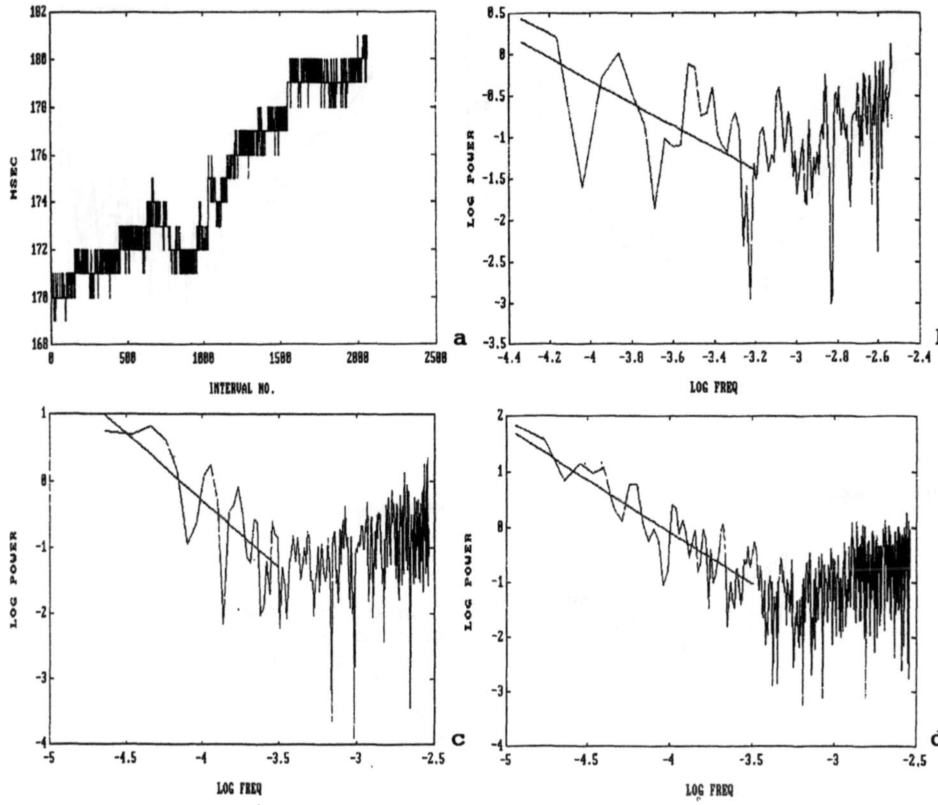

Fig. 2a-d. Time series of-inter beat intervals of an isolated perfused rat heart. Changes in scaling exponent of power spectrum of (**a**) by increasing number of intervals (**b-d**) 256, 512, 1024 intervals ($\alpha = 1.35$, 2.01, 1.87 respectively.)

days), this value can vary substantially. (Figs. 2-4) This is not remarkable in that it is probably reflective of a non-stationary process. It is useful to recall that an n-fold distribution $p(x(t_0), x(t_0+t_1), \ldots, x(t_0+t_n))$ is stationary if it does not depend upon the origin of the time axis. Practically this can be ascertained from the power spectrum of $x(t)$ if it follows a power law in the vicinity of f=0:

$$S_x(f) \propto f^{-\alpha}. \tag{3}$$

When α lies on (0, 1) the quantity is proportional to the Fourier transform of the correlation function [4]:

$$<x(t)x(t+\tau)> \propto \tau^{\alpha-1} (0 < \alpha < 1), \tag{4}$$

and

$$<x(t)x(t+\tau)> \propto \delta(\tau) (\alpha = 0). \tag{5}$$

For the condition that $\alpha = 0$, the time series is termed white noise;

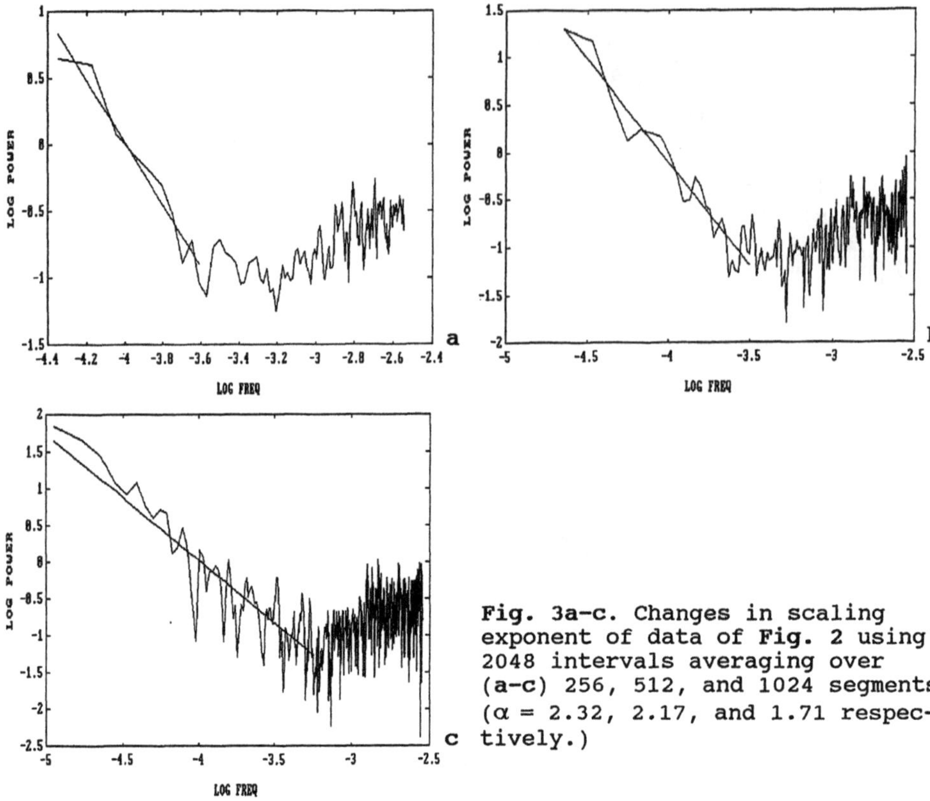

Fig. 3a-c. Changes in scaling exponent of data of **Fig. 2** using 2048 intervals averaging over (a-c) 256, 512, and 1024 segments ($\alpha = 2.32$, 2.17, and 1.71 respectively.)

for the condition $\alpha > 1$, equations (3,4) are not valid, and the time series is non-stationary.

Numerous models have been proposed for the generation of time series exhibiting power law scaling of the spectrum. A few with obvious possible relevance for physiologic systems are reviewed with particular emphasis on aspects of transiency:

2.1 Driven Oscillators

Although there have been many attempts to explain the underlying processes of 1/f scaling, in the present context, one suggested by Duong Van, et al. is that it results from the addition of several damped, driven oscillators of the form [5]

$$d^2\theta/dt^2 + (1/\tau)d\theta/dt + \omega_0^2 \sin(\theta) = 0\,, \tag{6}$$

where pendulum displacement is proportional to $\sin(\theta)$, τ is the damping, and ω_0 is a low amplitude natural frequency.

2.2 Multiple Delays

Alternatively, the hypothesis of Glass, et al. may be appropriate, whereby a control system can be modeled by multiple feedback loops [6]. With the addition of a time delay (τ) and control parameter, μ, an

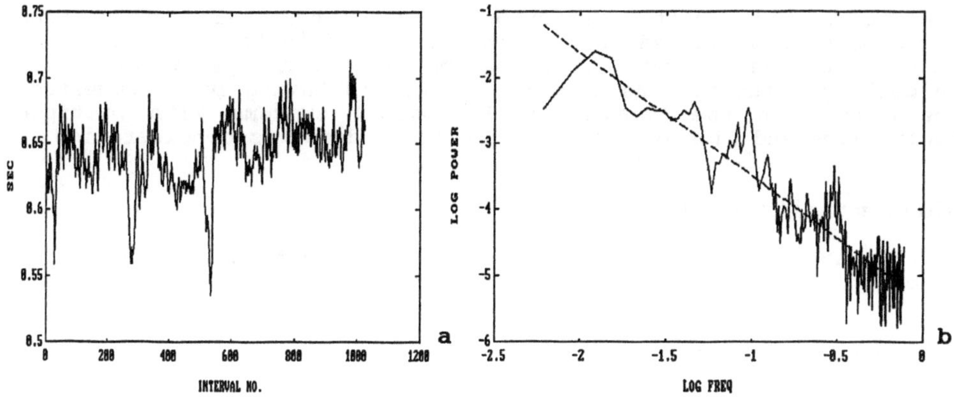

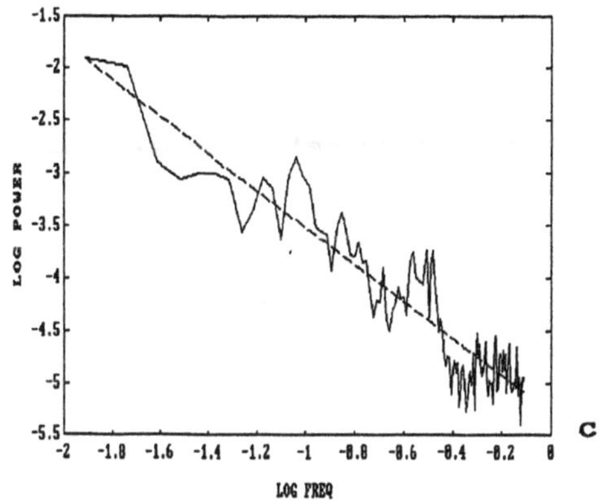

Fig. 4a–c. (**a**) Time series of inter-beat intervals for normal human male. (**b**) Power law scaling exponent using all 512 intervals ($\alpha = 1.88$). (**c**) Scaling exponent averaging 256 point segments ($\alpha = 1.76$.)

infinite dimensional differential delay equation is obtained [7]:

$$\frac{dx}{dt} = P(x(t), x(t-\tau), \mu_1, \mu_2, \ldots) - D(x(t), x(t-\tau), \mu'_1, \mu'_2, \ldots), \tag{7}$$

which can exhibit various oscillatory behaviors depending upon choice of parameters, and delay time. It is noted that such equations can exhibit exponential decay of the power spectrum, but that with noise such decay may obey a power law [8,9].

2.3 Poisson Process

Yet another proposition is that the controls may be a result of a Poisson process. Grüneis, et al. have demonstrated that neuronal spike trains of the mesencephalic reticular formation exhibit 1/f spectra which can be interpreted as a clustering Poisson process [10].

2.4 Self-Organized Criticality

Recently, the idea that many systems spontaneously evolve into dynamics characterized by power law distributions of scales and times has generated much interest. Bak suggests that large scale systems tend to become self-organized, and on the border of instability [11]. At the present time however, investigations into such systems are still rudimentary, with various regimes and models having been suggested; nonetheless, in the context of biological systems with multiple fluctuations and high dimensions, the model deserves consideration.

3 Experimental Observations

Numerous experiments have shown that under conditions of recognized pathology, the power law scaling can approach a white noise process:

3.1 Inter-Beat Intervals

Background. For the case of cardiac inter-beat intervals, it has been shown that the 1/f spectra do, in fact, have several resonances, possibly related to thermal, blood pressure, respiratory, and hormonal control, which may for large perturbation behave in a nonlinear manner [12,13]. (Fig. 5) It is also noted that the blood pressure resonance appears to be controlled by the sympathetic system and no delay, whereas the respiratory resonance is governed by the parasympathetic nervous system, and does have a delay.

Berger et al. have used a transfer function approach to these frequencies, and have found that the frequencies may respond as a low pass filter [14]. In this context it is noted that if a smooth dynamical system, $\vec{F}$, has power spectrum $S(f)$, then the power spectrum of the nth derivative of the time series, $x(t)$, is therefore $f^{2n}S(f)$. By Parseval's theorem:

$$\int_0^\infty f^{2n}S(f)df = < x^{(2n)}(t) > . \tag{8}$$

The highest order term dominates in the power spectrum, and for a fixed point, a power law roll-off for a low pass filter ensues for external noise. Essentially, this is the mechanism for recursive low-pass filters [8]. The suggestion is that the cardiovascular system behaves also as a low-pass filter, perhaps to remove certain levels of noise which may otherwise be deleterious.

The main resonances typically appear below 0.3 Hz, whereas the normal heart beat is in the range of 1-1.5 Hz. The implication is that the system is driven faster than the control oscillations can die out. This is confirmed by the fact that the resonances are rarely peaked, except in rather unusual control circumstances, with the result being rather broad peaks. (Fig. 5) Another possibility is that the individual oscillations do possess attractors, but that they annihilate each other, which results in transiencies as suggested Grebogi and Ott [15].

Ventricular Dilatation. Isolated perfused rat hearts have been progressively dilated with small balloons inserted in the ventricle to mimic pathologic overload states. The resultant power spectrum scaling of inter-beat intervals show a significant trend toward a white noise process [22]. (Fig. 6) The important observation here is that deformation of the excitable media (myocardium) is an important parameter to consider relative to excitation dynamics. The results have been indirectly supported by the observation that power law scaling of the ECG changes with cardiac dilatation and regional wall motion abnormalities [17].

Isolated Perfused Rat Hearts. Isolated perfused rat hearts demonstrated a wide range of oscillatory and chaotic phenomena which were modeled

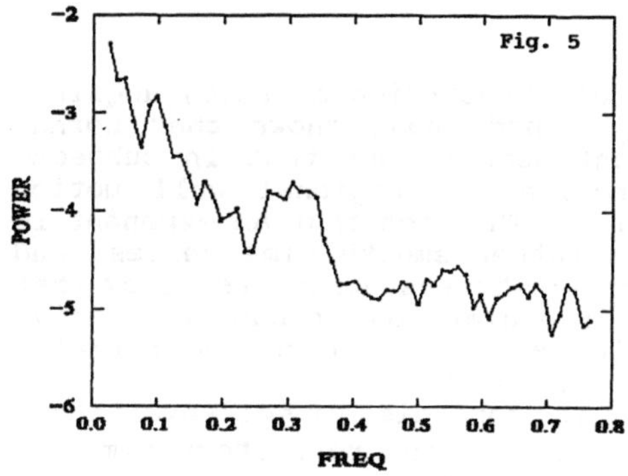

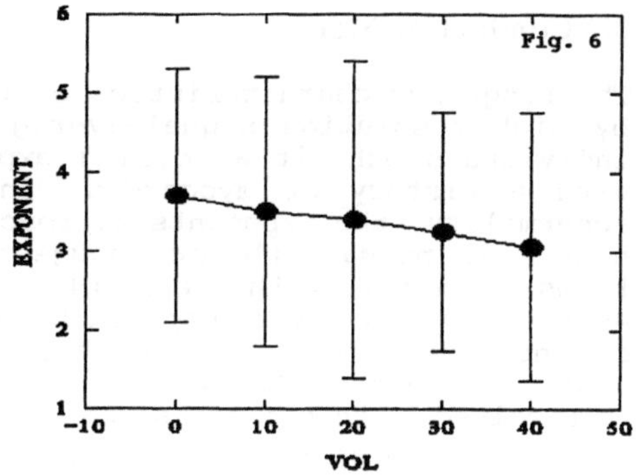

Fig. 5. Power spectrum for data of Fig. 4(a). Note peaks at 0.001, 0.01 and 0.25 Hz representing thermal, blood pressure, and respiratory activity respectively. (512 points averaged over 128 point segments.)

Fig. 6. Trend of scaling exponents of power spectrum with progressive dilatation of ventricle with 10 microliter aliquots.

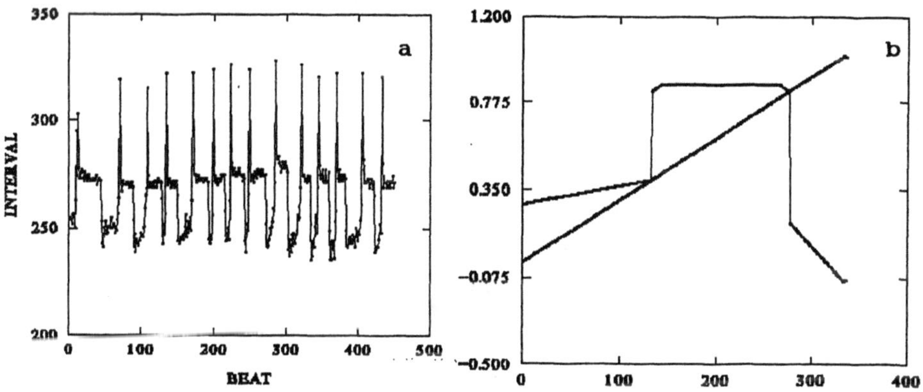

Fig. 7a,b. (a)Three state oscillations of isolated rat heart. (b) Piecewise linear map used to model heart beats of (a).

by a simple one-dimensional map. In many cases power law behavior of the intervals was demonstrated, although random-walk-like patterns devoid of scaling could also be demonstrated. This was interpreted as possibly resulting from "noised induced intermittency," or dynamics which change the slope of the map [23]. (Fig. 7) It is important to note, however, that the experimental hearts are isolated from the larger organism, and thus suggest only the intrinsic heart dynamics.

Volume Loading. In an experiment with patients with congestive heart failure (CHF) or cardiac transplantation, power law scaling approached a white noise spectrum with distention of the system by the acute administration of one and two liters of intravenous normal saline. Significant changes in important biochemical amines accompanied the administration. In one CHF patient barely visible broad-band oscillations developed in the spectra just prior to a brief run of ventricular tachycardia [24].

3.2 Conduction System

The frequency characteristics of the electrocardiogram (ECG) acquired by high resolution signal-averaging methods have shown that normal individuals exhibit a scaling exponent near 4, but that in subjects with a history of myocardial infarction, or regional wall motion abnormality, the exponents approach 3.5. The fact that an exponent is near 4 in normal subjects suggests a rather smooth time series, and takes it away from the realm of fractals; whereas exponents below that suggest waveforms which appear highly irregular, and perhaps indicative of increased "noise components," which may fractionate the normally highly correlated ECG signal [16,17]. (Figs. 8,9)

Involved in this process are two global properties of self-excitatory biologic membranes: latency to excitation, e, and refractory time, r. The ratio, $e/r = \mu$ is then a dissipative forcing term, governing the dynamics of inter-beat intervals [18]. Noisy depolarization dynamics, which may overtake the medium if sufficiently strong, or amplified, may make it difficult to respond adequately to the forcing term [17,19]. The noise itself may originate from a wide variety of sources including ionic changes, scarring and blocks [20,21].

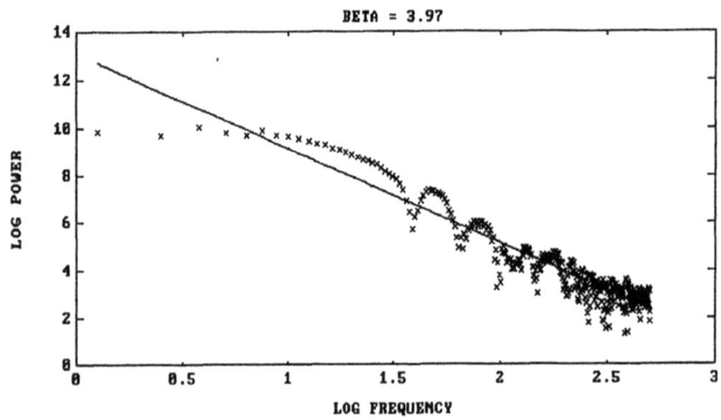

Fig. 8. Example of power law scaling for a signal-averaged ECG, using 800 points of a Euclidean norm for leads X,Y, and Z. (Beta:scaling exponent.)

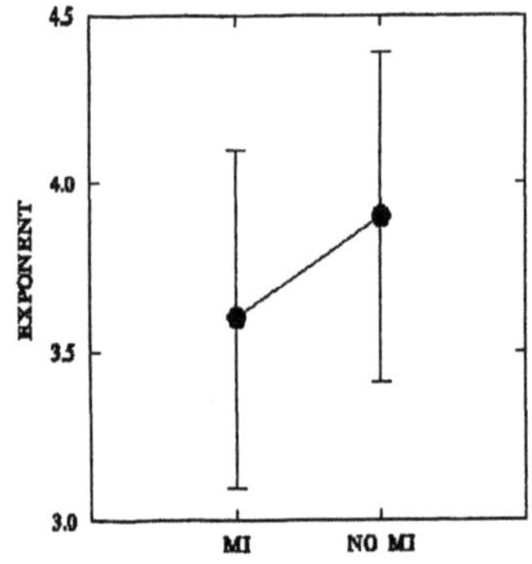

Fig. 9. Significant change in the scaling exponent in patients with a history of myocardial infarction (MI).

4 Pathologic Attractors and Entropy

An important point to consider in these power law observations is that very frequently the exponent of power law scaling is > 1, implying Brownian or other smooth processes. This is a critical observation in that it reflects the fact that the dynamics is not statistically stationary, whereas the typical 1/f behavior with exponent of one is at least quasi-stationary. Indeed, in many observations under circumstances of pathology, the dynamics approached a white noise process; i.e., invariance under time translations.

The implication of non-stationarity suggests a process which may, in fact, be a result of system transient relaxations interacting among the relevant systems. Calculation of global dimensions or entropies under such circumstances is not warranted.

Crutchfield has argued that the attractive hypothesis need not obtain in all explanations of turbulence, and indeed may not be relevant [25]. The idea is that deterministic systems can generate very long transients, whose geometry is organized by attractors which are never reached, but whose basins may be locally accessible. Information is not produced in such circumstances. This is especially true when one considers Ruelle's observation that most natural systems are not autonomous, at least for the observed (short) periods. Crutchfield also points out that if the local dynamics is periodic, when coupled to a large spatial system (cardiovascular system) the turbulence can exist for times much longer than the time of observation.

At the same time, nonetheless, attractor-like dynamics have been observed and modeled. In this context, however, it is noted that the circumstances are pathologic; e.g., an isolated, perfused rat heart which is "running down" in time. And other pathologic states, such as ventricular tachycardia, bigeminy, and severe congestive heart failure, may also be governed by attractors. The suggestion herein is that cardiac normalcy is governed by transiency; whereas pathology may occur as a result of "trapping" by an attractor (which may not necessarily be chaotic, and may be accessed by noise), with a reduced ability to respond to multiple "modes" of oscillation in phase space. Yet how and where does this occur?

An important observation in this context has been made by Mandell [26]. He describes ideal transformation matrices for cardiac inter-beat intervals based upon ideal theoretical asymptotic probability measures derived from an analogue, codimension-2 endomorphism:

$$M_1 \quad (mode\ locked\ bigeminy), \tag{9}$$

$$M_2, M_3 \quad (ergodic\ mixing), \tag{10}$$

$$M_4 \quad (ventricular\ tachycardia). \tag{11}$$

Furthermore, the entropy, h, of these processes is ordered as

$$h(M_3) > h(M_2) > h(M_1, M_4) = 0, \tag{12}$$

where h is the topological entropy [27]. And according to Shannon's coding theorem, the entropy of a receiving channel must be greater than the source. Consequently, if the beating of the heart approaches a low value (as in the case of tachycardia, or any other regularity), it becomes informationally isolated from the body's controls. (There is some debate regarding the status of fibrillation; i.e., whether or not it is an attractor [28]. Several studies suggest that fibrillation may be governed by a point attractor--the ultimate singularity [29,30].)

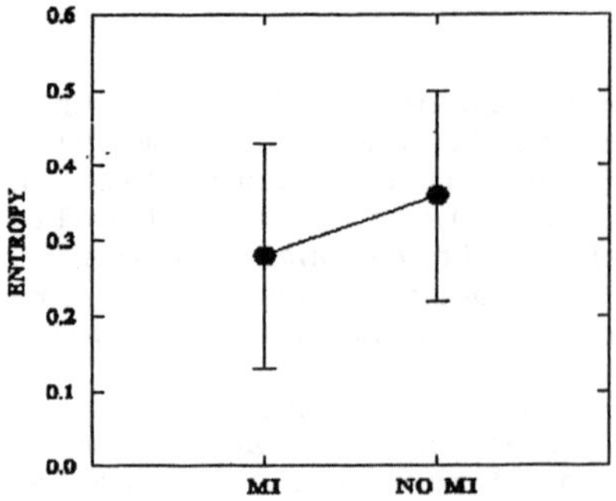

Fig. 10. Data from Ref. 32 demonstrating a decrease in the entropy of heart beat intervals in patients with a positive diagnosis of myocardial infarction.

Such a possibility has been demonstrated in the case of patients experiencing significant cardiac events, and was further demonstrated by Soviet cardiologists over a decade ago [31,32]. (Fig. 10) These studies, however, probably do not reflect asymptotic measures; instead, in the context of transients, such experimental observations suggest that sub-basins, "tendrils," and near-by dynamics may be just as important as attractors.

Further insight into the possibility that <u>local</u> dynamics are important can be obtained by considering local divergence rates for dynamical systems, which may provide the basis for a significant change in local entropy and is given by

$$\gamma(t) = \frac{d}{dt}\ln\left(\sum_{i=1}^{d}\delta x_i^2\right)^{1/2}. \tag{13}$$

An orthonormal set of vectors, $\delta_i^{(n)}$, the principal axes of an d-dimensional ellipsoid, define local divergence rates in various directions. The time averages of these are then the Lyapunov exponents [8,33]

$$\lambda_n = <\gamma_n(t)>_t = \lim_{t\to\infty}\frac{1}{t}\int_0^t \gamma_n(t)dt. \tag{14}$$

Local entropies can be derived from the local divergence rates, which may be important for the understanding of transient dynamics. Kaneko has analyzed cellular automata models for attractor entropy, and even relatively simple one-dimensional models can exhibit very long transients, with multiple attractors, and oscillations possibly governed by noise dynamics.[34]

Perhaps the most vulnerable component of the cardiovascular system is the heart's conduction system. A scaling exponent of 4 implies fairly long time correlation, which might be expected of a depolarization process which needs to be well coordinated in order to effect precise, efficient, contraction of the myocardium, yet it must be robust enough to withstand normal, physiologic "noise." Several models of cellular automata for spatio-temporal systems have been devised which can perform in this way [19,35-37]. Additionally, it has been suggested by Ott, et al. that, if chaotic attractors ensue, changes to periodic oscillations or other forms of attractors can easily be accomplished [38]. Nonetheless, if large enough portions of the myocardium are affected (critical mass?) deleterious results can ensue.

One way this can occur is through so-called "triggered activity" related to calcium oscillations. The nonlinear dynamics of the sarcoplasmic reticulum (SR) relative to its calcium oscillations may be an important factor with respect to the generation of noise and transients. The SR is important for the release of calcium in cardiac cells for the purposes of conduction and excitation-contraction coupling. Under normal conditions it has been reported to be an oscillator with a broad-band frequency, but a variety of behaviors ensues under overload conditions.

The implications for such oscillations have recently been delineated by Lakatta.[39] Not only can spontaneous oscillations of the SR trigger arrhythmias, it can lead to abnormal diastolic tonus, and limited systolic function as well, as is commonly seen in patients with congestive heart failure. And this, certainly, is one area which has not been adequately investigated; namely, the relationship between mechanical properties (including deformations) and conduction [40,41].

5 Diagnosis of Transients

The question remains, however, of devising a method which could demonstrate important local changes in divergence and entropy. Power law scaling unfortunately requires relatively stationary ensemble averages, and does not always elucidate non-linear processes. A graphical tool suggested by Eckmann, et al. [42] for use with non-autonomous systems, namely, recurrence plots, may provide a useful solution to the problem. It has already been used in the context of transients in spatio-temporal systems, speech signals, and heart beats.[23,43,44]

While commenting upon the difficulty of dealing with non-autonomous systems, Eckmann, et al. demonstrated that a time series suitably embedded in $\Re^n$ can be plotted on a symmetric matrix according to their distances, $x_i - x_j$: dots are placed on the matrix when a specified number of nearest neighbors are obtained within a ball of given radius from a point on the matrix. The resultant plot can demonstrate not only the existence of transients, and otherwise unobserved periodicity, but also is related to Lyapunov exponents and dimensions: areas which are "paler" away from the diagonal typically imply transients; Lyapunov exponents are related to the inverse of the length of short lines parallel to the diagonal. And dimensions are related by the pointwise mass function:

$$\frac{1}{T_{x_0}}(r) \sim B_{x_0}(r), \qquad (15)$$

where T is the recurrence time [45].

An application of this method was made to the CHF patient data mentioned previously. The algorithm was modified to utilize only the upper triangle of the matrix. In the case of the subject with tachycardia, clear patterns with increased recurrences were demonstrated just prior to the tachycardia. (Fig. 11) Implicit in this method is the realization that the entire attractor need not be reconstructed for meaningful information, nor does large embedding need to be accomplished, in that an embedding of D-1 is generally sufficient [38]. Only the area in a small neighborhood of points need be followed, since the appearance of sudden recurrences in relatively low embeddings suggests dynamics which are governed by lower dimensions, and perhaps with lower entropies [46]. The specific dynamics of such processes are not well understood, and although possible analytic models abound (Sec. 2), they are not typically useful experimentally. As a temporary solution, it is suggested that the Shannon entropy may be useful as a complexity measure for the quantification of recurrence plots: a totally random plot of recurrences would tend to exhibit maximum entropy [47].

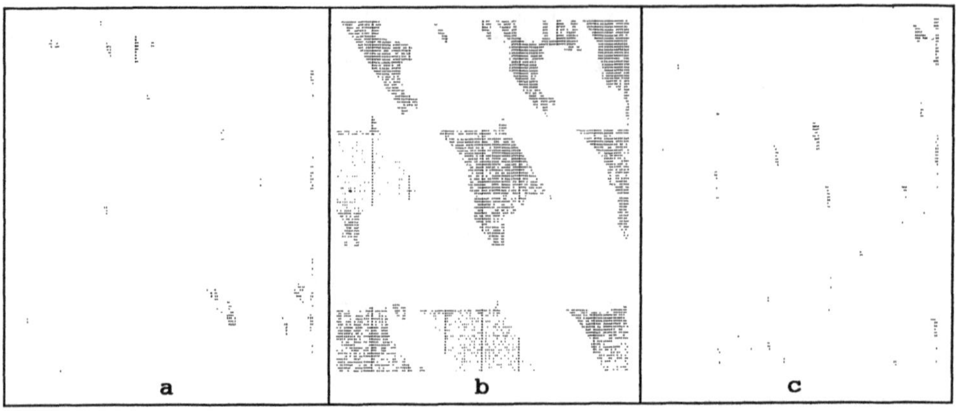

Fig. 11a-c. Use of recurrence plots in a patient experiencing ventricular tachycardia. **(a)**Baseline; **(b)**just prior to tachycardia and after the administration of one liter normal saline; **(c)**after two liters of normal saline.

6 Conclusions

In general it is proposed that: (i) the cardiovascular system is best modelled as a spatio-temporal system, which is governed by multiple inputs; (ii) these inputs are superimposed on one another, but their characteristic times are longer than the beats of the heart itself, which results in transients and inverse power laws; (iii) the coupling between these inputs varies, as do the scaling exponents (in general the most correlated systems are the most sensitive to noise perturbations); (iv) attractors (periodic or otherwise) may result from noise perturbation, and suggest a loss of physiologic responsiveness to the milieu if the associated entropy is less (as a communication channel) than the entropy of the governing system. Whether the dynamics gets stuck to a basin of attraction is dependent upon many factors, such as noise, strength of noise, strength of attraction, etc. Whether or not the attractor is reached depends upon the boundaries between multiple basins of attraction, and may explain why some arrhythmias may easily be terminated, whereas others may not, and may be related to relaxation times of transient components [38]. Certainly, it would seem that additional research is required to characterize the behavior of transients, and high dimensional complexity.

7 Acknowledgements

At various times, stimulating discussions and comments from Gottfried Mayer-Kress, Arnold Mandell, James Crutchfield, Mike Shlesinger, Celso Grebogi, and Charles Webber, Jr. have been greatly appreciated.

8 References

1. L. Glass and P. Hunter, Physica **43D**, 1 (1990).
2. D. Ruelle, Proc. Royal Soc. (Lond) A **413**, 5 (1987).
3. A. L. Goldberger, D. R. Rigney, B. J. West, Sci. Am. **262**, 40 (1990).
4. H. Takayasu, Fractals in the Physical Sciences (Manchester Univ. Press, Manchester, 1990).
5. M. Duong-Van, M. D. Feit, and G. W. Hedstrom, Phys. Lett. A **147**, 195 (1990).

6. L. Glass, A. Beuter, D. Larocque, Math. Biosci. **90**, 111 (1988).
7. M. C. Mackey and J. G. Milton, Comments on Theor. Biol. **1**, 299 (1990).
8. R. J. Deissler and J. D. Farmer, Los Alamos Preprint, LA-UR-89-4236 (1989).
9. A. Longtin, J. G. Milton, J. E. Bos, M. C. Mackey, Phys. Rev. A **41**, 6992 (1990).
10. F. Grüneis, M. Nakao, M. Yamamoto, T. Musha, H. Nakahama, Biol. Cybern. **60**, 161 (1989).
11. P. Bak, C. Tang, K. Wiesenfeld, Phys. Rev. Lett. **59**, 381 (1987).
12. J. P. Saul, News in Physiol. Sciences **5**, 32 (1990).
13. A. Selman, A. McDonald, R. Kitney, D. Linkens, Automedica **4**, 131 (1982).
14. R. D. Berger, J. P. Saul, R. J. Cohen, Amer. J. Physiol. **25**, H142 (1989).
15. F. Romeiras, C. Grebogi, E. Ott, Phys. Rev. A **38**, 463 (1988).
16. A. R. Osborne and A. Provenzale, Physica **35D**, 357 (1989).
17. J. P. Zbilut, J. Briller, N. Weinstein, W. Wiessner, in IEEE Computers in Cardiology (1990), to appear.
18. V. I. Krinskii, Biofizika **11**, 776 (1966).
19. C. H. Bennett, Physica **163A**, 393 (1990).
20. M. E. Josephson, L. N. Horowitz, A. Farshidi, S. R. Spielman, E. L. Michelson, A. M. Greenspan, Circ. **59**, 459 (1979).
21. D. J. Hearse and A. Tosaki, J. Mol. Cell. Cardiol. **20**, 213 (1988).
22. J. P. Zbilut, C. E. Lawless, M. F. O'Toole, G. Mayer-Kress, P. A. Sobotka, N. A. Voss, H. S. Loeb, Clin. Res. **37**, 893A (1989).
23. J. P. Zbilut, G. Mayer-Kress, P. A. Sobotka, M. O'Toole, J. X. Thomas Jr., Biol. Cybern. **61**, 371 (1989).
24. J. P. Zbilut, M. Koebbe, H. Loeb, G. Mayer-Kress, in IEEE Computers in Cardiology (1990), to appear.
25. J. P. Crutchfield and K. Kaneko, Phys. Rev. Lett. **60**, 2715 (1988).
26. A. J. Mandell, Physica **27D**, 235 (1987).
27. R. L. Adler, A. G. Konheim, M. H. McAndrews, Trans. Amer. Math. Soc. **114**, 309 (1965).
28. D. T. Kaplan: The Dynamics of Cardiac Electrical Instability. PhD Thesis, MIT, 1989.
29. G. Martin, J. V. Gimeno, J. Cosin, in IEEE Computers in Cardiology (1983), 213.
30. R. Dzwonczyk, C. G. Brown, H. A. Werman, IEEE Trans. Biomed. Eng. **37**, 640 (1990).
31. J. P. Zbilut and L. Lawson, Crit. Care Med. **16**, (1988).
32. E. Sh. Khalfen and B. M. Temkin, Kardiologiia (Moscow) **9**, 32 (1983).
33. J. S. Nicolis, G. Mayer-Kress, G. Haubs, in Stochastic Phenomena and Chaotic Behaviour in Complex Systems, edited by P. Schuster, (Springer, Berlin, 1984).
34. K. Kaneko, in Theory and Applications of Cellular Automata, edited by S. Wolfram (World Scientific, Singapore, 1986).
35. K. Kaneko, Physica **34D**, 1 (1989); K. Kaneko, Prog. Theor. Phys. Suppl. **99**, 263 (1989). Kaneko makes a case for the use of local Lyapunov exponents with transients. A persistent question, however, is how much data is required.
36. G. Mayer-Kress and K. Kaneko, J. Stat. Phys. **54**, 1489 (1989).
37. C. H. Bennett, G. Grinstein, Y. He, C. Jayaprakash, D. Mukamel, Phys. Rev. A **41**, 1932 (1990).
38. E. Ott, C. Grebogi, J. A. Yorke, Phys. Rev. Lett. **64**, 1196 (1990).
39. E. G. Lakatta, Perspectives Biol. and Med. **32**, 421 (1989).
40. M. J. Lab, Circ. Res. **50**, 757 (1982).
41. G. R. Heyndrickx, P. J. Vantrimpont, M. F. Rousseau, H. Pouleur, Amer. J. Physiol. **254**, H817 (1988).
42. J.-P. Eckmann, S. Oliffson Kamphorst, and D. Ruelle, Europhysics Lett. **4**, 973 (1987).
43. J. P. Crutchfield, Nuclear Phys. B (Proc. Suppl.) **5A**, 287 (1988).

44. G. Mayer-Kress and A. Hübler, in Quantitative Measures of Complex Dynamical Systems, edited by N. B. Abraham, (Plenum, New York, 1989).
45. J. Theiler, J. Opt. Soc. Am. A 7, 1055 (1990).
46. J. D. Farmer, Physica **4D**, 366 (1983). Farmer investigated the infinite dimensional delay equation and found that the entropy could remain constant despite increasing dimension. This, however, may be related to sample size.
47. P. Grassberger, in Measures of Complexity, edited by L. Peliti and A. Vulpani, (Springer, New York, 1988). See also C. Webber, Jr., this volume.

Part III

Respiratory Rhythms

The Respiratory System – Features of Modulation and Coordination

F. Raschke

Institute of Rehabilitation Research, Klinik Norderney,
Kaiserstr. 26-28, W-2982 Norderney, Fed. Rep. of Germany

Abstract. The respiratory system modulates and coordinates other physiological systems, being itself a modulated and coordinated system. Mutual dynamic interactions are the result, which are investigated in experiments on coupling mechanisms between cardiovascular-respiratory, and sensorymotor-respiratory systems.

1. Introduction

The specific human capability to intentionally use respiratory movement for phonation in verbal communication, in artistic vocal, instrumental, or motoric expressions, or to use respiratory techniques for optimization of athletic techniques and therapeutical processes, reveals a multidimensional action of respiratory muscles, ventilatory pattern, and airway system. Training by practising and intentionally guided training techniques may even lead to a specialization and adaptive improvement of those functions.

Coordinated activity of all respiratory muscle groups is an essential presupposition for these activities. However, differences in composition, intensity, and temporal order occur, according to the particular functional efficiency.

Generation of neural respiratory rhythm in the lower brain stem on a basic level is a self-sustained oscillatory process that is additionally governed by numerous feedback paths from various respiratory and nonrespiratory system functions. Rhythm generation and regulation, additionally to its autonomous function, reveals phase resetting, entrainment, or synchronization by diverse organismic functions. The mechanism of these feedback and feedforward mechanisms can be ascribed to mechanical, sensory-proprioceptive or enteroceptive, or central interneuronal coupling processes. These are described in a more general way in von Holst's terms as relative coordination [1]. This paper deals with both mechanisms: 1) respiratory rhythm as the modulator for many other organismic functions, and 2) respiratory rhythm as the dependent rhythm in terms of its modulated and coordinated activity. Our first step in approaching the complexity of these interactions was to gather data for a couple of synergistic activities within the autonomous and the vegetative systems, respectively. There follows a description of interaction between these and the sensory-motor system. Experiments with the respiratory system under natural conditions are given for locomotion, arm work, speech, and different functional states, such as relaxed wakefulness or sleep.

2. Respiratory System as a Modulator

Besides feedforward and feedback mechanisms that are involved in the respiratory pattern generation, the respiratory system itself is

Eds.: H. Haken and H.P. Koepchen

found to alter other systems' functions. In the literature plenty of experiments are concerned with the effect the respiratory rhythm exerts on sensory, motor, and autonomous, mainly cardiovascular system functions.

For the sensory system, the following have been described during the inspiratory phase: an increase in pupil diameter [2], an increase in patellar tension reflex [3], an increase in pain sensation and vigilance [4], and a decrease of acoustical reaction time [5] and pupil reflex [6], among others.

Motor functions and motor efficiency are modulated by means of the respiratory phase as known e.g. from sport. Investigations have shown that maximal force is increased during expiration [7]. Eyelid blinking [8] or tapping movements [9] preferentially occur during expiration.

Very pronounced modulations are found in systemic circulation, where hemodynamic variables (pressure, flow, volume) are modified with the respiratory cycle. They may cause effects on motor efficacy in terms of changing skeletal muscle perfusion or venous return from the lower limbs. Very many well-established investigations have been done on inspiratory cardiovascular changes: heart rate increase [e.g. 10], stroke volume and ventricle diameter increase of the myocard, right ventricular or arterial changes in blood pressure [e.g. 11], increase in skeletal vascular resistance [12,13], impairment of arterial baroreflex sensitivity [14], or increase in the occurrence of extrasystoles [15].

All these investigations reveal as a common characteristic some kind of modulation. Its direction, magnitude and phase depend on the ongoing respiratory activity. A widespread influence of respiration becomes evident, which in reality is even more complex and ubiquitous.

3. Respiratory System as a Modulated and Coordinated System

Numerous organismic functions exist where respiration becomes the dependent rhythm. The duration of the respiratory phases duration, the tidal volume, or the phase onset are modulated, being coordinated to other rhythmicities and distinct phases, e.g. those of the motor system. Weak or even strong respiratory phase resetting [e.g. 16] occurs during speech or swallowing.

On the other hand, coordinated activity is found in the ventilatory system itself. Chemosensibility of the respiratory system and intra- and extrathoracical airway patency are closely related to each other, as recent investigations have shown [e.g. 17,18,19]. Respiratory drive as assessed from both chemosensibility and neuromuscular activity considerably changes with time, leading to multioscillatory modulations of respiratory variables (tidal volume, minute ventilation, inspiratory/expiratory time, respiratory and airway muscle activity). Modulations with graded complexity have been found, which are compiled on the left side of Fig. 1. On the right side of the figure various mechanisms are suggested as coordinating respiratory system variables.

3.1 Modulation

With reference to Fig. 1, five different frequency bands are described with respect to modulation. The first type of modulation are respiratory rhythm synchronous active mechanisms keeping the airways patent. Their purpose is e.g. to prevent extrathoracical airways from collapsing during inspiratory negative pressure changes, or to pre-

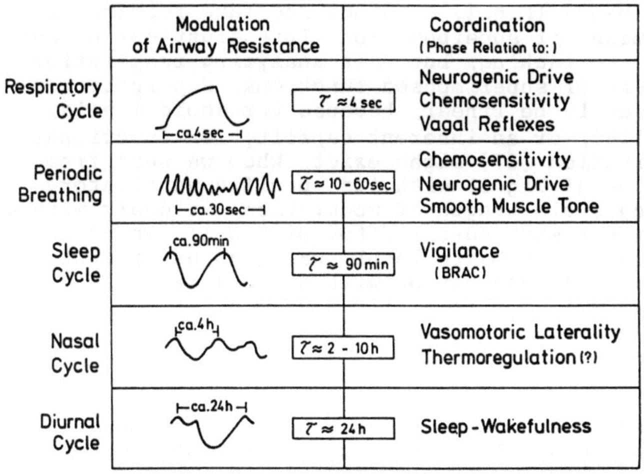

Fig. 1. Outline of five different frequency bands in the respiratory system where modulation is found in airway resistance or chemosensibility (left). Mechanisms explaining phase coordination to respiratory and nonrespiratory functions (right)

vent alveoli from collapse during expiration by increasing resistance to airflow during this phase [18]. Changes in aperture synchronous to the respiratory rhythm have been described separately for nasal [e.g. 20, 21], pharyngeal [e.g. 22], laryngeal [e.g. 23, 24], and tracheobronchial [25] segments of the airways. Proper functioning of these mechanisms essentially supposes coordinated or even synergistic neuromuscular activity between upper airway muscles, diaphragm, and intercostal muscles. Because coordinated activation of different muscle groups is realized, different cranial nerve nuclei must be involved as well as activation of respiratory motoneurons from pathways originating in the formatio reticularis of lower brain stem. There is a phase advance between dilating alae nasi [26] and inspiratory flow onset at the nares. It could be explained neither by lung compliance nor by delay in neural conduction time [18]. Myographic activity of the genioglossus muscle precedes that of the diaphragm by 70 ms [27], which likewise could be explained neither by feedback nor by conduction time. Only the assumption of a coordinated activity of the XIIth cranial nerve [18] can sufficiently explain this phenomenon.

A periodic breathing pattern [28] occurs in humans even under physiologic conditions, e.g. during exposure to high altitude or during sleep [29, 30]. Bronchial smooth muscle tone shows modulations with similar periodicities [25]. Recent investigations on overall airway resistance have shown that it changes synchronously with the periodicity of periodic breathing [31]. Reduced chemosensibility with sleep onset [32] is suggested as one of the mechanisms inducing this kind of modulation by impairment of loop gain [28]. It is not clear if an alteration in feedforward is involved [33], or if relative coordination [1] with interactions from the cardiovascular system [34] plays an additional role.

Chemosensibility [35] and airway resistance [36] vary with the sleep cycle [19]. Its duration of about 90 min is very stable. Nasal resistance exhibits a laterality cycle [20] with highly variable period length, greatly effecting the total airway resistance. Finally chemosensibility [19] and airway resistance [18] are modulated with a very pronounced 24 h rhythm.

These five frequency bands (Fig. 1) together modulate respiratory functions, probably by means of superposition. Investigations on such an integrative approach are lacking, but when analyzing respiration one should always be aware of superimposed functions. Superposition can, but need not necessarily be linear, because threshold levels, instationarity in time [18], or an inherent capacity for functional and temporal self-organization [37] might exist. When we went from the respiratory cycle (top of Fig. 1) with its specific ventilatory function to an increasing cycle length of modulation, we dealt with a more and more complex system structure (bottom of Fig. 1). The following issue will show on a more basic level, using an experimental approach, how coordination of respiration with other system functions is realized.

3.2 Coordination

Experiments are listed which show respiratory phases to be phase coordinated to a) the cardiovascular system under resting conditions or during sleep, b) locomotion during walking and jogging, c) sensory-motor system during arm work and speech.

3.2.1 Cardiovascular-Respiratory Coordination

Figure 2, with standardized cardiac cycle plotted on the x-axis and the onset of inspiration plotted on the y-axis as a histogram, shows that inspiratory onset starts preferentially at distinct phases of the cardiac cycle. Inspiration during wakefulness (W, just after awakening) is hardly phase coincident. Light sleep (1, 2) is characterized by systolic, stages 3 and 4 (deep sleep) by diastolic inspiratory onsets. Additionally the degree of coherence varies with the depth of sleep as derived from the height of the peaks. Maximum relative coor-

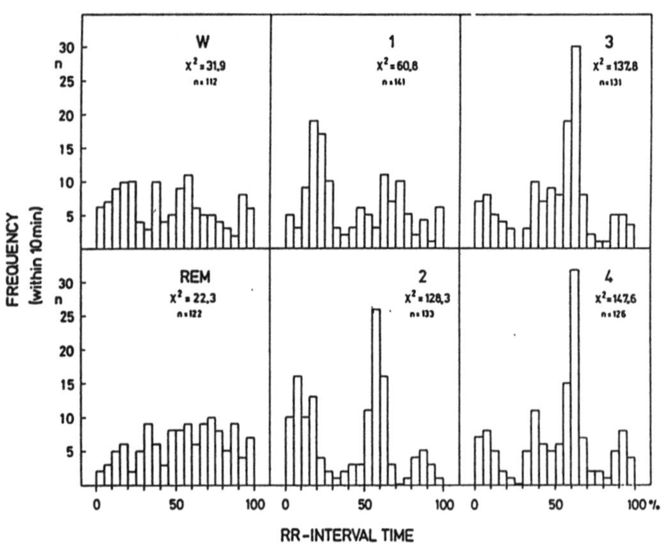

Fig. 2. Histograms of inspiratory onset during normalized cardiac cycle divided into 20 classes. 6 different sleep stages (W=awake, REM=rapid eye movement, stages 1, 2, 3, 4) of 10 minute records during each sleep stage of a healthy male subject. Chi^2 values indicate coupling rate

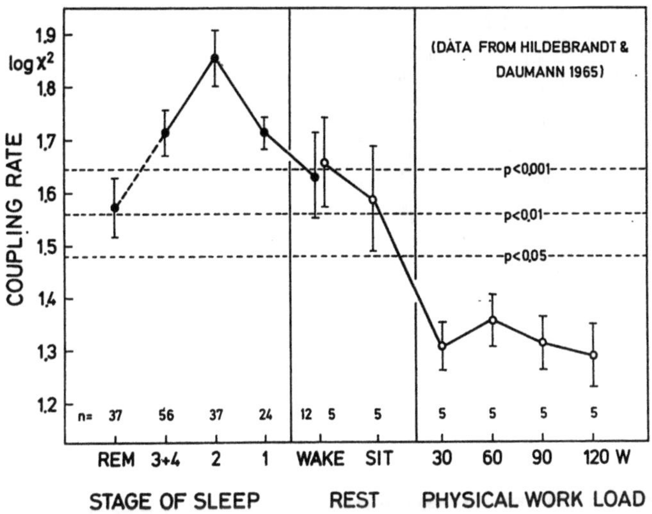

Fig. 3 Comparison of cardiorespiratory coupling rate (log chi^2, cf. fig. 2) in different sleep stages, awake but resting (supine and seated), and graded bicycle ergometer load (from [39]).

dination is found during sleep, as further analysis has shown [38], whereas wakefulness impairs the degree of coupling, which is even lost during physical strain. These results are shown in Fig. 3, where chi^2 values, taken as the deviation from a uniform distribution, have been calculated. A slight stress of 30 W leads to an abrupt complete loss of cardiovascular-respiratory phase coordination, as has also been found for psychovegetative strain [39].

3.2.2 Coordination of Locomotion and Respiration

Experiments on a treadmill were carried out in order to compare phase coordination of motor and respiratory phases in untrained and well-trained athletes. We investigated 10 subjects during walking (4-5 km/h) and jogging (7-9 km/h) with no restraints on breathing pattern, which was recorded with thermistor techniques without motor artifacts. Right heel strike as related to inspiratory or expiratory phase onset was recorded and plotted as a histogram in Fig. 4, normalized for stepping rate as done for heart rate in Fig. 2. The results, obtained from 5 minutes of locomotion in each subject and condition, are given as sum histograms totalling all subjects. Inspiratory phase coordination is shown at the top, expiratory phase coordination below. Arrows indicate heel strikes. Several characteristics become evident: 1) Inspiratory onset basically is more strongly coupled with locomotion than expiratory onset (chi^2-values). 2) During walking the untrained show a less distinct reference point for inspiration onset. 3) Untrained subjects preferentially show inspiratory and expiratory onset coinciding with the beginning or middle of the stance phase during jogging. 4) Inspiratory histograms of athletes reveal very distinct phase reference points.

In order to check if there is any advantage for distinct phase positions, we did the following supplementary experiment: Subjects were asked to start their inspiratory onset according to an acoustical signal which was phase delayed to heel strike at various interval times without their knowledge. Oxygen uptake was used as an indication

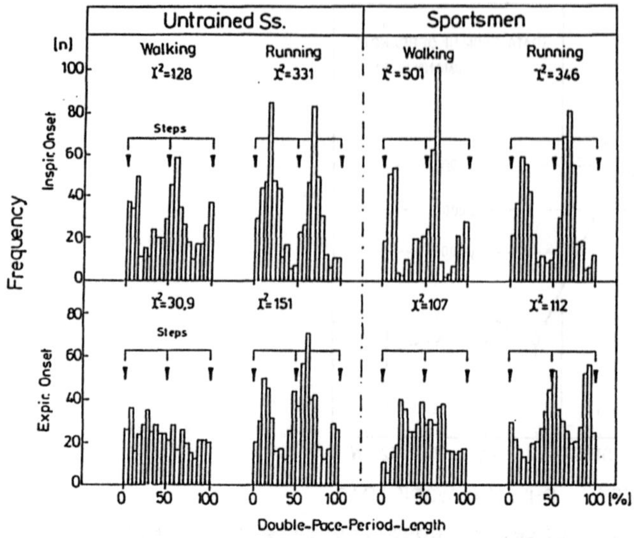

Fig. 4. Sum histograms of inspiratory and expiratory onset during normalized double pace period length during treadmill ergometry. Walking (4-5 km/h) and running (7-9 km/h) plotted separately for 5 untrained subjects and 5 athletes. Arrows indicate heel strike. Chi^2 values indicate coupling rate.

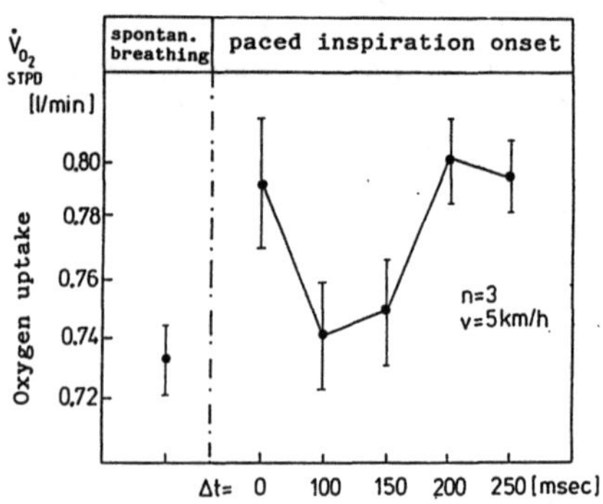

Fig. 5. Mean oxygen uptake during walking on a treadmill for 5 min during each condition. Spontaneous breathing on the left, paced inspiratory onset on the right with phase delays of 0, 100, 150, 200, and 250 ms to right heel strike. Frequency ratio breathing/pacing 1: 3. Mean +/- SE of an athlete on 3 different days.

of functional economy. The results are shown in Fig. 5, where the oxygen uptake of one trained subject was tested on 3 separate days under conditions of spontaneous breathing (left), and during paced breathing with varying phase references (right). Spontaneous breathing shows a minimum of oxygen uptake. Shifting phase reference over pace shows a phase delay between 100 and 150 ms after heel strike to come closest to spontaneous breathing conditions. Thus it can be concluded that, with regard to energy expenditure, an optimal matching is already realized under spontaneous conditions, which is reestablished when voluntary breathing is put into identical phase reference. (Mean athletes' stepping rate in Fig. 4 was 106/min, equalling 566 ms. 10

classes each of 56.6 ms gave peaks in the 2nd and 3rd class, which are equivalent to a phase delay from 113 to 169 ms between heel strike and inspiratory onset.)

3.2.3 Coordination of Sensory-Motor and Respiratory System

Motor coordination of limbs during locomotion [40], bimanual activity [41], and phase transitions of motor activity [42] have been investigated recently. Coordination of limb movements and respiration has been investigated for pedalling [e.g. 43, 44], abduction-adduction movements [45], and in a number of sports activities [7], where it is of great importance for a functional optimization [18]. But an analytical reference providing a rationale for training techniques in sports or medicine is still lacking. In order to obtain an insight into what coordination is like in everyday life, the following experiments on translatory arm movements and phonatory breathing have been arranged.

Fig. 6 gives the results of coordination between different translatory arm movements and spontaneous respiration under graded force exerted on a workpiece. From left to right, filing, sawing, forced filing, and pushing with maximal force were performed. It becomes quite obvious that with increasing force, coordination as calculated from chi^2 values (cf. Fig. 2) increases monotonically, switching from relative coordination [1] on the left to absolute coordination, i.e. synchronization, on the right. Additionally, the respiratory phase that reveals the stronger coherence to the motion switches from the inspiratory phase, for a mild load, to the expiratory phase for a heavy load.

Investigations on speech controlled breathing have shown that behavioral control of breathing overrides metabolic control [46]. The following experiment was performed to investigate how the timing of phonation interacts with the ongoing respiratory rhythm. In this experiment phonation of the greeting formula "hallo" was started at 13 different phases of the respiratory cycle. Respiratory flow and volume (pneumotachograph), O_2 and CO_2 concentration were recorded continuously for each respiratory cycle. Fig. 7 shows the result. Phonetic ventilation is shown in the upper volume curve as short downward

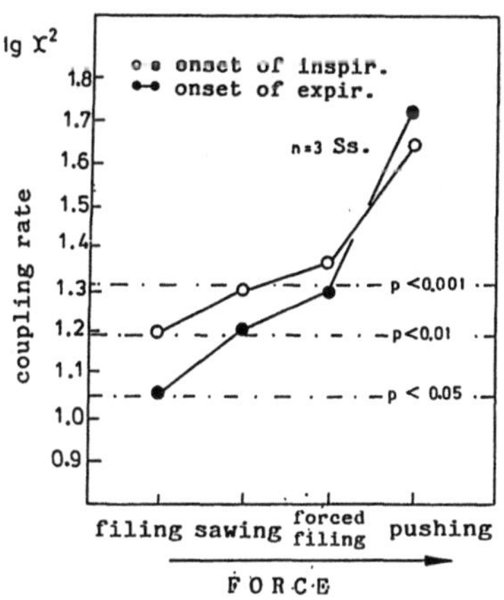

Fig. 6. Log chi^2 values of sensorymotor-respiratory coordination during different translatory arm movements with increasing force from left to right. Mean of 3 subjects. Separate calculations for inspiratory and expiratory onset.

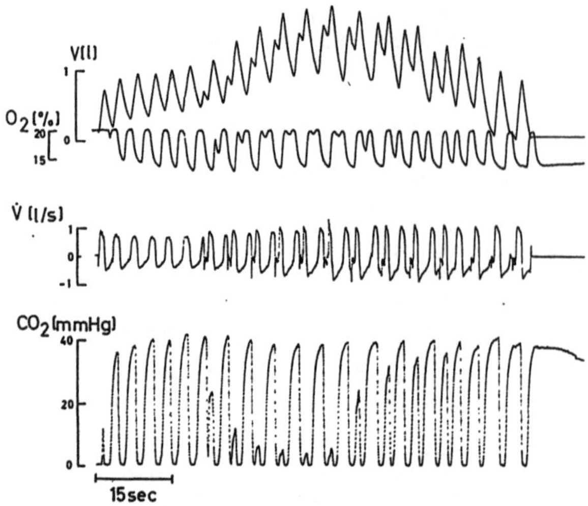

Fig. 7. Original pneumotachograph record of vocal breathing while exclaiming the greeting formula "hallo" in 13 different phases of the ongoing respiratory rhythm. From top to bottom: volume (lung capacity), O_2 concentration, flow, CO_2 concentration. Inspiration upwards in flow and volume curves.

movements, whereas inspiration goes upwards. On the left the greeting is started early in inspiration, changing to expiratory phonation on the right. The elevated level of this curve, returning to the initial level shows that there is a tremendous increase in functional residual capacity which is compensated during 'expiratory' phonation. Increase in end-expiratory O_2 concentration, frequency doubling of upward inspiratory flow movements, and decrease in end-tidal CO_2 concentration are together indicative of hyperventilation but insufficient gas exchange when phonatory resetting of respiratory rhythm occurs during inspiration. But onset of phonation during natural expiration brings the flow and volume patterns, and gas exchange back to normal values. A permutation of the sequence of phase reference was examined, giving the same result.

This investigation clearly demonstrates that an optimal matching between behavioral and autonomous control of respiration is an essential prerequisite for the metabolic balance, which can however be overridden by speech demands, but only for a very short time. Further investigations on speech controlled breathing have shown that during normal talking with moderate voice hypoventilation but sufficient gas exchange is realized, probably induced by altered mechanics of gas exchange during phonation. Future experiments are expected to contribute to the assessment of training rationales for phonetics, therapeutical use of physical therapy in respiratory medicine, or even the teaching of singing or playing wind-instruments.

4. Conclusions

We have considered a number of experiments where two capabilities of the respiratory system, autonomic and behavioral control, are exhibited. Dichotomous regulations are found as a peculiarity in this system. Easy switching from one system state to the other is an essential necessity. One realization is the interruption of the actual respiratory cycle, known as resetting. It is needed for speech, swallowing, coughing, laughing, sniffing, etc. On the other hand, longer lasting behavioral respiratory control, such as during locomotion, occupational work, or lecturing, requires the coexistence of both mechanisms, at least realized as 'relative coordination' [1]. Actually we are not able to describe or understand the underlying mechanisms in detail, but the theory of synergetics [37] has offered new analytical pathways as a means of approach. Compared to these complex mechanisms, control modalities become rather simple when we go to sleep, as shown for cardiovascular-respiratory coordination. In this case the 'only' interactions are modulation of cardiac cycle time by respiration (respiratory sinus arrhythmia, cf. [10]) and respiratory phase switching, triggered by cardiovascular events [39]. We have tried to separate these dynamic interactions, looking only at a quasi-stationary state. Numerous future efforts are required to completely analyze the dynamic autonomic processes as a whole.

References

1. E. v. Holst: Erg. Physiol. **42**, 228 (1939)
2. P. Borgdorff: Am J. Physiol. **228**, 1094 (1975)
3. W. Schmidt-Vanderheyden, L. Heinrich, H. P. Koepchen: Pflügers Arch. **317**, 56 (1970)
4. P. Engel: *Über Veränderungen der Reaktionszeit und ihrer Variabilität im Rhythmus der Atmung* (Med. Inaug.-Diss., Marburg 1963)
5. P. Engel, G. Hildebrandt: Pflügers Arch. **280**, 333 (1964)
6. K. Golenhofen & P. Petranyi: Biol. Kybernetik **4**, 55 (1967)
7. W. S. Farfel: *Bewegungssteuerung im Sport* (Sportverlag, Berlin 1977)
8. H. Engel: *Untersuchungen über die Koordination des Lidschlages mit Herz- und Atemrhythmus* (Med. Inaug.-Diss.; Marburg 1963)
9. E. E. Poole: Nature **189**, 579 (1961)
10. H. P. Koepchen: in *Autonome Innervation des Herzens - Medikamentöse Therapie bradykarder Rhythmusstörungen.* ed. by B. Brisse, F. Bender (Steinkopff, Darmstadt 1982)
11. M. De Burgh Daly: In: *Handbook of Physiology: The Respiratory System.* vol II, (Am. Physiol. Soc., Bethesda 1986)
12. H. P. Koepchen, H. Seller, J. Polster, P. Langhorst: Pflügers Arch. ges. Physiol. **302**, 285 (1968)
13. W. Jänig, H. Kümmel, S. Wiprich: in *Central Interaction between Respiratory and Cardiovascular Control Systems* ed. by H. P. Koepchen, S. M. Hilton, A. Trzebski (Springer, Berlin 1980)
14. D. L. Eckberg, Y. T. Kifle, V. L. Roberts: J. Physiol. **304**, 489 (1980)
15. J. M. Patrick, Z. Gibson, S. P. Hanley: Clin. Science **67**, 225 (1984)
16. D. Paydarfar, F. L. Eldridge: Am. J. Physiol. **252**, R55 (1987)
17. D. Weiner J. Mitra, J. Salamone, N. S. Cherniack: J. Appl. Physiol. **52**, 530 (1982)
18. F. Raschke: *Prinzipien der Atmungsregulation - Experimentelle Beiträge* Habilitations-Schrift Fachbereich Humanmedizin, Universität Marburg (1988)
19. F. Raschke, K. H. Möller: Pneumologie **43**, 568 (1989)
20. R. Eccles, R. J. Lee: Acta Otolaryngol. **9**, 257 (1981)
21. K. P. Strohl, C. F. O'Cain, A. S. Slutsky: J. Appl. Physiol. **52**, 1432 (1982)
22. O. P. Mathew, J. E. Remmers: in *Sleep and Breathing*, ed. by N. A. Saunders, C. E. Sullivan (Dekker, New York, Basel 1984)
23. D. Bartlett: in *Handbook of Physiol.* Sect. 3 *The Respiratory System* Vol. III (Am. Physiol. Soc., Bethesda 1986)

24. A. Brancatisano & L. A. Engel: in *Respiratory Function of the Upper Airway*, ed. by O. P. Mathew, G. Sant' Ambrogio (Dekker, New York, Basel 1988)
25. R. A. Mitchell, D. A. Herbert, D. G. Baker: J. Appl. Physiol. **58**, 911 (1985)
26. K. P. Strohl, M. J. Hensley, M. Hallett, N. A. Saunders, R. H. Ingram: J. Appl. Physiol. **49**, 638 (1980)
27. E. Önal, M. Lopata, T. D. O' Connor : J. Appl. Physiol. **50**, 1052 (1981)
28. M. C. K. Khoo, R. E. Kronauer, K. P. Strohl, A. S. Slutsky: J. Appl. Physiol. **53**, 644 (1982)
29. P. Webb: J. Appl. Physiol. **37**, 899 (1974)
30. A. Berssenbrugge, J. Dempsey, C. Iber, J. Skatrud, P. Wilson: J. Physiol. **343**, 507 (1983)
31. G. Warner, J. B. Skatrud, J. A. Dempsey: J. Appl. Physiol. **62**, 2201 (1987)
32. E. Önal, D. L. Burrows, R. H. Hart, M. Lopata: J. Appl. Physiol. **61**, 1438 (1986)
33. H. E. Hoff, C. G. Beckenridge: Arch. Neurol. Psychiat. **72**, 11 (1954)
34. G. Hildebrandt: Verh. dtsch. Ges. Inn. Med. **73**, 922 (1967)
35. N. J. Douglas, D. P. White, J. V. Weil, C. K. Pickett, C. W. Zwillich: Am. Rev. Respir. Dis. **126**, 758 (1982)
36. M. H. Sanders, S. E. Moore: Am. Rev. Respir. Dis. **127**, 554 (1983)
37. H. Haken: *Information and Self-Organization* (Springer, Berlin, Heidelberg 1988)
38. F. Raschke: in *Temporal Disorder in Human Oscillatory Systems* ed. by L. Rensing, U. a. d. Heiden, M. C. Mackey (Springer, Berlin, Heidelberg 1987)
39. F. Raschke: in *Cardiorespiratory and Cardiosomatic Psychophysiology* ed. by P. Grossman, K. H. L. Janssen, D. Vaitl (Plenum, New York 1986)
40. J. A. S. Kelso, D. L. Southard, D. Goodman: Science **203**, 1029 (1979)
41. J. A. S. Kelso: Am. J. Physiol. **246**, R1000 (1984)
42. G. Schöner: Biol. Cybern. **63**, 257 (1990)
43. J. Kohl, E. A. Koller, M. Jäger: Eur. J. Appl. Physiol. **47**, 223 (1981)
44. D. J. Paterson, G. A. Wood, A. R. Morton, J. D. Henstridge: Eur. J. Appl. Physiol. **55**, 530 (1987)
45. B. Raßler, S. Waurick, D. Ebert: Biol. Cybern. **63**, 457 (1990)
46. E. A. Phillipson, P. A. McClean, C. E. Sullivan, N. Zamel: Am. Rev. Respir. Dis. **117**, 903 (1981)

Phase Resetting of Respiratory Rhythm – Experiments in Animals and Models

F.L. Eldrige

Departments of Medicine and Physiology,
University of North Carolina at Chapel Hill,
Chapel Hill, NC 27516, USA

Abstract. Resetting of respiratory rhythm was studied in cats and in two models (Van der Pol and Fitzhugh-Bonhoeffer) of a limit cycle oscillator. All three behave similarly when perturbed by brief stimuli, when respiratory drive is changed or when stochastic processes are added. The findings support the idea that a limit cycle oscillator underlies mammalian respiration and show that a number of features of in vivo respiration can be mimicked by limit cycle models.

1. Introduction

The respiratory oscillator generates a rhythm within a larger system which constitutes a central pattern generator that produces the alternating inspiratory and expiratory neural activity that leads to breathing. Questions about the precise location of the oscillator, the neurophysiological basis of its operation and even whether it is a pacemaker or a neural network have not been resolved. Nevertheless, some attributes of the oscillator can be studied by phase resetting experiments.

In such experiments, the respiratory cycle is perturbed by various stimuli. When the oscilator recovers its initial rhythm after such stimuli, its phase may be reset relative to that of the control. The amount of resetting depends upon two factors, one of which is the strength of the stimulus and the other the time in the cycle at which it is given. The experimental paradigm is to present a brief perturbing stimulus to the ongoing rhythm. The effect of this stimulus on subsequent repeats of the rhythmic event is determined as a function of the phase of the cycle at which the stimulus was delivered. There are a number of ways to present the findings; we have chosen to present ours as topological phase resetting curves, in which times of resetting, cophases, i.e., times from the end of the stimulus to onset of rescheduled breaths are plotted against the time in the cycle at which the stimulus was delivered, the old phase, in our case the onset of inspiration. The data are normalized by assigning a value of one to the average period of control breaths preceding the stimulus. The plot is constructed from the resetting times after stimuli of a given strength are given at old phases throughout the respiratory cycle. The experiments are then repeated several times with stimuli of different strengths, ranging from very weak to very strong.

2. In Vivo Experiments and Models

In vivo experiments were conducted in anesthetized, paralyzed cats whose vagus and carotid sinus nerves had been cut, whose arterial Pco_2 was kept constant by means of a servo controlled ventilator and whose respiratory rhythm was monitored from integrated phrenic nerve activity. Two different brief perturbing stimuli were studied; one a facilitatory stimulus produced by an electrode placed in the mesencephalic reticular formation [1], the second an inhibitory stimulus produced by stimulating the superior laryngeal nerve electrically [2].

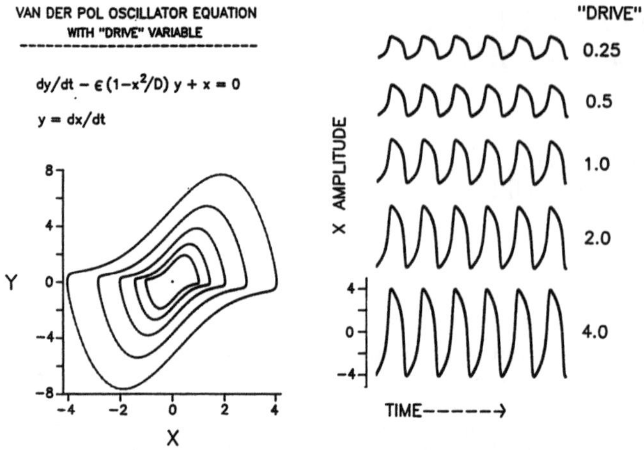

Fig. 1. Plots of limit cycles (left panel) of the Van der Pol oscillator with various "drive" (D) parameters, 0.25 (small), 0.5, 1.0, 2.0 and 4.0 (large). Right panels show rhythms and amplitudes for these limit cycles.

One model experiment used the Van der Pol (VDP) limit cycle oscillator equation [3] which had been modified to include a term for respiratory "drive" (D) [4],

$$dy/dt = E \cdot (1 - x^2/D) \cdot y - x \qquad \text{and}$$

$$y = dx/dt \; .$$

Parameter E is the model's coefficient of relaxation, which defines the shape of the limit cycle and rate of the oscillator's return to the limit cycle after a perturbation [3]. We chose a value of 2.0 because it produces a fairly rapid (approx. 0.5 of a cycle) return (see Fig. 10). Parameter D, or "drive", defines the size of the VDP oscillator in numerical terms. Increasing D has the effect of increasing the size of the limit cycle in both directions without changing its shape. The equation was incorporated into a digital computer program that generated the limit cycles. The program could be run without "noise" but, when desired, pseudorandom noise from the computer's random number generator could be added to produce small variable changes of x and y at each of the 7616 times they were computed in single cycle.

Limit cycles generated by the VDP model at different values of D are shown in Fig. 1, along with wave forms. At the center of the limit cycle is a phaseless set (x=0, y=0). On theoretical grounds, if a perturbation of appropriate strength were given at the critical old phase and were to shift the state point from the limit cycle to the phaseless set, the rhythm would be annihilated [5]. In practice, ambiguities in the digital computer make it impossible to reach the phaseless set during running of the model.

The second model was the Fitzhugh modification [6] of the Bonhoeffer-Van der Pol oscillator, the equations for which are

$$dx/dt = E \cdot (y + x - (x^3/D)/3 + z) \qquad \text{and}$$

$$dy/dt = -(x - a + by)/E,$$

where $1-2b/3 < a < 1$; $0 < b < 1$; and $b < E^2$.

In our experiments, E was 3, a was 0.7, b was 0.8 and z was −0.4 or −0.6. Figure

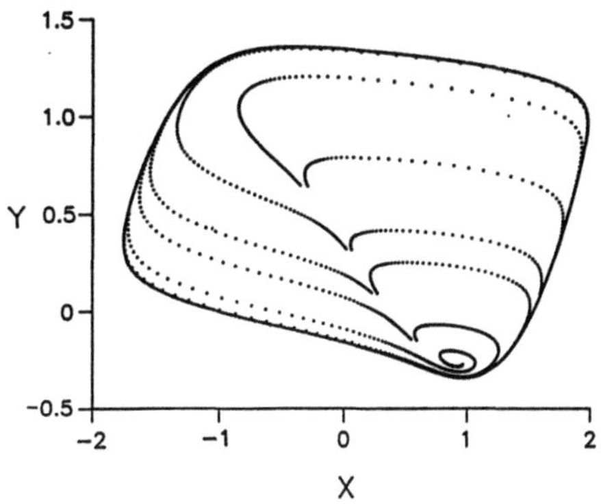

Fig. 2. Plot of Fitzhugh-Bonhoeffer oscillator with D of 1.0. Dotted lines represent trajectories of return to limit cycle after nine separate perturbations.

2 shows a limit cycle generated by the Fitzhugh model ($z = -0.4$) and the trajectories of return (dotted curves) to the limit cycle of nine perturbations of different strengths and old phases. The opposite directions of return for each of the four pairs of perturbations, matched for only small differences of strength, define a saddle-like singularity rather than a point singularity. When z is not zero, this model does not contain a point singularity and the rhythm cannot be annihilated by a single perturbation. "Noise" could be added to the Fitzhugh model in a manner similar to the VDP model.

The experimental protocol for both models was the same as for the <u>in vivo</u> experiments. First, a control cycle was generated; then a perturbation ("stimulus") was given at some time (old phase) in the next cycle and the program allowed to complete three more cycles. Old phases, times to succeeding cycles (cophases) and periods were calculated. The perturbation for the VDP model involved a single step change of y. The "stimulus" for the Fitzhugh model involved very small changes of x and y given at each iteration of the computations for a finite time, usually about 20 to 25% of the cycle. This type of stimulation mimicked the <u>in vivo</u> experiments more closely than did a single step stimulus.

3. Phase Resetting

3.1 Phase Resetting <u>in vivo</u>

With stimulation of the midbrain of the cat, we found continuous phase resetting patterns [1]: Type 1 for weak stimuli [Fig. 3, panels I (10 Hz) and II (20 Hz)] as defined in the figure; Type 0 resetting for strong stimuli [panels IV (30 Hz) and V (100 Hz)]; and unpredictable or ambiguous phase resetting [panel III (25 Hz)] when intermediate strength stimuli were given at an old phase of approximately 0.4 of the cycle, at the transition inspiration to expiration, i.e., a phase singularity. A three-dimensional plot made from the resetting patterns generated by various strength stimuli depicted a helicoid surface whose axis represents the oscillator's phase singularity (Fig. 4, left panel).

With stimulation of the superior laryngeal nerve in the cat, which inhibited respiration, the findings were similar except that the axis of the helicoid occurred at the transition from expiration to inspiration [2] rather then at the I-E transition.

These findings indicated that respiratory rhythm in the cat is generated by a continuous limit cycle oscillator, developing Type 0 resetting patterns with strong stimuli and a phase singularity with an intermediate strength stimulus,

PHASE RESETTING PATTERNS

TYPE 1--WEAK RESETTING
Resetting time (cophase) has net change of one cycle as stimulus time (old phase) varies through one full cycle.

AMBIGUOUS--PHASE SINGULARITY
Resetting time (cophase) becomes highly variable for stimuli of one intermediate strength given at unique time (old phase) in the cycle.

TYPE 0--STRONG RESETTING
Resetting time (cophase) has net change of zero cycle as stimulus time (old phase) varies through one full cycle.

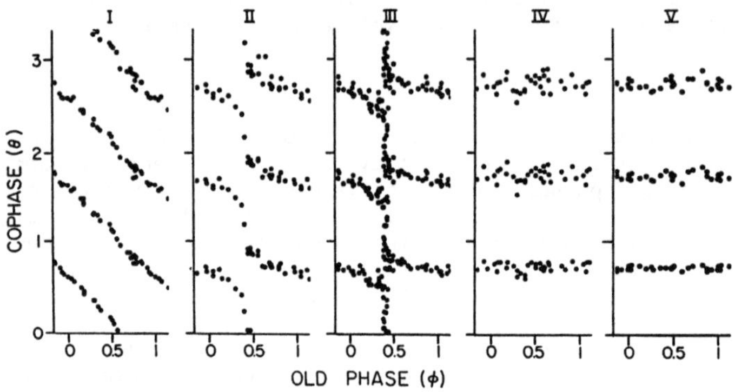

Fig. 3. Definitions of phase resetting patterns (top). Plots of resetting of respiratory rhythm in a cat (bottom); increasing strengths of midbrain stimuli from left to right, from weak (panel I, Type 1 resetting) to strong (panel V, Type 0 resetting). Panel III shows ambiguous resetting with intermediate strength stimuli given at old phase of 0.4. (Reprinted from Ref. 8).

and not by discrete on-switch, off-switch mechanisms which would show neither of these features [7].

3.2 Phase Resetting in the Models

With perturbations of various strengths both limit cycle models showed resetting patterns that were similar to those of the cat. Panels I and II of both the VDP (Fig. 5) and the Fitzhugh (Fig. 6) show Type 1 resetting with small perturbations, panels IV and V show Type 0 resetting with large perturbations, and both panels III show ambiguous resetting at the singularity when strength of the perturbations is intermediate. Figure 4 (right panel) shows the helicoid surface of resetting patterns from the VDP model; its similarity to that of the cat is obvious. The Fitzhugh model also develops a similar helicoid surface (not shown). The overall similarity of resetting patterns in the limit cycle models to those of the cat supports the idea that a continuous limit cycle process underlies mammalian respiratory rhythm generation.

3.3 Effect on Phase Resetting of Changes of Respiratory "Drive"

We observed in the experiments in cats that a change of respiratory drive, for example an increased respiratory activity produced by hypercapnia, could cause a

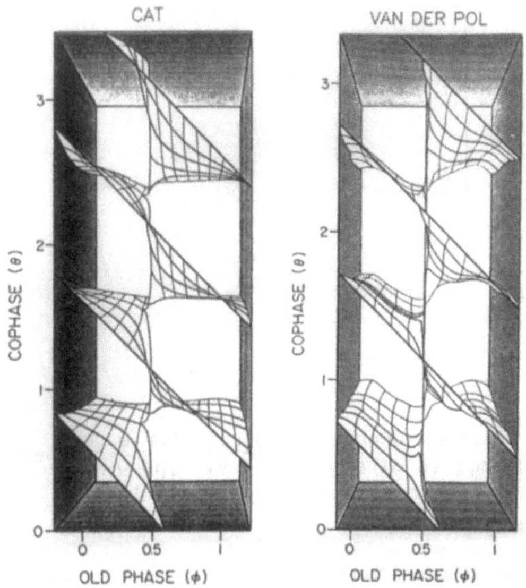

Fig. 4. Three-dimensional surfaces recontructed from resetting plots of cat (Fig. 3) and Van der Pol model. Both surfaces depict a helicoid whose boundary follows a corkscrew path around the vertical axis, which represents the singularity. (Reprinted from Ref. 9).

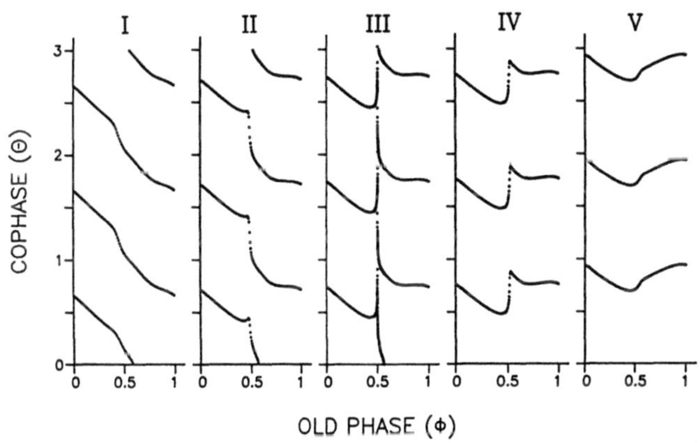

Fig. 5. Plots of phase resetting in the Van der Pol oscillator model, with weak (panel I, Type 1 resetting) to strong (panel V, Type 0 resetting) stimuli. Intermediate strength stimuli (panel III) lead to an ambiguous phase reset-ting pattern (Reprinted from Ref. 9).

change of phase resetting pattern, even though parameters of the perturbing stimulus remained constant. We therefore performed more studies in cats in which we kept the strength of stimulus constant but 1) systematicallly increased respiratory drive by increasing arterial Pco_2 or by electrically stimulating a carotid sinus nerve (CSN) and 2) decreased drive by cooling the intermediate areas of the ventral medulla [4]. Phase resetting patterns were generated before and after each change of drive, which was indexed by amplitude of integrated phrenic activity.

An example of the effect of increasing respiratory drive by hypercapnia is shown in Fig. 7. The effect of increase of drive and respiratory activity, with

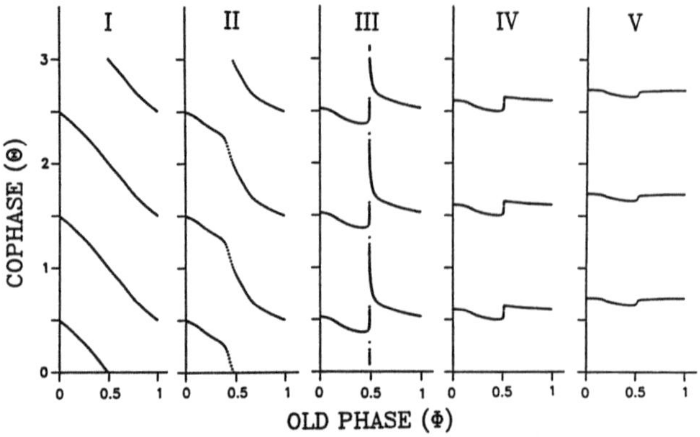

Fig. 6. Plots of phase resetting in the Fitzhugh-Bonhoeffer oscillator model, with weak (panel I, Type 1 resetting) to strong (panel V, Type 0 resetting) stimuli. Intermediate strength stimuli (panel III) lead to an ambiguous phase resetting pattern.

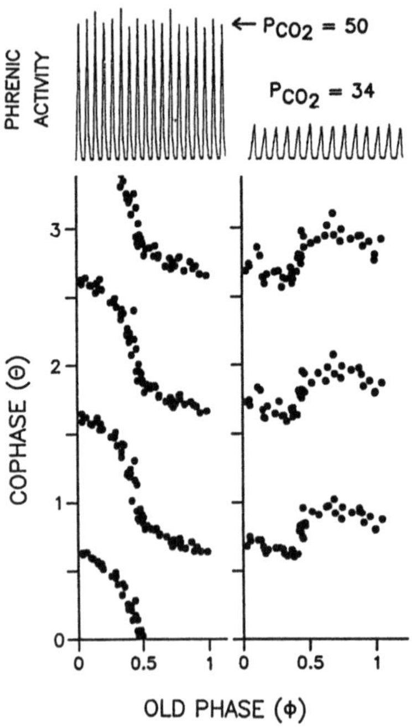

Fig. 7. Effect on phase resetting of increasing respiratory drive by hypercapnia. Right panel shows low Pco_2, a low level of respiratory (phrenic) activity and Type 0 (strong) resetting pattern. Hypercapnia and a high level of respiratory activity are associated with a weak (Type 1) resetting pattern, despite a constant perturbing stimulus strength. (Reprinted from Ref. 8).

perturbing stimulus strength kept constant, was to convert a Type 0 (strong) resetting pattern at end-tidal Pco_2 of 34 mm Hg to a Type 1 (weak) pattern at Pco_2 of 50 mm Hg. An increase of respiratory drive by means of carotid sinus nerve stimulation, while end-tidal Pco_2 remained constant, led to a similar change of resetting pattern from Type 0 (strong) when respiratory activity was

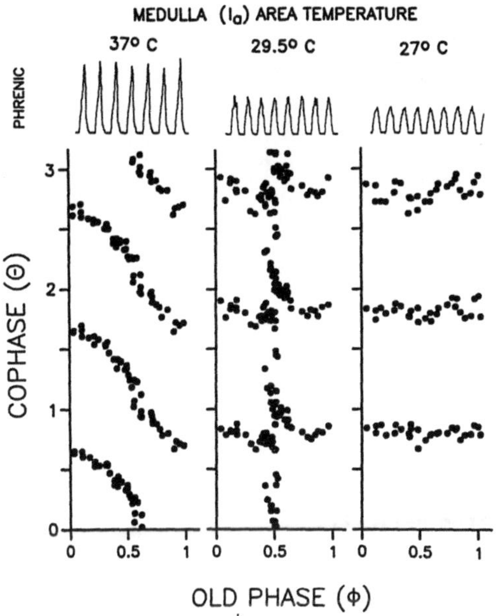

Fig. 8. Effect of phase resetting, at constant Pco_2, of changing respiratory drive by cooling intermediate areas of medulla. Constant mesencephalic perturbing stimulus. Left: Type 1 (weak) resetting when medullary temperature is 37.5 °C and phrenic activity high. Right: Type 0 (strong) resetting when temperature is 27 °C and phrenic activity low. Center: ambiguous resetting pattern consistent with phase singularity, when medullary temperature is 29.5 °C and phrenic activity is intermediate. (Reprinted from Ref.4).

low to a Type 1 (weak) pattern when respiratory activity was high [4]. This finding shows that the change with hypercapnia was not due to a direct effect of CO_2/H+ on neurons of the oscillator.

Figure 8 is an example of a cat whose respiratory drive was decreased by localized cooling of intermediate areas of the ventral medulla at a constant end-tidal Pco_2. The left panel shows the Type 1 (weak) resetting pattern found when Pco_2 (49 mm Hg) and respiratory drive were high and medullary temperatures normal (37.5 °C). When respiratory drive had been reduced by cooling of intermediate areas to 27 °C, with no change of Pco_2, the same perturbing stimulus now acted as a strong stimulus causing a Type 0 resetting pattern (right panel). Raising the local medullary temperature to 29.5 °C led to an increase of drive and phrenic activity (middle panel) and the same perturbing stimulus now led to an ambiguous resetting pattern with a wide scatter of cophases, characteristic of a singular response at an old phase of 0.5.

The studies show that a change of respiratory drive, evoked by several different types of mechanisms, affects the ability of a perturbing stimulus of constant strength to cause phase resetting of respiratory rhythm; the same stimulus that acts functionally as a strong perturbation at low drive acts functionally as a weak perturbation at high drive.

Effects on phase resetting patterns produced by changing the drive (D) parameter of the VDP model are shown in Fig. 9. All five plots were generated with the same perturbing stimulus (a change of y of 2.6063) which was chosen to produce an ambiguous pattern when drive was 1.0. There was a progressive change of pattern from Type 0 (right) when drive was small to Type 1 (left) when drive was large, even though perturbing stimulus was the same.

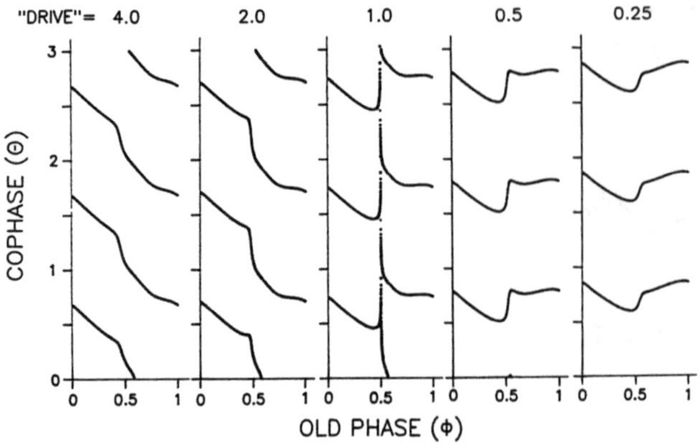

Fig. 9. Phase resetting plots generated with the Van der Pol model with constant perturbing stimulus (change of y = 2.6063) and various drive (D) parameters. As drive increases, phase resetting pattern changes from Type 0 (right) to Type 1 (left) with an ambiguous response at intermediate drive of 1.0 (center). (Reprinted from Ref. 4).

The resetting patterns with changing drive were similar in both oscillator models and were qualitatively similar to the <u>in vivo</u> experiments. Since the stimulus was provably constant in the model, the findings support a conclusion that increasing respiratory drive, by increasing the size of the limit cycle of the oscillator, leads to a change of functional effectiveness of the same stimulus and to the resulting changes of resetting patterns. Increased limit cycle size should affect functional effectiveness of the perturbation. This is shown in Fig. 10 for the VDP model. Top panels show a limit cycle with a small D of 0.25 and the effects of five perturbations of equal strength (a change of y of 3.0) given at various times in the cycle. Because of the small size of the limit cycle, all initial points after perturbation are shifted so that they do not enclose the phaseless points. Type 0 resetting therefore results [7]. Bottom panels show a limit cycle with a large D of 4.0. Because the limit cycle is large, the initial perturbed points, although shifted the same absolute magnitude as with the smaller drive, do enclose the phaseless point; a Type 1 resetting pattern is therefore generated.

Thus, phase resetting of an oscillator depends not only, as noted above, on the absolute strength of a stimulus and the phase at which it given, but also upon the size of the limit cycle. In phase resetting experiments, where effects of a perturbing stimulus are being studied, it is important that drive and the size of the limit cycle be kept constant; this will be especially important for respiration where drive can easily change. The findings also suggest that the stability of respiratory rhythm will be greater at a high level of drive than at a low level because any random perturbing effect will have a smaller functional effect. Finally, the findings have an implication for studies of entrainment because of the stronger functional effect of a perturbation when drive is low.

3.4 Effect of Stochastic Processes on Phase Resetting

Although we found continuity in the resetting plots in some animals, we also found in some animals plots that appeared to be discontinuous, particularly at the intermediate stimulus strength and at the old phase that should have yielded a continuous pattern at the singularity. An example is shown in Fig. 11, where weak stimuli led to Type 1 and strong stimuli to Type 0 responses, but the intermediate strength led to apparent discontinuities with clustering of cophases around certain values despite a number of stimulus trials at the same old phase of approximately 0.4 of the cycle.

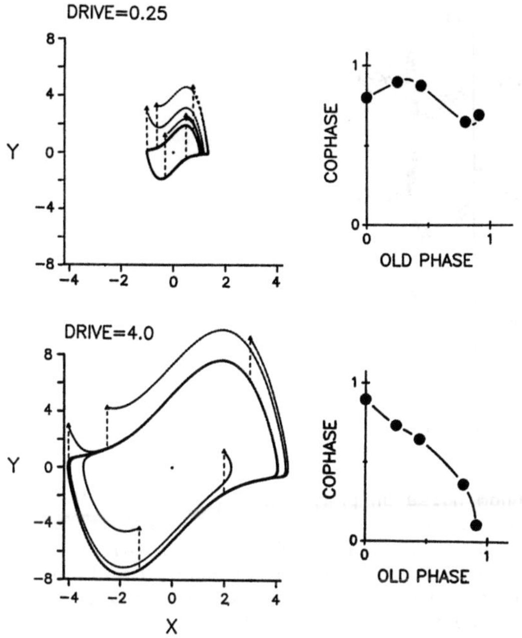

Fig. 10. Top panels: limit cycle generated by the Van der Pol model at low drive (left). There are also five separate perturbations (old phases 0, 0.25, 0.44, 0.8, 0.91) to points indicated by triangles and five subsequent trajectories of return. Because the perturbations are all on one side of the phaseless set, a Type 0 (strong) resetting pattern results (right). Bottom panel: limit cycle generated at high drive (left) and five perturbations of same magnitude as at top. Because these perturbations are on opposite sides of the phaseless point, a type 1 (weak) resetting pattern results (right). (Reprinted from Ref. 4).

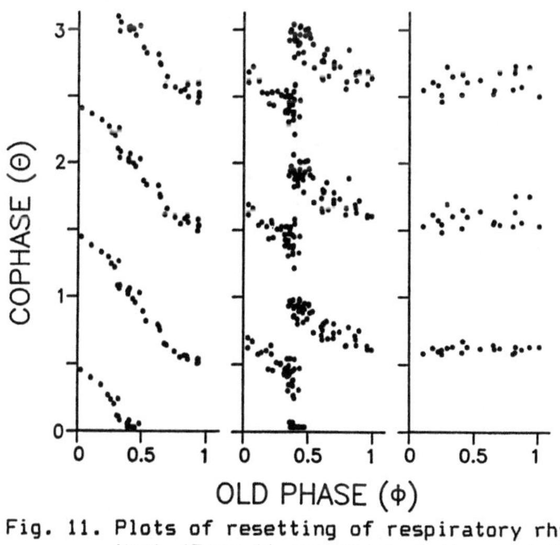

Fig. 11. Plots of resetting of respiratory rhythm in cat with midbrain stimuli. Weak (Type 1) resetting pattern at left, strong (Type 0) pattern at right. Intermediate strength stimuli yield an indeterminant pattern with apparent discontinuities of cophase despite a number of trials at old phase of 0.4. (Reprinted from Ref. 9).

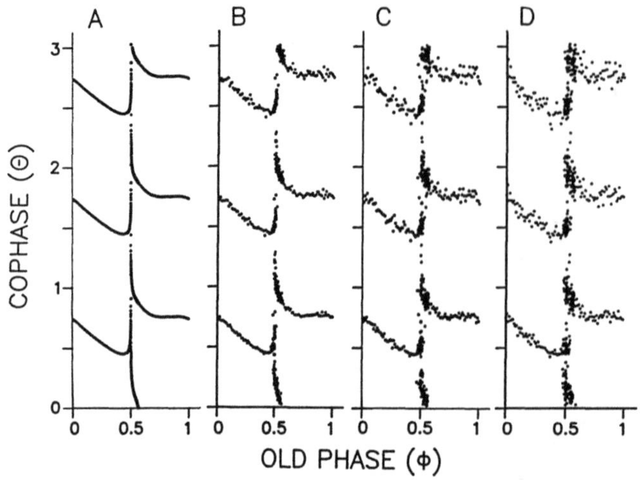

Fig. 12. Demonstration of effect of random noise on phase resetting in the Van der Pol model. All plots were generated with same variables except for addition of noise. A: no noise. B: small amount of noise. C: twice noise of B. D: three times noise of B. The continuous phase resetting pattern of the noiseless model is converted by noise to one in which clustering of cophases occurs at certain time domains.

We felt that such apparent discontinuities might be due to stochastic processes inherent in the biological preparation rather than contradicting the continuous limit cycle hypothesis. We therefore studied the effect of "noise" on phase resetting in our models. The VDP model without noise in Fig. 12A shows the resetting plot for a stimulus strength that yielded a continuous curve at the singularity (same as panel III of Fig. 5). When a small amount of random noise was added, the plot developed clear clustering of cophases (Fig. 12B). When two times (Fig. 12C) and three times (Fig. 12D) the noise were added, the clustering of cophases and appearance of discontinuity became more marked. The Fitzhugh model had the same behavior.

A quantitative analysis of this effect in both models showed that while noise causes clustering of cophases in preferred domains, it does not lead to true discontinuity. Fig. 13 graphs the cophases of the first cycle after each of 100 stimulus trials which were performed at an old phase of 0.503, near the singularity. Without noise (left panel), all 100 cophases were the same. With noise, cophases for the 100 trials at the same old phase became dispersed (right panel); although all cophases are represented, they clearly cluster at certain preferred times. The Fitzhugh model showed the same behavior.

Thus, addition of noise to the models can convert the continuous pattern of phase resetting at the singularity to one in which clusters of resetting times, or cophases, occur in certain time domains. There is, nonetheless, no true discontinuity (Fig. 13) for cophases of most durations are represented for the first cycle after the stimulus. Because of the clusterings, however, a large number of model stimulus trials was necessary to demonstrate the continuity. We suggest that in biological preparations, where stochastic processes are ubiquitous and only a small number of stimulus trials is usually possible, the probability of obtaining all cophases is small. There is thus the appearance, but not the reality, of discontinuities in the resetting plot. The apparent discontinuities in biological preparations, rather than contradicting the limit cycle mechanism, are probably explained by the underlying form of the limit cycle which, under the influence of stochastic processes, makes more likely access to certain trajectories of return to the limit cycle after a perturbation, and by the limited number of experimental points that can be obtained during an experiment [9].

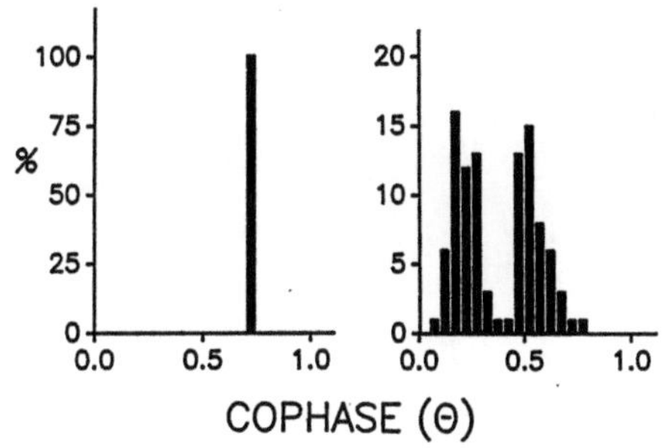

Fig. 13. Left panel: plot of cophase incidence for 100 resetting trials in the Van der Pol model when no noise is used. All cophases are the same. Right panel: plot of cophase incidence for 100 trials when noise (same as that in Fig. 12C) is incorporated into model.

Acknowledgement

This work was supported by USPHS Grant HL-17689.

References

1. D. Paydarfar, F.L. Eldridge: Am. J. Physiol. **252**, R55 (1987)
2. D. Paydarfar, F.L. Eldridge: Am. J. Physiol. **250**, R721 (1986)
3. B. Van der Pol: Phil. Mag. 2, 978 (1926)
4. F.L. Eldridge, D. Paydarfar, P.G. Wagner, R.T. Dowell: Am. J. Physiol. **257**, R271 (1989)
5. A.T. Winfree: The Geometry of Biological Time, (Springer-Verlag, New York 1980)
6. R. Fitzhugh: Biophysical J. **1**, 445 (1961)
7. L. Glass, A.T. Winfree: Am. J. Physiol. **246**, R251 (1984)
8. F.L. Eldridge: In Modeling and Parameter Estimation in Respiratory Control, ed. by M.C.K. Khoo, (Plenum Press, New York 1990), pp. 125-135
9. F.L. Eldridge: In Respiratory Control: a Modelling Perspective, ed. by G.D. Swanson, F.S. Grodins, R.L. Hughson, (Plenum Press, New York 1989), pp. 379-388

Rhythmogenesis of Deterministic Breathing Patterns

C.L. Webber, Jr.

Department of Physiology, Loyola University of Chicago,
2160 South First Avenue, Maywood, IL 60153, USA

Abstract. The physiological act of breathing functionally can be conceived as being governed by separate central rhythm and pattern generators within the brainstem/spinal cord which are subject to nonlinear afferent feedback and suprapontine feedforward inputs of varying coupling strengths. This complex network sculptures the spatiotemporal configurations and movements of the diaphragm and chest wall to effect the exchange of gases between alveolar space and the environment. As is common for many intrinsically stable physiological systems, the respiratory controller is characterized by nonstationarities and transients, the presence of which may preclude the accuracy of otherwise useful ergodic techniques in the analysis of complex systems. From the geometric perspective, however, recurrence plots are robust against dynamical transients and system drifts, and they are proving to be very helpful in assessing the dynamics of breathing patterns. Their general applicability to numerous other rhythmical phenomena in the physiological realm (cardiovascular, central neural, gastrointestinal systems, etc.) is intuitive. Finally, Shannon entropy, a gauge of randomness and information, and percent recurrence are introduced as possible measures of complexity for biological systems in quasi-steady states or transition phases.

1. Introduction

As life is synonymous with rhythms, so physiological organisms are best equated with complex systems. This complexity extends both to time and spatial domains which, in synergetic combination, result in rich repertoires of behavior, enabling life forms to adapt to and survive in a wide range of environmental situations and challenges. Due to inherent nonlinearities in biological systems, the intrusion of perturbing noise of external (environmental) or internal (dynamical) origin, and the ubiquitous presence of nonstationarities and discontinuous transients on different length and time scales, meaningful investigation of living dynamical systems becomes a formidable task. Despite these pragmatic difficulties, the study of chaotic dynamics in biological systems is gaining momentum [1]. Recent advances, for example, suggest that nonlinear biological systems are governed in essence by transients, and that strong attractors may be pathologic [2].

2. Applicability of Nonlinear Analyses to Biological Systems

At the outset one might ask whether the powerful mathematical tools of nonlinear dynamics are applicable to biological and physiological systems in general, or to the respiratory system specifically [3,4]. At first glance this hunch seems appropriate insofar time series and spatial structures of living systems possess superficial dynamical and architectural similarities with mathematical constructs [5]. Mathematical systems are comprehended with a fair degree of quantitative sophistication (fractal dimensions, entropy, Lyapunov exponents) [6], but the same cannot be stated for living systems. On closer examination one realizes that although biological and mathematical systems share internal nonlinear characteristics, living systems are faced with added complexities such as environmental noise and state changes which induce incessant nonstationarities and system drifts [7,8].

Estimations of dimensions and Lyapunov exponents for mathematical systems such as the logistic difference equation or Hénon strange attractor presume stationarity in the system under study [9]. Practically speaking, this means that the first several hundred iterations of the deterministic equations must be discarded in order to move the initial dynamics out of basins of attraction and onto the stationary attractors themselves. To perform similar ergodic computations on living systems necessarily presumes system stationarity which, if not present, would be expected to distort any forthcoming complexity estimation to some unknown degree. It remains unclear just how robust such mathematical estimations are for biological systems in the presence of long time scale drift and short time scale noise. Also, even if stationarity can be assured, calculation of the correlation dimension according to the algorithm of Grassberger-Procaccia has been shown to possess systematic errors for embedding dimensions greater than 5 [10]. Clearly, caution is called for in computation of estimates describing complex systems characterized by adiabatically fluctuating parameters [11] and multiple negative feedback loops, the latter of which may actually stabilize the dynamics [12].

3. Dynamical Breathing Patterns

Spontaneous breathing patterns were recorded in unanesthetized, unrestrained rats which were instrumented for the chronic recording of intrapleural pressure more than a week earlier. Under chloral hydrate anesthesia, intrathoracic balloon catheters of high compliance were implanted across the fifth rib interspace using sterile surgical techniques. Figure 1 shows the X-ray image of a well-positioned balloon catheter close to the diaphragmatic surface and below the heart. This placement was very important to minimize cardiac artifacts in respiratory fluctuations of thoracic pressure.

Chronic rodent breathing patterns were studied for two distinct physiological states as illustrated in Figure 2. During the isolated-quiet state, rats rested in solitude in their resident cages. Breathing movements were undisturbed, relatively slow and shallow. To encourage a disturbed-active state, an intruder rat was suddenly placed into the resident's cage which resulted in prompt acceleration of the breathing frequency and increased depth of the pattern. Stimulating interactions were allowed to continue for about 30 minutes which were essentially free from fighting. To quantify the data, VCR-recorded

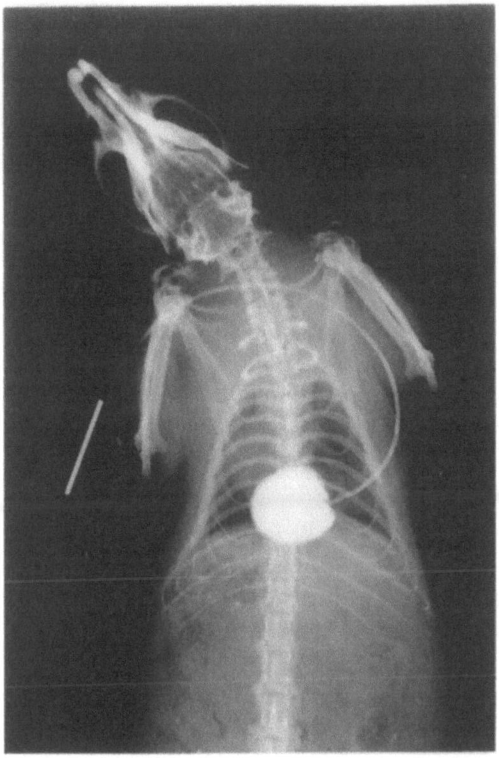

Fig. 1. X-ray image of barium sulfate filled balloon-catheter system in a rat. The 1 mL balloon within the thoracic space is centered on the diaphragm well below the heart to avoid mechanical cardiac artifacts. Calibration bar = 20 mm.

thoracic pressure changes were digitized at a rate of 500 Hz, but only discrete respiratory periods (rhythm function) and changes in thoracic pressure (pattern function) were written out to harddisk files on a cycle-by-cycle basis using a custom algorithm. Up to 4500 consecutive respiratory cycles were digitized by this means.

One important type of nonlinear analysis of breathing patterns is the representation of respiratory dynamics in the phase plane. Representative phase-plane portraits are shown in Figure 3 for the two selected physiological states of quiet and active breathing. These plots are generated by phase-delaying the thoracic pressure signal by 15 msec (digital delay unit) and plotting it against the undelayed version of the same signal. The delay necessarily opens the dynamic into loop structures (analogous to Lissajous figures), all of which are inscribed in counter-clockwise trajectories. Technically, to reveal the maximal profile of any attractor's geometry, the selection of the delay time should match the first minimum in the autocorrelation function or the first minimum in the minimal mutual information content [see 13]. The thoracic pressure loops displayed are representations of attractor-like processes which follow predictable or unpredictable paths, respectively, depending on whether the dynamic is recorded during the quiet state or active state. The quiet patterns resemble the stable limit-cycle behavior of the respiratory oscillator [14], but the active patterns

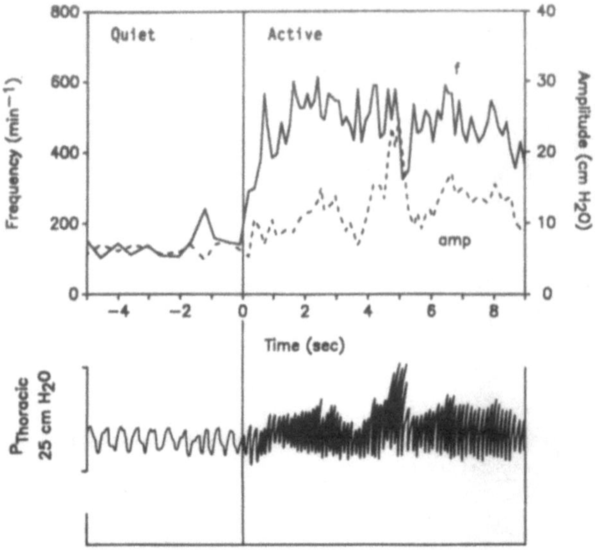

Fig. 2. Resident-intruder model of respiratory dynamics. The polygraph recording shows the quiet breathing pattern of an undisturbed rat. At time 0, introduction of an intruder rat induces an active breathing pattern. Discrete respiratory variables of frequency and amplitude are measured for each state.

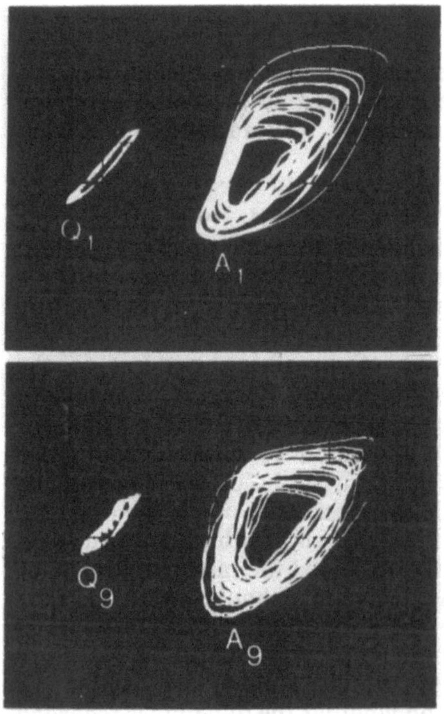

Fig. 3. Phase-space plots of respiratory dynamics. Continuous thoracic pressure fluctuations are phase-delayed (15 msec) and plotted (vertical axes) against the undelayed version of the signals (horizontal axes). During quiet breathing in two rats (Q_1 and Q_9) subsequent loops superimpose in stable limit-cycle behavior. During active breathing (A_1 and A_9), however, the loops exhibit variability and present as unstable (stochastic or chaotic) limit cycles. All loops inscribe in the counter-clockwise direction. Calibration bars = 15 cm H_2O.

show much more variability, reminiscent of the chaotic dynamics exhibited on the Rössler strange attractor [15].

4. Recurrence Plots of Dynamical Systems

Eckmann et al. [16] have introduced a graphical tool for the diagnosis of dynamical systems that performs well on short data sets beset by transients and nonstationarities. Recurrence plots reveal real, but subtle time correlation information from discrete or continuous time (or amplitude) series. This geometric technique may prove to be more useful than ergodic computations which require data points numbering in the neighborhood of ten to the power of the system's dimension. For example, a system of dimension 6 demands a data set length around one million elements, a formidable barrier against which most biological phenomena fail because they repeat only at relatively low frequencies. Even if one million events were recorded, the biological preparation most likely would have experienced several state changes which would defeat the numerical techniques just the same. To increase the digitization rate on continuous data is no solution either, for such an approach generates false correlations in the scaling region of interest and is therefore categorically disqualified. Recurrence plots bypass all these problems and are capable of revealing periodic structures and clustering properties in the reconstructed signals that are not apparent in the original time (or amplitude) series [17].

The recurrence plot methodology has been successfully applied to rodent breathing patterns as is herein described. Figure 4 catalogues this methodology for generating recurrence plots using discrete respiratory periods, although the mathematics works well for any contiguous data set. Strictly speaking, there are slight modifications from Eckman's original description, but our C-code works well on 286 and 386 microcomputers with EGA or VGA color graphics adapters. In this particular example, sequential periods are delayed (lag=1 cycle), embedded (ED=4), and averaged by the computation of the Euclidean Norm (EN). In essence, the EN symbolizes a moving time average through the data set, each vector of which is represented by a relative EN value and distinguished by a unique index number. The relative EN simply forces the average vector values to a mean of 100 without disturbing the original relative variance (coefficient of variation remains unchanged). There are alternative norms that could be used (minimum norm, maximum norm), but they were not selected for the present analyses. After performing relative EN computations, absolute distances (DIST) between all possible vector pairs (rel EN_I and rel EN_J) are then computed. Plotting DIST values at the I,J coordinates generates three-dimensional recurrence plots. Plotting points at I,J coordinates for DIST values falling within a cutoff region generates two-dimensional recurrence plots. In either case, whenever any vector is compared against itself ($I_{index} = J_{index}$) the DIST value is by necessity zero.

Three-dimensional recurrence plots produced from quiet and active rodent data (respiratory periods) are illustrated in Figure 5. The graphs are presented in spike format (left) and topological surfaces (right) to show how renditions of dynamical systems in time can be mapped as geometrical "landscapes" in space. Vector distance in the vertical dimension is plotted for each I,J vector pair being compared in the x,y plane. A line of identity extends from "southwest" to "northeast" where vectors compared with themselves have zero distances. Also,

METHOD OF GENERATING RECURRENCE PLOTS
FROM DISCRETE PERIOD DATA

Embedding Dimension (ED) = 4

Lag = 1 cycle

Euclidean Norm (EN) = $SQRT(t_1^2 + t_2^2 + t_3^2 + ... + t_{ED}^2)$

t = discrete periods of sequential respiratory cycles (sec)

ED = 1	2	3	4	EN	Rel EN	Index
0.726	0.732	0.898	0.710	1.541	102.098	0
0.732	0.898	0.710	0.548	1.465	97.063	1
0.898	0.710	0.548	0.840	1.522	100.839	2
0.710	0.548	0.840				
0.548	0.840					
0.840						

I_{index}	J_{index}	DIST = ABSOLUTE(Rel EN_I - Rel EN_J)
0	0	0.000
0	1	5.035
0	2	1.259
1	0	5.035
1	1	0.000
1	2	3.776
2	0	1.259
2	1	3.776
2	2	0.000

Recurrence Plots:
3-Dimensional Plot of I, J, DIST (x,y,z)
2-Dimensional Plot of I, J (x,y) for DIST <= cutoff

Fig. 4. Recurrence Plot Method. Discrete data (periods or amplitudes) are lagged and embedded to detect the presence of subtle periodic structure and transients in complex dynamical systems in N-dimensional space. The final data (I, J, DIST) can be used to construct two- or three-dimensional recurrence plots. See text for further important details.

the plots are symmetrical across this imposed diagonal due to the expression of distances as absolute differences between I and J vector pairs. In this example, 21 by 21 vectors are compared yielding a total of 441 vector distances which are scaled over the range from 0 to 100 relative units (ru). Because relative EN values are used in the computations, plots from different states are directly comparable and show a rougher topography for the active state ("mountains") as compared to the quiet state ("plains"). In order to navigate along this geometrical construct of a physiological dynamic, however, an interpretive "map" is needed. As will be shown, two-dimensional recurrence plots have the ability to reveal low-distance deterministic dynamical features not intuitively obvious in their three-dimensional counterparts.

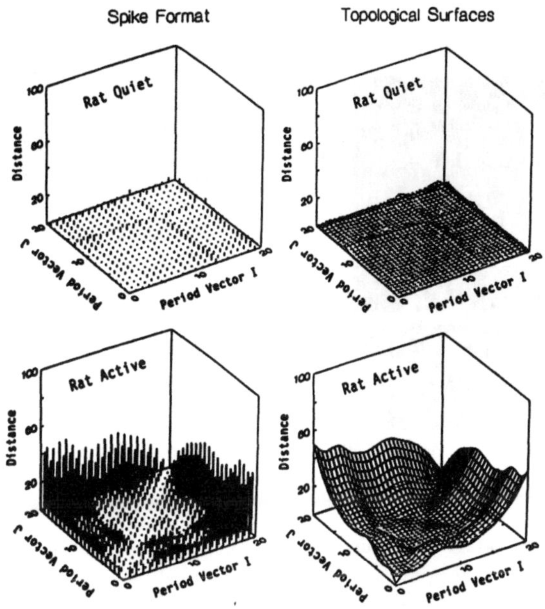

Fig. 5. Three-dimensional recurrence plots. Vector distances are plotted as amplitude functions at specific I,J coordinates either as spikes (left) or topological surfaces (right). The original input data is from respiratory periods of a rat in a quiet state (top) or active state (bottom). The change in states results in dramatic alterations in the topological representation of the global dynamic.

5. Differentiation of Deterministic Behavior from Transients

Theoretically, three distinct modes are available to any dynamical system during the synergetic, time-space evolution of observable outputs. State-dependent selection of these options include processes that are: 1) perfectly stochastic and indeterminate; 2) periodic and determinate; 3) aperiodic and deterministic. The last case represents ordered (rule-obeying) chaos which stands in stark contradistinction to true stochasticism. Chaotic systems remain unpredictable, of course, not for lack of deterministic rules as for random processes, but because initial conditions can never be known with infinite precision. Also, the important role of noise in these dynamical alternatives is not fully appreciated and remains an important topic for dynamical systems research [18,19]. The possibility of superimposition of random noise on deterministic systems underscores the importance of multimode unions and intersections to produce intermediate patterns of unique (complicated) quality.

The value and power of recurrence plots is their ability to detect transients and embedded determinism in dynamical systems as distinguished from stochastic noise. The interpretive framework (our "map") to locate these dynamical structures is now discussed. First, a two-dimensional recurrence plot for a completely periodic signal interrupted by two discontinuities is diagrammed in Figure 6. The DIST cutoff is selected so low that only vector distances equal to zero score as recurrences (black pixels). The presence of transients at cycles 10 and 15 produce the typical vertical and horizontal banding at corresponding I and

Cycle	Period (sec)
1	0.5
2	0.5
3	0.5
4	0.5
5	0.5
6	0.5
7	0.5
8	0.5
9	0.5
10	1.5 *
11	0.5
12	0.5
13	0.5
14	0.5
15	1.5 *
16	0.5
17	0.5
18	0.5
19	0.5
20	0.5
21	0.5
22	0.5
23	0.5
24	0.5
25	0.5
26	0.5
27	0.5
28	0.5
29	0.5
30	0.5

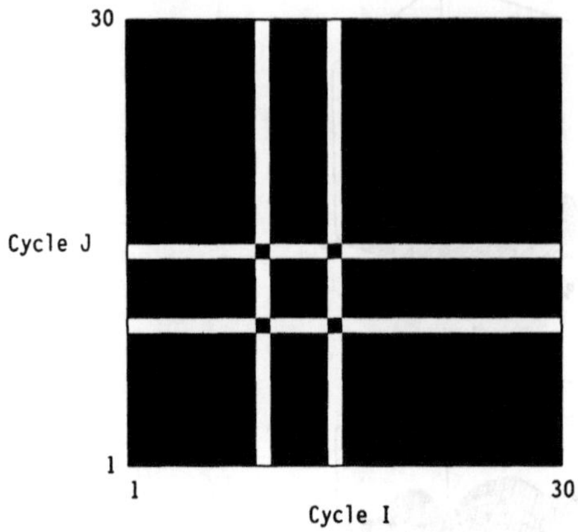

788 black / 900 total pixels = 88% recurrence

Fig. 6. Theoretical two-dimensional recurrence plot for a periodic pattern with transients. Key parameters are defined as: delay = 1; ED = 1; cutoff = 0. Recurrences are plotted as black pixels for vector distances equal to the cutoff. All cycles show recurrence with each other, except for distances involving the discontinuities at cycles 10 and 15. The two discontinuities (asterisks) produce horizontal and vertical banding at corresponding I,J positions, decreasing the percent recurrence.

J coordinates where the vector distances are much greater than zero (white pixels), save when the vectors are compared with themselves. The degree of recurrence can be quantified by calculation of percent recurrence as defined in the figure. For the respiratory system, one would anticipate that apneic pauses (long periods) or deep sighs (large pressure changes) in otherwise seeming regular data would lead to the expected horizontal and vertical patternings. Likewise, signal drifts present in nonstationary dynamics should decrease the percent recurrence and give broad horizonal and vertical regions of no recurrence. The key here is that transients can be studied by raising or lowing the selected cutoff. The use of color coding to stratify DIST values within the window becomes extremely useful in such cases.

Secondly, a two-dimensional recurrence plot of an aperiodic signal with three repeating cycles is shown in Figure 7. Again, the cutoff is selected such that recurrences are marked only when vector distances are zero (black pixels). The deterministic structure of the repeating dynamic is easily detected and plots as short diagonals to the main reference diagonal amidst a field of chance recurrences exhibiting no pattern. The diagonals arise from pattern repetitions at different times down the dynamic and become the hallmark for deterministic behavior in dynamical systems. Increasing the embedding parameter permits the detection of deterministic structure in higher dimensional systems. Also, the length of the short linear segments is, theoretically, inversely related to the

Cycle	Period (sec)
1	0.3
2	0.8
3	0.5 *
4	0.9 *
5	0.6 *
6	0.3 *
7	0.2
8	0.7
9	0.4
10	0.2
11	0.7
12	0.1
13	0.3
14	0.2
15	0.5 *
16	0.9 *
17	0.6 *
18	0.3 *
19	0.4
20	0.3
21	0.8
22	0.1
23	0.7
24	0.5 *
25	0.9 *
26	0.6 *
27	0.3 *
28	0.9
29	0.4
30	0.8

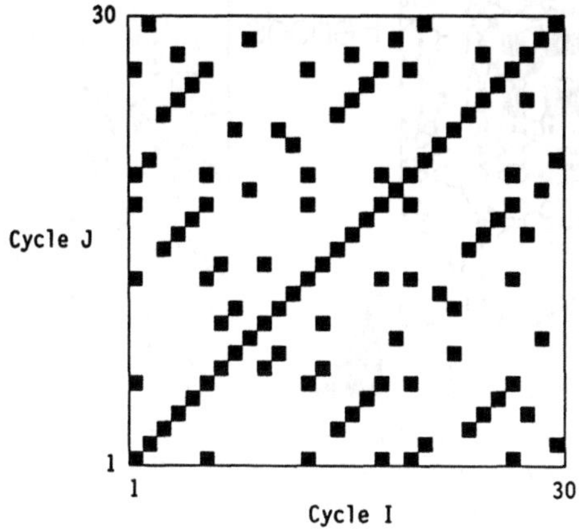

110 black / 900 total pixels = 12% recurrence

Fig. 7. Theoretical two-dimensional recurrence plot for an aperiodic pattern with repeating structure. Key parameters are defined as: delay = 1; ED = 1; cutoff = 0. Recurrences are plotted as black pixels for vector distances equal to the cutoff. Only those cycles that have identical periods show recurrence with each other. The repeating periods (asterisks) produce three short diagonal lines each four units long parallel to the main diagonal, increasing the percent recurrence.

largest positive Lyapunov exponent [16] which, when positive, signals the presence of chaotic dynamics. For example, recurrence plot analysis of either the Hénon strange attractor or the logistic quadratic difference equation in its chaotic regime results in sharply defined short linear segments. The question of interest now centers on the fine structure of two-dimensional recurrence plots of actual physiological data. What evidence is there for transient behavior and deterministic structure in the functional respiratory controller of the nervous system which, for convenience sake, can be viewed as consisting of separate rhythm and pattern generators [20,21]?

Two-dimensional recurrence plots from rodent breathing patterns are presented in Figures 8 and 9 for respiratory periods and pressures respectively. These variables can be carefully compared for quiet (top) and active (bottom) states. These high resolution graphs are generated from 480 reference cycles using a lag of 1 cycle, embedding dimension of 10 cycles, and cutoff of 1.0 ru. Next to each recurrence plot is the histogram distribution of the 230400 vector distances (480 by 480 cycles). In general, when the state changed from quiet to active there was a decrease in percent recurrence (paling of plots) for both period data (21.02% to 3.24%) and pressure data (13.10% to 5.47%) due to a wider distribution of vector distances above the fixed cutoff in the active state. Both sets of recurrence plots show evidence for the presence of transients (horizontal and vertical banding) and deterministic behavior (short diagonal line segments

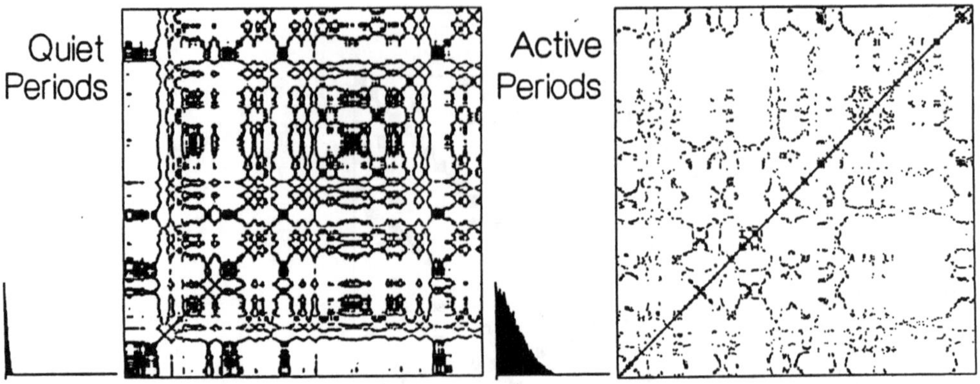

Fig. 8. Recurrence plots of respiratory periods for quiet (top) and active (bottom) states. Each plot represents data from 480 reference respiratory cycles which combine to generate 230400 vector distances. The distribution of vector distances is presented in histogram format with horizontal scaling from 0 to 150 relative units and vertical scaling from 0 to maximum counts per bin.

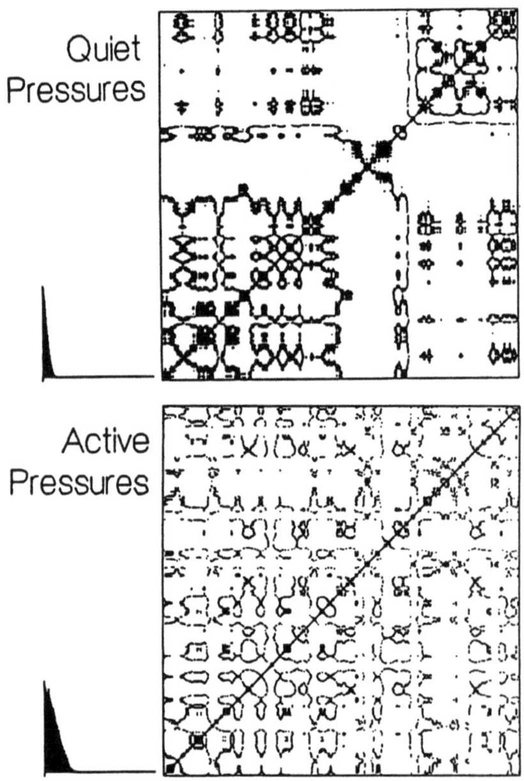

Fig. 9. Recurrence plots of respiratory pressures for quiet (top) and active (bottom) states. Each plot represents data from 480 reference respiratory cycles which combine to generate 230400 vector distances. The distribution of vector distances is presented in histogram format with horizontal scaling from 0 to 150 relative units and vertical scaling from 0 to maximum counts per bin.

parallel to the main diagonal) which leads to two conclusions regarding the respiratory controller. First, the respiratory dynamic, even in seemingly stable steady states, resides within a field of transients. Second, rhythm generators (period) and pattern generators (pressure) for breathing possess deterministic structuring despite subtle buffeting by noise processes (experimental, dynamical, environmental). Expanding on this latter issue, noise induces jitter in both timing and amplitude functions such that the short diagonal segments exhibit either minor broadening or occasional gaps.

6. Shannon Entropy and Physiological State

Classically, Shannon entropy values are used as measures of randomness [22]. In the context of information theory, increased entropy is associated with increased information and decreased predictability [23]. Shannon entropy measurements were applied to the histogram distributions of vector distances in an attempt to quantify recurrence plot information. As demonstrated in Figure 10, Shannon entropy values (H) depend on the profile of vector distances over an N-bin histogram. Specifically, the number of events per bin determine individual bin

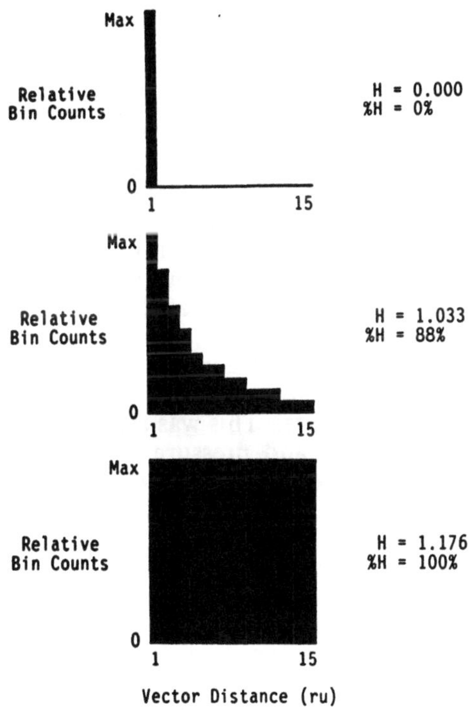

$$H = -\sum_{b=1}^{N} P_b * \log(P_b) \quad \text{and} \quad H_{max} = -\log(1/N)$$

where: P = histogram bin probability
b = bin number
N = maximum number of histogram bins (15)

Fig. 10. Calculation of Shannon entropy values (H) as measures of randomness. Theoretical vector distances are distributed into 15-bin histograms. If distances all fall within a single bin, entropy is minimized (top); if distances distribute evenly across all bins, entropy is maximized (bottom); any other histogram profile will result in entropy values between these two extremes. Percent entropy values (%H) are computed for comparison of data sets.

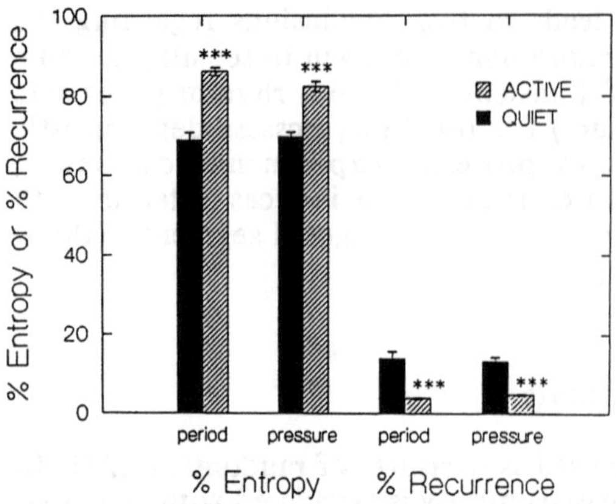

Fig. 11. Average percent entropy and percent recurrence values. Both respiratory period data and pressure data from a group of 10 rats show significant state-dependent changes in two measures of dynamical complexity. Comparing active states with quiet states, percent entropy values significantly increase and percent recurrence values significantly decrease. Statistics were performed with a paired Student's t test, the triple asterisks denoting P values less than 0.0005 (highly significant).

probabilities (P_b) which when mathematically transformed (P_b*log[P_b]) and summed compute the absolute entropy. For the 15-bin histogram shown, if all vector distances fall into a single bin, absolute H is 0.000 units. Alternatively, if the vector distances are equally distributed across all bins, the absolute entropy maximizes at 1.176 units. Between these two extremes reside non-zero, sub-maximal entropy values depending upon the specific distribution profile. If bin resolution and the maximal number of histogram bins are kept constant it is possible to compute percent entropy values (%H), which are useful for entropy comparisons between different data sets. For respiratory data, vector distances were distributed over 1500-bin histograms with a bin resolution of 0.1 units. For the data previously presented (Figs. 8 and 9), percent entropy values were always higher for the active state as compared to the quiet state. This was true for both respiratory period data (%H = 86.91% vs. 59.40%) and pressure data (%H = 94.53% vs. 66.16%). These results indicate that there is more randomness in respiratory variables during the active-disturbed state, the source of which is currently unknown.

For statistical verification of these concepts, recurrence plot analysis was performed on 10 rats during states of quiet and active breathing. As summarized in Figure 11, the percent entropy values for period and pressure data were significantly higher in the active state. Accompanying these changes were significant decreases in percent recurrence due to the greater variability in vector distances in active breathing states. In sum, these results substantiate the use of percent entropy and percent recurrence as measures of complexity of dynamical systems, but close examination of the recurrence plots themselves yields more detailed information on the transients and deterministic behavior in the dynamics of interest. New quantitative criteria for locating deterministic structure at fine scales (low cutoff levels) must be established as a scientific goal for future work in this area.

7. Conclusions

Argument has been put forth herein that complex dynamical systems in the physiological domain are better assessed by geometric than ergodic procedures. This is so because the presence of real-world discontinuities and nonstationarities tend to "float" dynamical systems, blurring numerical results. Recurrence plots appear to possess utility in detecting transients and deterministic structure in dynamics, and that from relatively short data sets. The art work of Figure 12, a child's view of chaos, is presented as a reminder that a dynamical system evolving in time can be represented as a complex spatial tapestry which possesses both beauty and, on close examination, non-random internal structuring. Armed

Fig.12. Graphical portrayal of dynamical chaos as expressed by the artistic talents of a child. Despite the global complexity of the total picture, local structures are prevalent within smaller scaling regions. Likewise, recurrence plots have the ability to focus on local dynamical behavior and are robust against transients and discontinuities occurring at larger scales.

with an interpretive framework, recurrence plot technology may contribute to the problem of distinguishing normal from abnormal dynamics. This graphical approach is pertinent to any rhythmical system in the body (heart, lungs, brain, etc.) and may prove to be of importance in the context of dynamical diseases [24]. An advanced application of the technique might be to update recurrence plots at a fast rate, transforming single-frame "stills" into dynamical "movies". The time-course of diseases could be studied in this manner and, optimistically, prove to have predictive value in human medicine.

8. Acknowledgements

Numerous and valuable discussions on nonlinear dynamics with Joseph P. Zbilut are greatly acknowledged. The X-ray picture in Figure 1 was obtained with the help of the late Paul D. Davis. Also to be thanked are John L. Silva, Jr. for his assistance with the experimental preparations, Stephanie A. Webber, daughter of the author, for her chaos art work in Figure 12, and Sue Ming Yang for her professional skills in photography.

References

1. L.F. Olsen, H. Degn, Quart. Rev. Biophys. 18, 165 (1985).
2. J.P. Zbilut, This Volume.
3. C.L. Webber, Jr., J.P. Zbilut: in Cardiorespiratory and Motor Coordination, edited by H.-P. Koepchen and T. Huopaniemi (Springer-Verlag, Berlin, Heidelberg, 1991).
4. C.L. Webber, Jr., J.P. Zbilut, Proc. IEEE Eng. Med. Biol. Soc. 12, 1863 (1990).
5. A.L. Goldberger, D.R. Rigney, B.J. West, Scientific Amer. 262, 42 (1990).
6. L.-S. Young, Ergod. Theor. and Dynam. Systs. 2, 109 (1982).
7. A.J. Mandell, Am. J. Physiol. 245, R484 (1983).
8. R. Lydic, FASEB J. 1, 6 (1987).
9. P. Grassberger, I. Procaccia, Physica 9D, 189 (1983).
10. I. Dvorák, J. Klaschka, Phys. Lett. A 145, 225 (1990).
11. D. Ruelle, Proc. R. Soc. Lond. A 413, 5 (1987).
12. L. Glass, C.P. Malta, J. Theor. Biol. 145, 217 (1990).
13. G. Mayer-Kress, J. Holfuss: in Temporal Disorder in Human Oscillatory Systems, edited by L. Rensing, U. an der Heiden, M.C. Mackey (Springer-Verlag, Berlin, Heidelberg, 1987).
14. F.L. Eldridge, D. Paydarfar, P.G. Wagner, R.T. Dowell, Am. J. Physiol. 257, R271 (1989).
15. O.E. Rössler, Phys. Lett. A 35, 397 (1976).
16. J.-P. Eckmann, S. Oliffson Kamphorst, D. Ruelle, Europhys. Let. 4, 973 (1987).
17. G. Mayer-Kress, A. Hübler: in Quantitative Measures of Complex Dynamical Systems, edited by N.B. Abraham (Plenum, New York, 1990).
18. J. Guckenheimer, Nature 298, 358 (1982).
19. B.J. West, M. Shlesinger, Am. Scientist 78, 40 (1990).

20. J.L. Feldman, J.C. Smith, D.R. McCrimmon, H.H Ellenberger, D.F. Speck: in Neural Control of Rhythmic Movements in Vertebrates, edited by A. Cohen (John Wiley and Sons, Inc., New York, 1988).
21. J.L. Feldman, J.C. Smith, H.H. Ellenberger, C.A. Connelly, G. Liu, J.J Greer, A.D. Lindsay, M.R. Otto, Am. J. Physiol. 259, R879 (1990).
22. C.E. Shannon, Bell Syst. Tech. J. 27: 379 (1948).
23. H. Haken: Synergetics: An Introduction, (Springer-Verlag, Berlin, Heidelberg, 1983).
24. M.C. Mackey, J.G. Milton, Ann. NY Acad. Sci. 504, 16 (1987).

Part IV

Motor Coordination

Synergetic Dynamics of Biological Coordination with Special Reference to Phase Attraction and Intermittency

J.A.S. Kelso, G.C. DeGuzman, and T. Holroyd

Program in Complex Systems and Brain Sciences, Florida Atlantic University, Boca Raton, FL 33431, USA

Under conditions in which absolute phase and frequency synchronization are neither essential nor attainable for biological functioning, some form of relative coordination is still possible. Attraction toward certain phase and frequency relations remains, but phase slippage as well as occasional skips and jumps occur as the component units adjust spatially and temporally. We establish the connection between this less rigid form of coordination and intermittency, a generic feature of dynamical systems near tangent bifurcations. Intermittency provides a mechanism for entering and exiting mode-locked states, endowing the system with a vital mix of flexibility and coherence. In the intermittent régime close to critical points, the system possesses a 'predictive' or 'anticipatory' property. The identified dynamics are level-independent and may be essential to a number of different biological functions.

1. Introduction

Over the last decade or so, synergetic concepts [1]: a) of collective variables (or order parameters) characterizing coordinated states; b) of abrupt changes in coordination pattern corresponding to phase transitions or bifurcations; c) of multistability and loss of stability leading to pattern change, etc. have been demonstrated to provide a plausible basis for understanding coordination in biological systems on a number of levels of description. Under well-defined experimental conditions, coordination patterns often appear in the form of frequency- and phase synchronization among interacting components. Although this work originated in studies of rhythmic tasks in humans [2,3], remarkable parallels exist with research on central pattern generators in invertebrate and vertebrate neuronal networks [4]. In contrast, however, to work aimed at finding general construction principles for central pattern generators [5], the focus of the synergetic approach is on *instabilities*, both as a means of empirically differentiating one pattern from another (thereby opening a path into modeling) and as a mechanism for flexibly switching among multiple coordinated states under nonspecific parametric influences.

Since the effective degrees of freedom of many high dimensional, dissipative dynamical systems often reduce to the space of coupled, nonlinear oscillators perhaps it is not so surprising that frequency and phase synchronization are

Eds.: H. Haken and H.P. Koepchen

ubiquitous behaviors in natural systems. But whereas the physicist sees the hint of *universality* in such phenomena – that the same dynamics may underlie cooperative behavior (coordinated patterns) regardless of origin – it is only recently that the neurobiologist has come to recognize that the same patterns may be produced by very different "mechanisms" [e.g. 6,7]. It is a surprise to many, for example, that cells in the mammalian visual cortex [e.g. 8,9] display the frequency and phase synchronization characteristic of central pattern generators. Yet from another perspective, phase and frequency synchronization may be an elementary, even fundamental way to functionally coordinate spatially separate components whose behavior evolves in time.

The behavioral physiologist Erich von Holst [10] coined the term *absolute coordination* when two or more components oscillate at the same frequency and maintain a fixed phase relation. Elsewhere, we have shown that the phenomenon of absolute coordination can be understood in detail using the concepts of synergetics and the mathematical tools of nonlinear dynamical systems [e.g. 11]. In quite a large number of cases, relative phase, ϕ, was identified as an order parameter or collective variable capturing the ordering relations among the active components. Multistability and transitions among phase-locked states were observed at a critical value of a continuously changing control parameter, usually (but not always) movement frequency. *En route* to these transitions, theoretically predicted features of synergetic systems such as enhancement of fluctuations, critical slowing down of the order parameter and the switching time distribution were observed in experiments. Once the pattern dynamics were found by mapping observed patterns onto attractors of the order parameter dynamics it was possible to derive them by nonlinearly coupling the individual oscillatory components. The latter were mapped onto limit cycle attractors whose functional form was again determined by detailed experiments (for detailed references and a review, see e.g. [12, 13]).

Our aim here is to build upon and extend this somewhat 'ideal' picture in which absolute coordination is characterized by mode-locked attractors, to accommodate experimental observations of: a) periodically driven systems, in which loss of entrainment occurs (Section 2); and b) multifrequency coordination in which a conflict is created between sustaining a constant frequency ratio between the components and an intrinsic tendency on the part of these components to synchronize in-phase or antiphase (Section 3). In such situations, the system solves the problem adaptively by balancing spatial and temporal features of the pattern. Thus, we observe, for example, phase slippage between the components, the insertion of additional cycles and hopping in and out of frequency- and phase-locked patterns. In contrast to experiments in which special efforts are made to pin the system into a particular phase- and frequency-locked pattern, we find *tendencies* toward certain frequency ratio values and *tendencies* toward in-phase or anti-phase attraction. Von Holst (1939/73) coined the terms Magnet or M-effect and *relative coordination* to describe such tendencies. We establish the connection between this less rigid, more flexible form of coordination and *intermittency*, both in a continuous flow form of the collective variable, relative phase dynamics (Section 2) and a discrete version of the dynamics which we call the *Phase Attractive Circle Map* (Section 3). Intermittency is a generic feature of dynamical systems near tangent or saddle node bifurcations. It refers to a region around a destabilized saddle node which contains both a stable (attracting) and an unstable (repelling) direction. Typically, the system spends some time near the attractive fixed point (a frequency- and phase- or mode-locked state in our case) and then escapes through the unstable direction before returning via a reinjection process (see, e.g. [14]).

By living *close* to mode-locked attractors but not in them, an optimal form of coordination results because both coherence *and* flexibility are given by the dynamics in the intermittent régime. Intermittency provides a mechanism for coming in and going out of phase- and frequency attractive states, thus preventing the biological sytem from getting stuck. As Mandell has repeatedly argued [e.g. 15], intermittency may be a quite crucial feature of neuropsychobiological dynamics. Recent experimental and modeling work on visual cortex, for example, stresses short term, but long range frequency- and phase-synchronization [8, 9, 9a]. On the other hand, it may be just as important for vision to have a mechanism for entering and exiting coherent patterns, and even for staying out of resonant mode-locked states. A similar argument applies to neuronal circuits for central pattern generation where flexibility as well as reliability is desired. Finally, we note also that by staying near critical points the system possesses a property that could be called *predictive* or *anticipatory*. This feature is due, in the present case, to the essential nonlinear nature of the underlying dynamics.

2. Relative Coordination I: Driving a Rhythm

When an external stimulus periodically "drives" the response of the subject we may consider the relation between stimulus and response as a pattern [16]. The resulting dynamics may be understood in terms of pattern formation in nonequilibrium systems, entirely analogous to previous work (e.g. [11]). In an experiment on hand-metronome coordination the subjects' task was to synchronize peak flexion of the index finger with the metronome in either of two modes of coordination: (1) on the beat or (2) off the beat. For each mode the pacing frequency was monotonically increased (or in another condition, decreased) in fixed steps. The key points of the synergetic strategy are: a) to identify a collective variable (relative phase, ϕ, between hand and metronome in this case); and b) to map the pattern formation process onto attractor dynamics of the relative phase. Limiting the expansion of the equation of motion to second order (for details see [16]) we find

$$\phi = \delta\omega - a\sin\phi - 2b\sin 2\phi + \sqrt{Q}.\xi_t, \qquad (1)$$

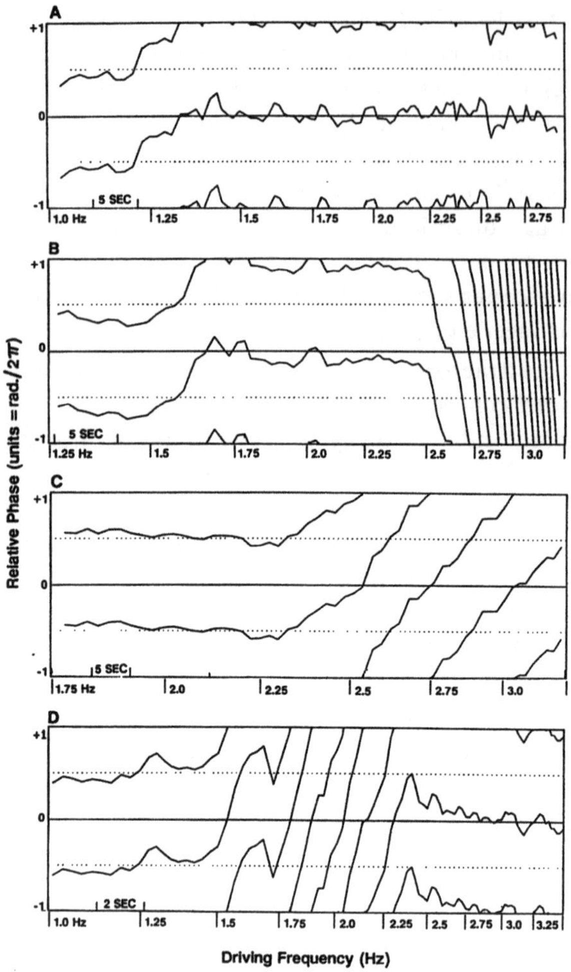

Figure 1: Representative plots showing relative coordination between the hand and a metronome (y-axis) as a function of the driving frequency in Hz (x-axis). Relative phase is normalized to [0,1] and the plot is duplicated in the interval [-1,0]. Tendency for phase attraction persists even though the oscillatory components exhibit different periodicities. The characteristic features are interspersed intervals of strict mode-lockings, occasional phase wanderings, and longer epochs of phase wandering as metronome frequency is increased (from [16]).

where the parameters a and b are related to environmental requirements (e.g. $\mid a/4b \mid$ is proportional to the stimulus frequency) and $\delta\omega$ allows for slight desynchronization between metronome signal and limb. For reasons discussed elsewhere [18] a Gaussian white noise, ξ_t, may be added as a stochastic influence of strength Q.

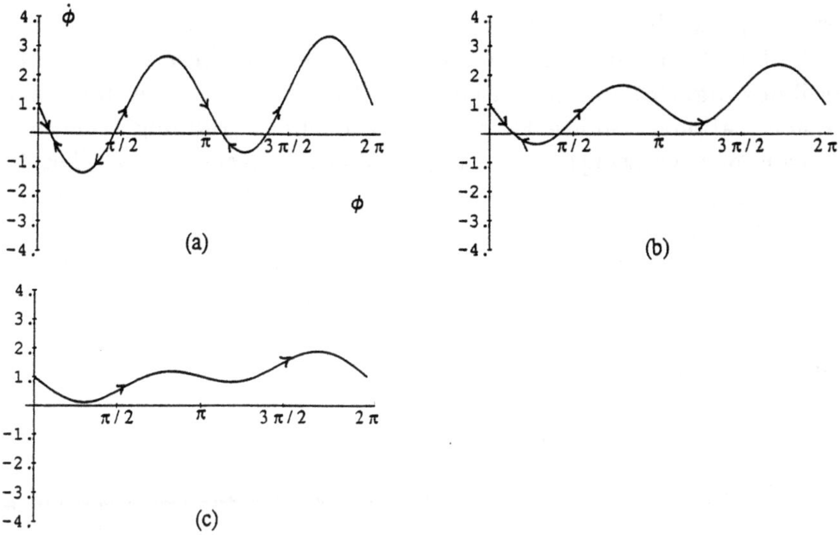

Figure 2: Plot of $\dot{\phi}$ vs. ϕ as given in Equation (1) for $a = 0.5$ Hz, $\delta\omega = 1.0$ Hz and $b = 1.0$ Hz (a), $b = 0.5$ Hz (b) and $b = 0.25$ Hz (c). The arrows indicate the flow of $\dot{\phi}$. Zeros of $\dot{\phi}$ are stationary solutions that are stable if the slope is negative and unstable if the slope is positive at the stationary state (adapted from [16]).

Equation (1) captures the observed pattern behavior including: (1) stability and loss of stability of the patterns, the latter in the form of a transition from the off-beat to on-beat mode when metronome frequency is increased; (2) hysteresis; and (3) loss of entrainment (phase and frequency desynchronization) similar to what von Holst [10] called *relative coordination* and others call *sliding coordination.* Relative coordination refers to conditions where a tendency for phase attraction persists even though the oscillatory components exhibit different periodicities. The characteristic features of (3) are: (a) interspersed intervals of strict mode-lockings (in phase or antiphase) and occasional phase wanderings, and; (b) longer epochs of phase wandering as metronome frequency is increased (see Figure 1).

This empirical behavior is captured by eq. (1) and occurs when, by increasing the detuning parameter, $\delta\omega$, the original fixed points of the relative phase dynamics disappear (see figure 2). As we shall see (Section 3), similar phenomena occur in dynamical systems when a low dimensional map loses fixed points near tangent bifurcations.

A good way to intuit the solutions of eq. (1) is to plot the right hand side of this equation as a function of ϕ for various parameter values. The system has stationary states where $\dot{\phi}$ crosses the ϕ-axis. These are stable states if the slope of $\dot{\phi}$ is negative and are unstable states if the slope is positive at this point. The arrows in figure 2 indicate the direction of flow. For $\Delta\omega \ll a$ and $\Delta\omega \ll b$, two stable states exist, one close to in phase and one close to antiphase (fig. 2a). If, for example, b is decreased, the amount of attraction to these stable states decreases until one (fig. 2b) and ultimately both (fig. 2c) cease to be stationary solutions. When stationary solutions are no longer found, "running" solutions (ϕ ever increasing or decreasing), or, if the periodicity convention is applied to relative phase, "wrapping" solutions occur. There is no longer any phase or frequency-locking, a condition called loss of entrainment.

Due to the frequency detuning parameter in eq. (1), the symmetry of the relative phase dynamics under the operation $\phi \to -\phi$ is broken. (Parenthetically, all the interesting spin-offs of the original Haken, Kelso & Bunz model, e.g. intentional switching, learning, multicomponent coordination (for a review, see [17]) arise when the symmetry of the dynamics is broken or lowered.) Thus the bifurcation structure is no longer a pitch fork (two stable fixed point solutions emerging from one and vice-versa at a critical parameter value) but of saddle-node type. This is shown in figure 2(b) where the dynamics contain a stable fixed point near $\phi = 0$ and an unstable fixed point near $\phi = \pi$ above the line. Similarly, in figure 2(c) *both* fixed points are unstable: the previously stable fixed point around $\phi = 0$ drifts and loses stability when the function leaves the line. One obtains a similar conclusion by examining the qualitative behavior of the potential (Fig. 3). Here, the local minima serve to trap the system into one of the intrinsic phase states depending on the initial condition (Fig. 3a). By decreasing the parameter b, the localized wells are made shallow, allowing the system to escape after some time (Fig. 3b-3c).

An important prediction from this theoretical formulation concerns loss of stability, which is at the origin of the observed transitions. Such stability loss can be observed through measures such as the relaxation time of relative phase, the time it takes to relax to the pattern after a small perturbation is applied, or the amount of fluctuations of relative phase, measured in terms of phase variability. When a transition is approached and the system is prepared in the state that loses stability these measures increase strongly. One can see this intuitively in figure 2. As a coordinative state loses stability, the slope of $\dot{\phi}$ at that state becomes flatter, indicative of *critical slowing down.* The system takes longer and longer to return to the initially prepared state. The corresponding broadening of the attractor lay-out corresponds to *enhancement of fluctuations.* For transitions among attractors, a further measure is the length of the transient of the actual switch – the so-called switching time – that can be calculated theoretically given the stability measures [18]. All of these theoretically predicted features have been observed in a number of experimental systems (e.g. [19-22]).

A final and most interesting aspect emerges when the so-called 'running' solutions of eq. (1) corresponding to Fig. 2c are examined. Even though there are no stable fixed points, these running solutions have a fine structure, i.e. they spend more time at those relative phase values at which the force $\dot{\phi}$ is minimal. This is exactly the phenomenon of *relative coordination* discovered

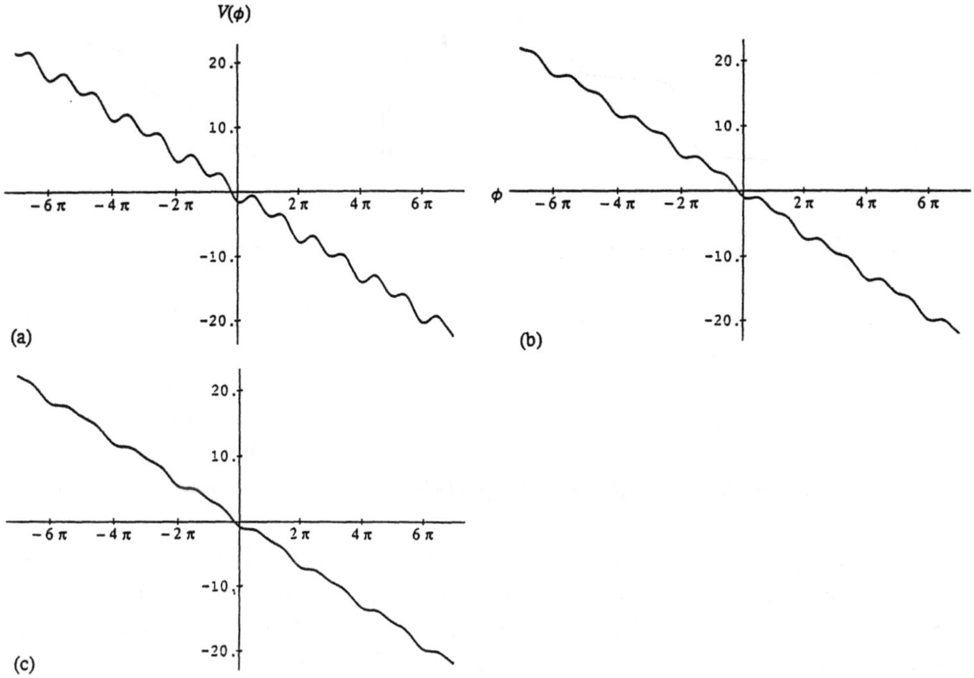

Figure 3: Potential corresponding to the right hand side of Equation (1) for the three sets of parameter values in Figure 2. The region around each minimum acts like a local well that weakly traps the system into an inphase state. We thank H. Haken for suggesting this representation (see also [11]).

first by von Holst (1939/73). There remains an attraction to certain phase relations even though the relative phase itself is unstable. Figure 4 shows a numerical simulation of Eq. (1) in this parameter régime. The observed periodic modulation of relative phase is a signature of relative coordination and can vary considerably due to noise, illustrating that this periodicity is not, in itself, a stable state.

We suggest that orderly but nonstationary behavior is a general feature of coordination in real biological systems (see figure 1). A casual survey of time-dependent processes on many levels of description in many different experimental systems (cell membrane, heart cells, respiratory neuron firings, locust flight patterns, physiological rhythms....) suggests that absolute phase and frequency synchronization may be the exception rather than the rule. Rigid entrainment (absolute coordination) which reflects asymptotic convergence to a well-defined frequency and phase relation state (a mode-locked attractor) may be more a feature of biological oscillator *models* than of reality. Discrepancies between ideal models using oscillators that can "only do it right all the time" versus real biological systems that "do it right occasionally" yet still function, may therefore be more than finding the appropriate oscillator and coupling functions.

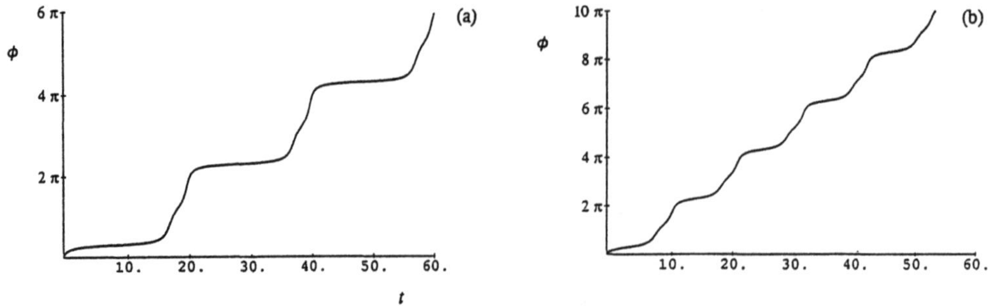

Figure 4: Numerical simulation of Equation (1) in the regime in which only running solutions exist. (a) $\delta\omega = 1$ Hz, $a = .5$ Hz, $b = 0.30$ Hz; (b) $\delta\omega = 1$ Hz, $a = 0.5$ Hz and $b = 0.25$ Hz. The slight shortening of the plateau when $b = 0.25$ Hz is qualitatively understandable since the basic frequency of oscillation (α to $\mid a/4b \mid$) is slightly higher and therefore the plateau durations are shorter (adapted from [16]).

In short, biological systems clearly exhibit *attraction* in this flexible régime where the individual components vary in period and amplitude, but strictly speaking there are no *attractors*, at least in the sense we have defined the term, i.e. as attractors of the collective variable dynamics. Below we make this point even more explicit by establishing the connection between phase slippage or relative coordination in multifrequency experiments and *intermittency* in a discrete map analog of eq. (1), which itself is due to abrupt slips away from a previously stable fixed point (cf. Fig. 2). The ubiquity of relative coördination may be explained by intermittency, which is a generic feature of dynamical systems near a saddle node or tangent bifurcation.

3. Relative Coordination II: Multifrequency Processes and the Phase Attractive Circle Map

One of the characteristic features of coordination is that a set of events in one signal is correlated somehow with a set of events in another. Experimentally, the signals themselves may be quite different in nature; however, often they come together at easily recognizable points. For example, it is easy to see in typical records that rather different electrical events in the cat's visual cortex – single unit activities and local field potentials – are tightly coordinated for short periods of time [e.g. 8,9]. As a measure of coordination, provided that events such as onsets, peaks or valleys are identifiable, other features of the signals may be ignored. This, along with other evidence, suggests that biological coordination holds only at some points in the signal "trajectories" of the individual components: in between, trajectories may vary spatially (e.g. characteristic amplitudes) and temporally (e.g. characteristic periods).

That the individual components may be coupled only at certain points also comes out as a dominant feature of experiments in which the frequency relations between rhythmically moving components are experimentally manipulated. In initial experiments [23, 24] the frequency and amplitude of one component (Left Hand) were determined passively by a torque motor while the other component (the Right Hand) actively maintained a base rhythm. Six different frequency ratios were studied: f(L):f(R)=4:3, 3:2, 5:3, 2:1, 5:2 and 3:1 over long (90 sec.) time series. More recently, we scanned the frequency ratio in ascending and descending order from just below 1:1 to just above 2:1 and *vice versa* in steps of 0.1 Hz. In none of the experiments was the initial phase between the components fixed, yet there was still a strong tendency for phase attraction to occur.

Collectively, our multifrequency experiments revealed the following features: 1) phase attraction to in phase and antiphase even in situations when absolute coordination is not easily attained; 2) differentially stable multifrequency patterns (as measured by the variability of the frequency ratio) depending on the required frequency ratio (2:1, 3:1, and 3:2 being least variable); 3) short-term spontaneous jumps from less stable to more stable frequency ratios, as revealed through analysis of power spectra; 4) transitions between in phase and antiphase modes *within* a frequency ratio (e.g. 2:1); and 5) amplitude modulation of the driven (LH) component and the insertion of occasional *extra* cycles of LH in order to maintain an in phase relation when the frequency ratio was not 1:1.

This modulation arose when the driver was allowed some flexibility (by lowering the torque from the driving motor), indicating that a two-way interaction between the components was present, of sufficient strength that an otherwise perfect harmonic trajectory was modified. These modulated cycles appeared rarely and randomly near 1:1 but more frequently near 2:1 where sometimes even trains of alternate low and high amplitudes were seen (see figure 5). The modulation is due to the cross-coupling between the components causing a back reaction of the LH on the input torque signal.

To characterize coordination in multifrequency situations in which the components synchronize at only certain event times and may not have a well defined frequency relationship, we need a local measure that characterizes time relationships between events. An often quoted measure of coordination is the relative frequency with which these events occur, such as the frequency ratio. But frequency related measures are with respect to events in components taken individually, not between them, and are therefore inadequate. A better measure is the *point estimate of relative phase* calculated experimentally as

$$\phi_n = 2\pi \frac{\tau_n}{T_n}, \tag{2}$$

where T_n is the peak-to-peak period of component 1 beginning at time t_0, and τ_n is the interval between that event and the peak onset of component 2.

To study how the components are coordinated in time, we examine the evolution of the phases ϕ_n. The utility of (2) for expressing coordinated behavior in

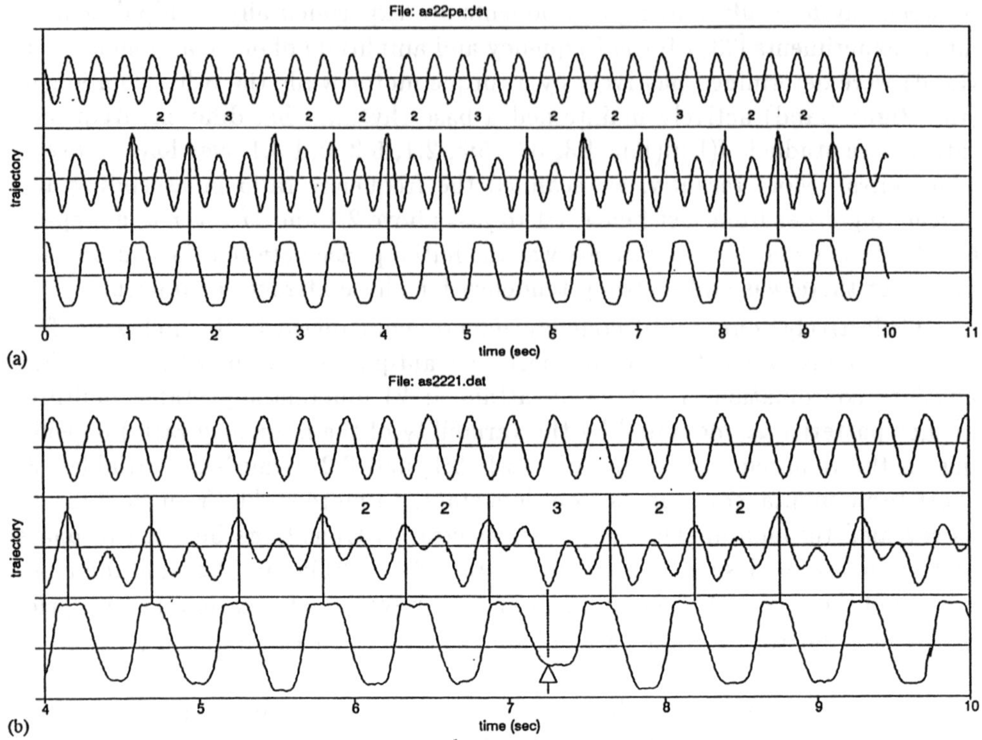

Figure 5: (a) Trajectories of the input signal used to drive the torque motor (top), the actual motion produced by the driven left hand (middle) and the "free" hand (bottom) when the required frequency condition is near 2:1. Enhancement of the peaks near inphase regions points to the discrete nature of the coordination, i.e. intermediate portions of the left hand trajectories are adjusted to sustain the natural tendency to be inphase with the right hand (indicated by the vertical lines). Occasional slips occur when the position of the enhanced peak extends beyond the broadened peak of the right hand (3 steps instead of 2). (b) Similar plot near 2:1 from another time interval showing the slippage effect on the right hand.

terms of the evolution of ϕ_n at discrete times τ_n depends on the actual dynamics of the system. Phase transition experiments and consequent theoretical modeling show that the relative phase dynamics, with observed fixed point attractors (in phase and antiphase), can be derived from nonlinearly coupled autonomous nonlinear oscillators. We can readily extend this continuous form of the phase dynamics to a discrete circle map of order 1 which has the property that in the 1:1 case not only are $\phi = 0$ and $\phi = 1/2$ fixed points but also that either one or both are stable (angle ϕ is now normalized to the interval [0,1]). We put the map in the form

$$\phi_{n+1} = \Omega + f(\phi_n) \tag{3a}$$

$$= \phi_n + \Omega + g(2\pi\phi_n) \mod 1, \tag{3b}$$

where Ω is an added constant term corresponding to the initial frequency ratio and g is a smooth periodic function of period 1 in ϕ (for complete details, see [24]).

Performing a Fourier series expansion of g, the lowest order expansion that can satisfy e.g. locally stable fixed points at in phase and antiphase is

$$g(2\pi\phi) = a_1 \sin(2\pi\phi) + a_2 \sin(4\pi\phi). \tag{4}$$

Transforming (4) to the new parameters (K,A) by letting $a_1 = -K/2\pi, a_2 = -KA/4\pi$ and using (4) in (3b) we obtain the *phase attractive circle map*

$$\phi_{n+1} = \phi_n + \Omega - \frac{K}{2\pi}(1 + A\cos 2\pi\phi_n)\sin 2\pi\phi_n \quad \mod 1. \tag{5}$$

For the standard circle or sine circle map which characterizes a bewildering array of coupled multifrequency systems in nature (see [25] for a review) K is a measure of the amplitude strength of an external driving force, or, in general a nonlinear coupling parameter. Also for a general $g(x)$ in (3b), the dominant parameter that scales the overall magnitude of g may be given this interpretation. The "intrinsic parameter" A expresses the bistability of relative phase and behaves as the ratio 4 b/a in the original Haken, Kelso & Bunz [11] model (Eq. 1 with $\delta\omega = 0$). Since A is the ratio of the coefficients for the sin ϕ and sin 2ϕ terms in the Fourier expansion of g, A is roughly inversely proportional to the basic frequency of oscillation. For a given frequency ratio, either both inphase and antiphase affect the overall performance, or only the inphase is important depending on how fast the basic frequency is. Thus the parameter A is a measure of the relative importance of the intrinsic phase states 0 and π.

The relative strength of A, representing the intrinsic phase attractive dynamics and K, the nonlinear coupling term, determines many of the observed experimental features of multifrequency behavior mentioned above. The mechanism for pattern *selection*, in fact, may be seen as a *competition* between these terms. For example, it is easy to see why only a few, low-integer frequency lockings are seen in biological systems and why the simple frequency ratios in our experiments are the most stable (least variable). The reason is that the widest mode-locked regions or *Arnold tongues* (e.g., 1:1, 2:1, 2:3 etc.) corresponding to absolute coordination are the most stable and attractive. If the experimental noise level is low, it is possible to observe other locked states, but the presence of noise can easily kick the sytem into nearby, more stable attractors (e.g. from 5:2 to 2:1). In addition, theoretical features that are not readily observable in biological experiments, such as chaotic transients and transitions from one chaotic region to another can be found in the map [26]. Such random switching back and forth between two chaotic states is deterministic, not due to noise.

Our main goal here, however, is to show how the phase attractive map can explain relative (not just absolute) coordination. A close look at the trajectories shown in figure 4 reveals 'slips' and sometimes even 'jumps' in cycles of the LH driven component. For example, when a peak of the LH is just at the edge of a flattened peak of the RH, either the LH jumps over a cycle or shortens the next one. An analogous situation happens in the example of a father walking along with his small child. Since their intrinsic frequencies are different, synchronization is difficult unless either one or both adjust their periods. If the mean frequencies are constant for time scales not too large compared to the larger of the two periods (the father's) a less rigid form of coordination may be observed. Like our data (cf. Fig. 5 and von Holst [10] for many more examples), in order to keep pace with each other, the child sometimes skips a step or the father slows down and we observe *phase slippage*, followed by more or less regular cycles again.

A generic or universal feature of dynamical systems (and, of course, of our map) near a tangent bifurcation is the phenomenon of *intermittency*, i.e. abrupt slips away from a previously stable fixed point. Phase slippage in a circle map occurs because of an unstable phase direction while return to more regular cycles is due to the stable direction. That is, the order parameter relative phase is *hyperbolically stable*, attraction taking place in one direction and divergence in another. The surface containing convergent trajectories is called the *stable manifold* while the surface containing the divergent trajectories is called the *unstable manifold*. Figure 6 shows how the phenomenon of intermittency occurs in the map. In figure 6(a) a pair of fixed points are shown, one stable (slope

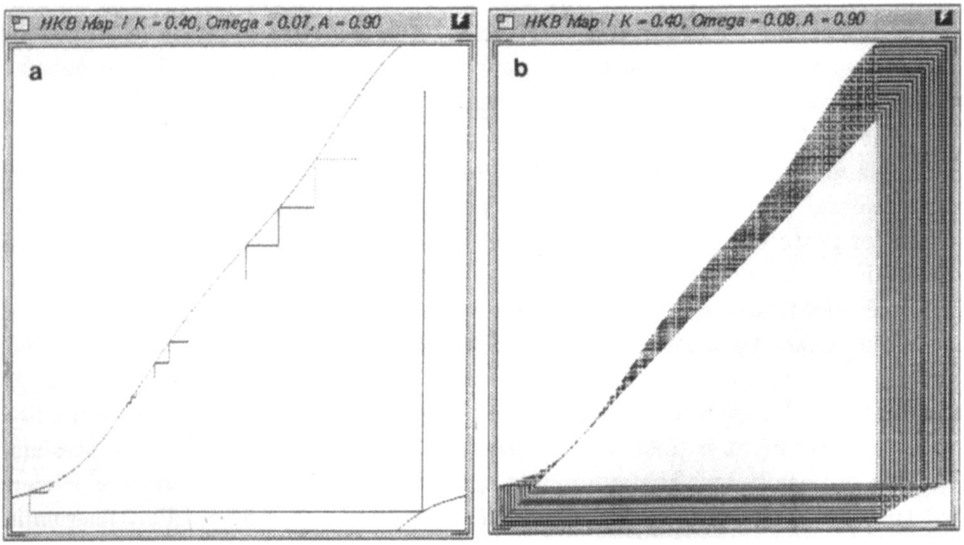

Figure 6: Return maps (Eq.5) of ϕ_{n+1} (y-axis) against ϕ_n (x-axis) (a) Mode-locked behavior at 1:1 showing a pair of fixed points, stable and unstable, close to the origin. (b) Intermittency occurs when the graph is lifted slightly above the diagonal (see text for details).

< 1) the other unstable, i.e. there is frequency-locking. When the slope of the function is exactly 1 the graph is just tangent to the diagonal, a single fixed point subsists. When the graph leaves the diagonal, the iterates pass through the channel, go round the circle and enter from the other side. Phase slippage occurs between the two quasi-regular oscillatory periods and there is no frequency locking (Figure 6b) because the fixed points have disappeared.

The appearance of phase slippage means that between two channel crossings one of the oscillators gains a period with respect to the other: exactly the phenomenon of *relative coordination.* Note that close to the tangent bifurcation the phase concentrates (*phase attraction persists*) and slows. The reason, of course, is that the iterations are compressed in the channel. The relative phase, ϕ is plotted as a function of time (x-axis) in figure 7 as the system approaches a fixed point (1:1 phase- and frequency-locked state). As one proceeds from Figure 7(a) to 7(c), with the frequency ratio, Ω, decreasing, the time spent in the channel (revealed by the flat portion of the curve) gets longer and longer. Abrupt slips in the phase occur following these pauses: the phase leaves the channel and reenters after going once around the circle where it visits once more the 'remnant' or 'ghost' of the fixed point. Only the relative frequency of occurrence of phase slippage is affected as the tangent point is approached; the shape of the orbit does not change. Much more exotic, bursting behavior is shown in Figure 7(d) where the map parameters are in the chaotic regime, i.e. above the critical surface where the map is noninvertible. Mixed with slow phases near the fixed point are 'turbulent' bursts of unpredictable duration, rather like the spontaneous firing of neurons. For non-1:1 frequency ratio, one observes slippage of the orbits instead (see Kelso & deGuzman [23] Fig. 9 for an example in the case of 2:1). Intermittency in this case may be formulated in terms of the behavior of an appropriate iterated point, i.e. second iterate in the case of 2:1. In Figure 8 is plotted the orbit near a tangent bifurcation at 2:1 using the first (a) and the second (b) iterate. In Figure 8b, the pattern of jumps between the two regions of concentration is qualitatively very similar to the phase wandering and phase locking found in the experiment.

We note that the appearance of phase gathering near mode-locked solutions has an *anticipatory* quality about it. A good way to see this is through the corresponding bifurcation diagrams shown in Fig. 9(a) and 9(b). In this figure, the fuzzy area reflects quasiperiodic motion because the frequency ratio between the components is irrational. Notice, however, that the progressive darkening "anticipates" the upcoming stable solution (the single line) which indicates the system is trapped or mode-locked 1:1 (Fig. 9(a) or 2:1 (Fig. 9(b)). The map (5) is invertible if the function $f(\phi)$ is monotonic or the slope $f'(\phi)$ is never zero. The boundary in the parameter space separating invertible and noninvertible regions is called the *critical surface.* Mode-locked regions, parameter regions of absolute coordination in our language, are found below the critical surface of the map. For $A = 0$ and $K < 1$ the map is still invertible. Much richer

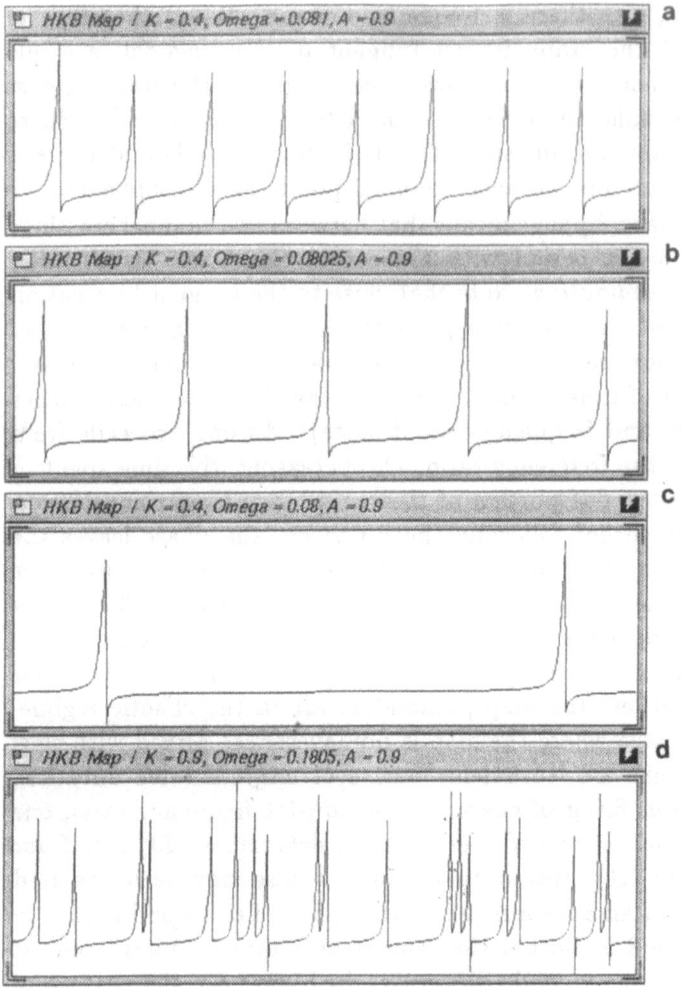

Figure 7: (a)-(c) Plots of the time series ϕ_n for decreasing values of Ω, showing the slowing down of the bursts. For fixed K and A, the frequency of the burst depends on how far Ω is from the fixed point. (d) Bursting in the chaotic régime of the map.

dynamics, of course, are possible beyond the critical surface where the map loses invertibility. Figure 9(c) shows a bifurcation diagram as K is increased *along* the 1:1 Arnold tongue, i.e. a constant distance C from the tongue, instead of a straight cut through the $[K, \Omega]$ parameter space as in Fig. 9(a). Virtually no stable solutions exist along the tongue except for very narrow bands of overlap. Nevertheless, even though the phase is unstable, a good deal of orderly structure remains as the dark band shows.

Our main points are: (i) that relative coordination – a ubiquitous, but largely unexplained phenomenon in biological systems – is to be understood as

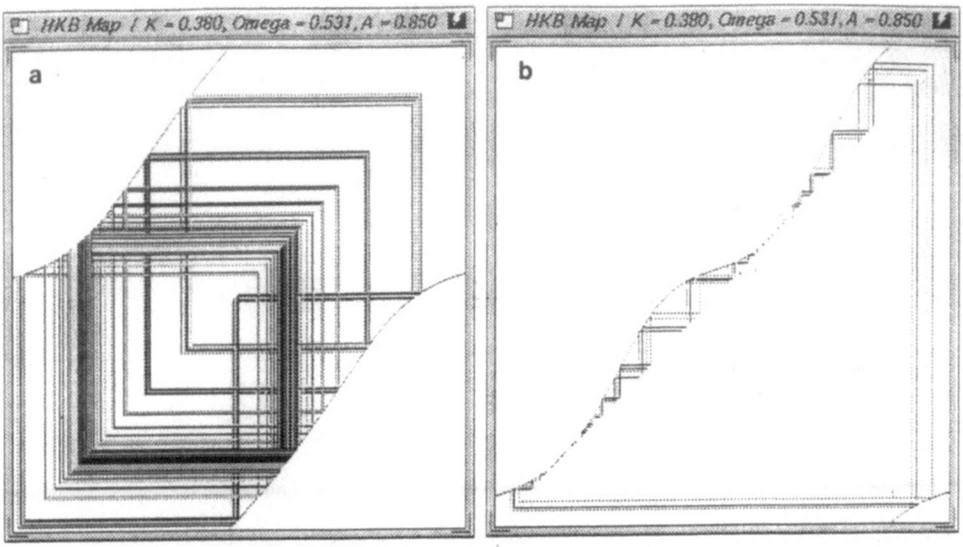

Figure 8: Intermittency in the phase attractive map near 2:1. Return maps are plotted using the first iterate (a) and the second iterate (b). Parameters are identical in each case.

a generic property of dynamical systems (flows or maps) near tangent bifurcations which leads naturally to intermittency. One may speak of a "dwell" time, the average time the system stays near the same point (or the same orbit as in a non-1:1 case) before running away. Because the frequency ratio among the components is not necessarily 1:1, the fixed point states of the intrinsic dynamics need not be realized. An optimal synchronization would then be measured by how close the system gets, overall, to these intrinsic phase points; and (ii) the growth of phase densities and the resulting slowing is *predictive* of upcoming critical points, what one might call an *anticipatory dynamical system* [27]. Anticipatory dynamical systems stay in contact with the future by living *near* but not *in* mode-locked regions, thereby retaining an optimal mixture of flexibility and stability (see also [28]).

4. Conclusions

If experimental systems can be prepared so as to correspond neatly to low dimensional coupled oscillators, coordination can be well captured in terms of mode-locked states. However, asymptotic convergence to mode-locked orbits (absolute coordination) may be the exception in biological systems: an, albeit useful, idealization upon which to develop a deeper understanding. Building upon previous theoretical and experimental studies of coordination we explore the consequences of 'softening' or 'loosening up' the boundary constraints on the system's behavior. Then a much less rigid, more flexible form of relative

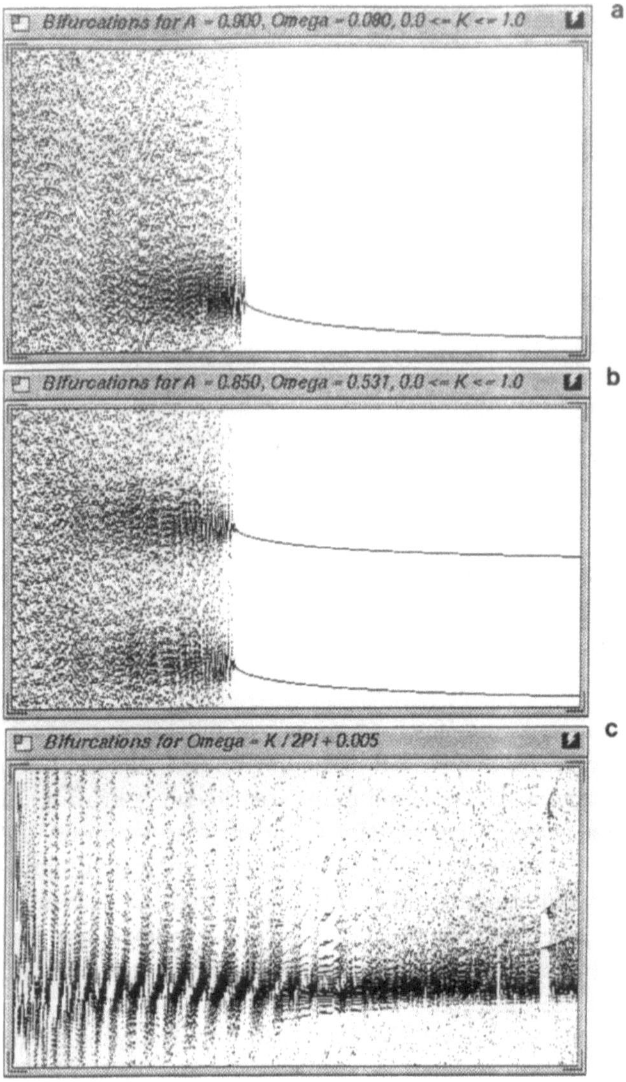

Figure 9: Bifurcation diagram showing the phase behavior (y-axis) when crossing the Arnold tongues at 1:1 (a) and 2:1 (b) for fixed A and increasing K. (c) Phase behavior along a path just outside the 1:1 tongue and parallel to its boundary.

coordination emerges. Instead of rigid mode-locked states, we see that component behavior is still strongly coupled at certain points (phase attraction) but occasional skips and jumps occur. We establish the connection between relative coordination and intermittency both in a continuous differential form of the collective variable dynamics and a discrete phase attractive circle map. The former has the advantage that it can be explicitly derived from the component level under certain assumptions. The latter is easier to work with and is consistent with observations that the 'trajectories' of components play a minor role compared to the temporal coordination of certain spatial events. It has been hypothesized [29] that the brain (and probably biological systems in general) is a self-organizing system that operates close to instability points, thereby allowing it to switch flexibly from one state to the other. A related possibility is that biological systems are intrinsically unstable (or marginally stable) operating close to stable fixed points. The latter, in some sense, is a reflection of the former. Regardless, our studies indicate that intermittency is quite crucial, affording the possibility of phase- and frequency-synchronization but also providing the necessary mechanism for going in and out of coherent states as well as for exploring the full state space. We intuit that mind, brain and behavior need this mechanism if they are to retain the vital mix of coherent yet flexible function.

Acknowledgments

The research discussed here was supported by NIMH (Neuroscience Research Branch) Grant No. MH42900, ONR contract N00014-88-J1191 and BRS Grant RR07258. We thank Hermann Haken and Arnold Mandell for discussion and encouragement. We dedicate this paper to our friend and colleague, A.J.M.

References

1. Haken, H. (1977/1983). *Synergetics, an introduction: Non-equilibrium phase transitions and self-organization in physics, chemistry and biology.* Berlin: Springer-Verlag, 3rd edition.
2. Kelso, J.A.S. (1981). On the oscillatory basis of movement. *Bulletin of the Psychonomic Society, 18*, 63.
3. Kelso, J.A.S. (1984). Phase transitions and critical behavior in human bimanual coordination. *American Journal of Physiology: Regulatory, Integrative and Comparative Physiology, 15*, R1000-R10004.
4. Cohen, A.V., S. Rossignol, & S. Grillner (Eds.) (1988), *Neural control of rhythmic movements in vertebrates.* New York: John Wiley.
5. Kopell, N. (1988). Toward a theory of modelling central pattern generators. In A.V. Cohen, S. Rossignol, & S. Grillner (Eds.), *Neural control of rhythmic movements in vertebrates* (pp. 369-413). New York: John Wiley.
6. Marder, E. (1989). Modulation of neural networks underlying behavior. *Seminars in the Neurosciences, 1(1)*, 3-4.
7. Selverston, A.I. (1988). Switching among functional states by means of neuromodulators in the lobster stomatogastric ganglion. *Experientia, 44*, 376-383.
8. Gray, C.M., Konig, P., Engel, A.K., & Singer, W. (1989). Oscillatory responses in cat visual cortex exhibit inter-columnar synchronization which reflects global stimulus properties. *Nature, 338*, 334-337.

9. Eckhorn, R., Bauer, R., Jordan, W., Brosch, M. Kruse, W., Monk, M. & Reitboeck, H.J. (1988). Coherent Oscillations: A mechanism of feature linking in the visual cortex? *Biological Cybernetics, 60*, 121-130.

9a. Barinaga, M. (1990). The mind revealed? *Science*, 249, 856-858.

10. von Holst, E. (1939/1973). Relative coordination as a phenomenon and as a method of analysis of central nervous function. Reprinted in: *The collected papers of Erich von Holst.* Coral Gables, Florida: University of Miami Press.

11. Haken, H., Kelso, J.A.S., & Bunz, H. (1985). A theoretical model of phase transitions in human hand movements. *Biological Cybernetics, 39*, 139-156.

12. Jeka, J.J. & Kelso, J.A.S. (1989). The dynamic pattern approach to coordinated behavior: A tutorial review. In S.A. Wallace (Ed.), *Perspectives on the coordination of movement* (pp. 3-45). Amsterdam: North-Holland.

13. Schöner, G., & Kelso, J.A.S. (1988a). Dynamic pattern generation in behavioral and neural systems. *Science, 239*, 1513-1520.

14. Bergé, P., Pomeau, Y. & Vidal, C., 1984. *Order within chaos: Towards a deterministic approach to turbulence.* New York: John Wiley.

15. Mandell, A.J. (1983). From intermittency to transitivity in neuropsychobiological flows. *American J. Physiol., 245*, R484-R494.

16. Kelso, J.A.S., DelColle, J.D., & Schöner, G., 1990. Action-perception as a pattern formation process. In M. Jeannerod (Ed.), *Attention and Performance XIII* (pp. 139-169), Hillsdale, N.J.: Erlbaum.

17. Kelso, J.A.S. (1990). Phase transitions: Foundations of behavior. In H. Haken & M. Stadler (eds.). *Synergetics of Cognition*, Berlin: Springer-Verlag, pp. 249-268.

18. Schöner, G., Haken, H., & Kelso, J.A.S. (1986). A stochastic theory of phase transitions in human hand movement. *Biological Cybernetics, 53*, 442-452.

19. Kelso, J.A.S., Scholz, J.P., & Schöner, G. (1986). Non-equilibrium phase transitions in coordinated biological motion: critical fluctuations. *Physics Letters, A118*, 279-284.

20. Kelso, J.A.S., Buchanan, J.H. & Wallace, S.A. (in press). Order parameters for the neural organization of single, multijoint limb movement patterns. *Exp. Brain Res.*

21. Scholz, J.P., & Kelso, J.A.S. (1989). A quantitative approach to understanding the formation and change of coordinated movement patterns. *Journal of Motor Behavior, 21*, 122-144.

22. Schmidt, R.C., Carello, C. & Turvey, M.T., (1990). Phase transitions and critical fluctuations in visually coupled oscillators. *J. Exp. Psychol. Human Perc. & Perf., 16*, 227-247.

23. Kelso, J.A.S. & DeGuzman, G. (1988). Order in time: How cooperation between the hands informs the design of the brain. In H. Haken (ed.), *Neural and synergetic computers.* Berlin: Springer-Verlag, pp. 180-196.

24. DeGuzman, G.C. & Kelso, J.A.S. (in press). Multifrequency behavioral patterns and the phase attractive circle map. *Biological Cybernetics.*
25. Glazier, J.A. & Libchaber, A. (1988). Quasiperiodicity and dynamical systems: An experimentalist's view. *IEEE Transactions on Circuits and Systems, 35*, 790.
26. Kelso, J.A.S., DeGuzman, G.C. & Holroyd, T. (in press). The self organized phase attractive dynamics of coordination. In A. Babloyantz (Ed.) *Self-organization, Emerging Properties and Learning,*. New York: Plenum.
27. Kelso, J.A.S. (in press). Anticipatory dynamical systems, intrinsic pattern dynamics and skill learning. *Human Move. Sci.*
28. Beek, P.J. (1989). *Juggling Dynamics.* Amsterday: Free University Press.
29. Haken, H. (1983). Synopsis and introduction. In E. Basar, H. Flohr, H. Haken & A.J. Mandell (eds.), *Synergetics of the Brain.* Berlin: Springer-Verlag.

Synchronization of Rhythm in Motor Actions

M. Stadler[1], *S. Vogt*[2], *and P. Kruse*[1]

[1]Department of Psychology, University of Bremen,
W-2800 Bremen 33, Fed. Rep. of Germany
[2]Max Planck Institute for Psychological Research,
W-8000 München 40, Fed. Rep. of Germany

Abstract. Motor rhythms studied on the level of behavioral phenomenology are surveyed in this paper and alternative views to the concept of motor programming are presented. First, v. Holst's concept of coupled oscillators and the Fourier coding model are discussed. The experimental data presented do not confirm frequency-specific processing if dissociated from the phase angle information. Then physical and cognitive principles of self-organization in the motor domain are investigated. Following the results of experimental studies the principle of exploitation of kinetic energy seems to be an organizing factor in motor actions. Furthermore the results of serial reproduction experiments show that the cyclical movement rhythms converge to the same eigen-structure of simple harmonic relations. Finally it is concluded that there is a transition in the development of goal-directed action from goal-adjustment to goal-attraction.

1. Introduction: Behavioral phenomenology as the level of description

Motor actions in animals and men are composed of processes on at least four levels of description (see figure 1). On all levels - the neural, muscular and behavioral level - we find rhythmic activity as a basic phenomenon: The neural impulses, the contractions of the muscle fibers, and the cyclic activity of body movements. In our survey we shall pre-

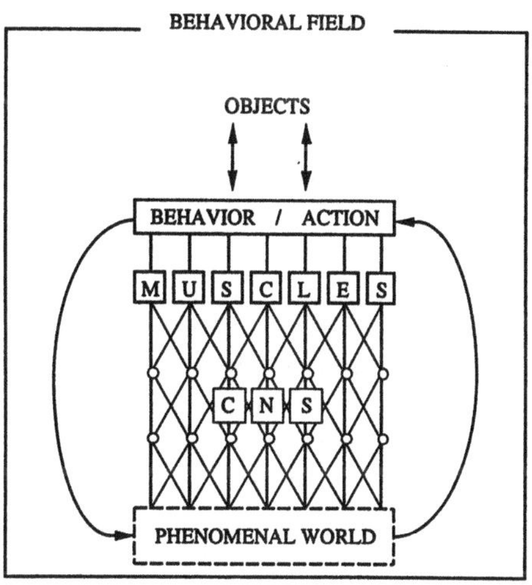

Fig. 1. Levels of description of motor processes

Springer Series in Synergetics, Vol. 55 **Rhythms in Physiological Systems**
Eds.: H. Haken and H.P. Koepchen

sent data and results mainly on the behavioral level and focus on the determination of actions by the behavioral field in the sense of Koffka [1].

For a long time motor scientists tried to explain behavioral performance by motor programs, considered to be elaborated in the brain to control hundreds of motor units in order to achieve certain muscle tensions, muscle lengths, and joint angles which finally result in an adequate topology and metrics of overt behavior. This approach, however, involved insuperable theoretical difficulties: especially the problem of countless degrees of freedom of biological motor systems and the phenomenon of motor equivalence, i.e. the fact that even in a constant environment one and the same action can be realized by a number of different motor solutions. The former approach, which favoured the hierarchic composition of independent motor elements has been repeatedly opposed by alternative views of motor action that postulate the synergy of elements on all levels of the motor system, the interaction between these levels, and the importance of the phenomenology of behavior in motor organization. "No single level has ontological priority over any other", as Kelso put it [2]. At least four theoretical sources of ideas converging in that direction can be identified:

- Erich von Holst's concept of coupled oscillators and the nature of order processing in the CNS [3];
- Nicolai Bernstein's idea of the image of achievement as a central determinant of specific complex movements and his introduction of the method of Fourier analysis to the description of motor processes [4];
- the Gestalt-theoretical view that motor organization obeys the fundamental Gestalt principles of non-additivity, transposability, and 'Prägnanz' (economy) [5];
- the approach of synergetics [6], which identifies motor actions as collective variables on a macroscopic level that enslave the elementary processes from which they have emerged. This approach allows a formal mathematical basis of self-organization processes in motor control to be elaborated for the first time [7].

In the next sections some basic concepts and experimental results of these alternative approaches will be surveyed, which may be the basis of further theoretical and mathematical analyses of biological motor systems.

2. Coupled oscillators and the magnet effect

Ever since the work of the German neurobiologist Erich von Holst it has been widely accepted that the locomotion of animals is guaranteed by endogenous rhythmicity. In that view, locomotor actions of the extremities are caused by certain pacemakers in the nervous network that autonomously generate rhythmic signals. These so-called oscillators, responsible for the different muscles in the locomotion system of an animal, may be distributed over the whole brain. However, as von Holst found out, they are coupled in a variety of ways. Let us have a look at the rhythmic movements of the pectoral and tail fins of a medulla fish studied by von Holst (figure 2). Every motor rhythm studied in that way has a certain stability - a maintenance of its periodicity over some time interval, which can be modulated by the attractive forces of coupled oscillators. The so-called **magnet effect** (entrainment) may result in a change of frequency as well as in phase shifts to achieve harmonious relations of different rhythms. These frequency or phase shifts may happen slowly or in a sudden phase transition. The latter has been studied in detail by Haken and Kelso by means of applying synergetic methods [7].

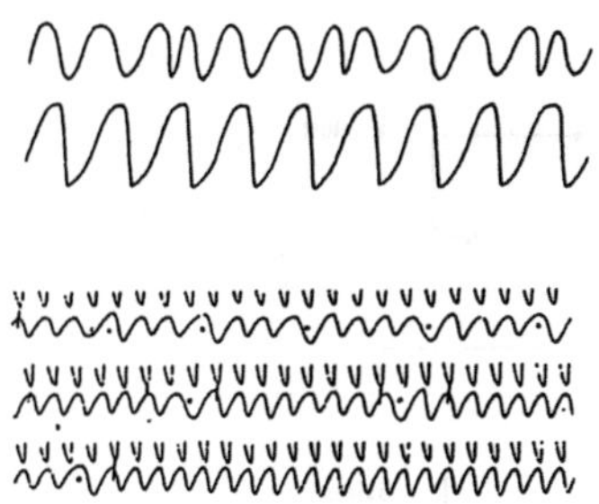

Fig. 2. The magnet effect in the medulla fish (from v.Holst [8]). Above: Relative coordination of the pectoral and dorsal fins; below: Absolute coordination of the pectoral and tail fins.

According to von Holst the oscillators controlling the various extremities are subgrouped by oscillators controlling the single muscles and motor units. Each subgroup acts on every other subgroup and is influenced reciprocally. The synchronization and superimposition of these oscillators result in a collective process of perodic rhythmicity - the behavior of the organism - that has a backward effect on the elementary oscillators responding in a resonant way.

For the experimental study of the magnet effect it has turned out to be useful to define an independent and a dependent motor rhythm by external constraints. In general the dependent rhythm is characterized by a greater flexibility (a smaller number of effective constraints), thus allowing synchronizations with the independent (more strongly constrained) rhythm. In a complex task context, such attraction processes were demonstrated in a study performed in our laboratories [9]. Human subjects had to perform a tracking task (paced contour-tracking) with the right hand (independent rhythm). They were given the additional task of tapping simultaneously the rhythm of a simple melody (dependent rhythm) using their left hand. Figure 3 shows the slow synchronization of the two motor actions in the course of a number of performances. The obvious entrainment-effect is a useful model for a complex motor learning process. The reverse possibility of decoupling contralateral rhythms is much more difficult as every musician knows.

3. Frequency analysis and synthesis as a model of perceptual-motor organization

In addition to the magnet effect Erich von Holst identified a second basic principle of synchronization between two oscillations in the motor system: In his experiments with medulla fishes, the independent rhythm of one fin sometimes modulated the dependent rhythm by **superimposition** [10]. Also in the thirties, without having knowledge of these investigations, the Russian physiologist Nicolai Bernstein thoroughly studied rhythmical human movements (such as walking, striking with a hammer, filing or piano playing) and found that the spatio-temporal trajectories of most of them may be interpreted with an accuracy of within a few millimeters as the sum of three or four harmonic oscillations (Fourier trigonometrical sums) [11]. Thus, while von Holst described superimposition as a cross-talk phenomenon between two spatially separate rhythms, Bernstein used superimposition and decomposition methodologies in order to study the characteristics of individual complex movement patterns.

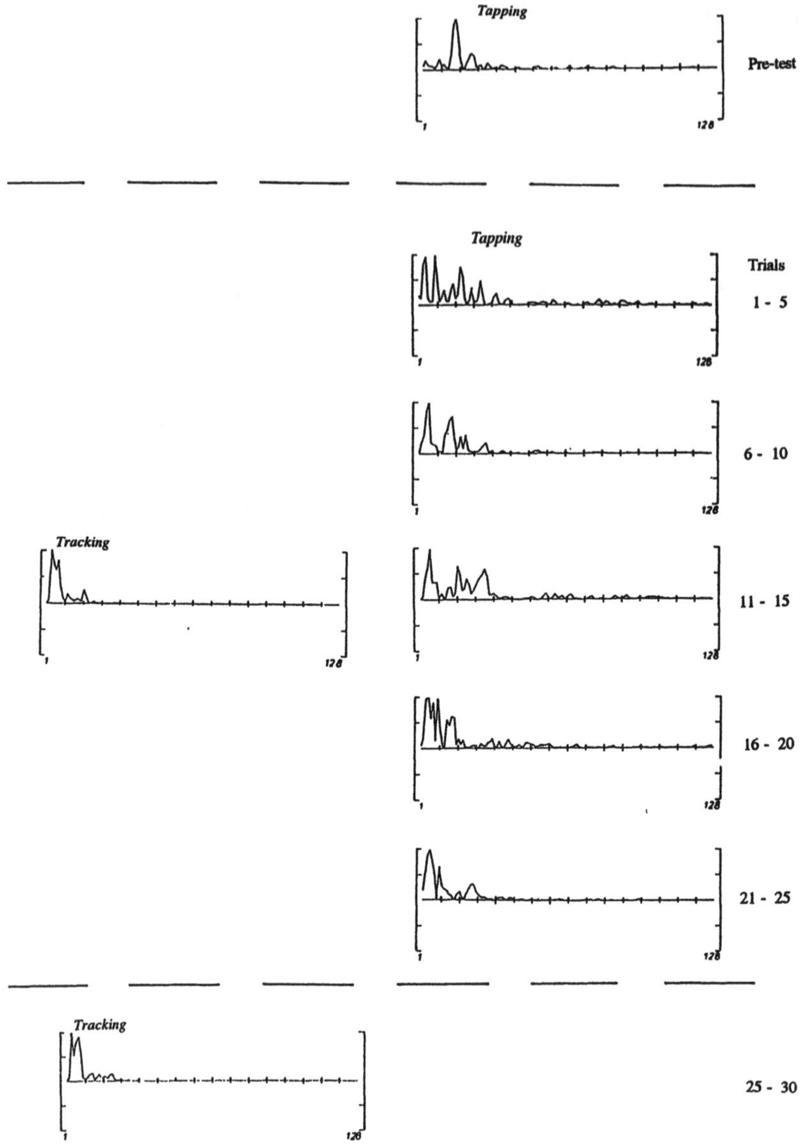

Fig. 3. Power spectra of an independent tracking task (left side), and a dependent melody tapping task (right side). Between the dashed lines these motor tasks are performed simultaneously. The tapping rhythm is entrained by the tracking frequencies in the course of the learning process (from Wehner & Stadler [9]).

The consecutive step from descriptive method application to a psychological model of trajectory representation was taken by Pribram and later by Gallistel [12]. Gallistel favored the idea that - following the Fourier theorem - any complex sequence of movement can be generated by superimposing a very few sinusoids in certain phase relations and that, by a variation of amplitudes and phase angles, a large number of different spatio-

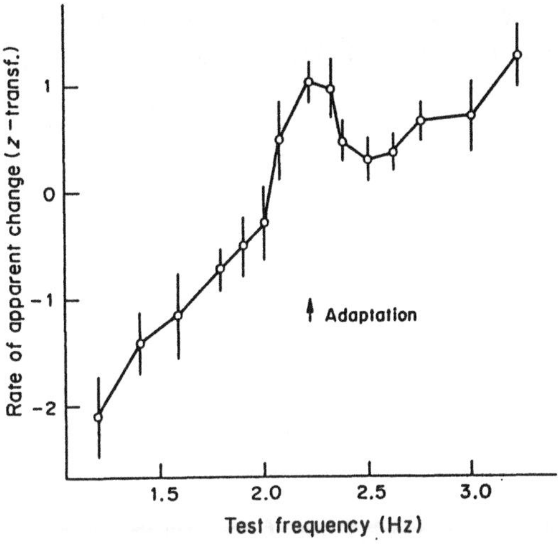

Fig. 4. Frequency selective channel in the visual system measured by selective adaptation of stroboscopic alternative movement (from Kruse, Stadler & Wehner [15]).

temporal patterns of movement can be obtained. Oscillators might be, Gallistel concluded, "the alphabet with which central programs are written" (p.80). Finally, Pribram, Sharafat, & Beekmann [13] found frequency selective reactions in a range of half an octave by recordings of single cells in the motor cortex of the cat during periodic passive movements of the forepaw.

It is worthwhile to synoptically consider these issues concerning the motor system with corresponding results in the perceptual domain. It has long been known that frequency analysis of complex soundwaves exists in the auditory system; but in the last decades research interest has focussed, also, on the visual system. All animal extremities are suspended at the trunk like multiple pendula. Thus, most of the translatory motions of animals and men (and even of plants being moved in the wind) consist of simple or superimposed sinusoids as can be visualized by the patch-light technique of Gunnar Johansson [14]. Therefore, frequency selective channels can be assumed also for visual motion perception. This was confirmed by our group [15] in an investigation of long-range apparent movement (see figure 4).

Considering these results together with the old hypothesis that memory might be a process of distributed macroscopic reconstruction of former brain states by means of resonance, as was discussed by the gestaltists [16], one could put forward a global model of perceptual-motor organization: The stimulus input of biological motion is analyzed by resonant channels in the sensory system. The resulting frequency distribution is stored in the CNS by means of power-spectral analoga. Simultaneously the system composes its motor output by superimposing sinusoidal oscillations. In this model the central representation of external (biological) motions as well as the general features of the image of movement are expressed as power-spectral analoga. Given that biological movements require only a small number of sine wave descriptors, the model allows an enormous compression of the information which would be necessary to describe or store all the coordinates of movement trajectories in the temporal domain. Furthermore, the power spectra allow reliable estimations of the past and the future of the window of a motion trajectory

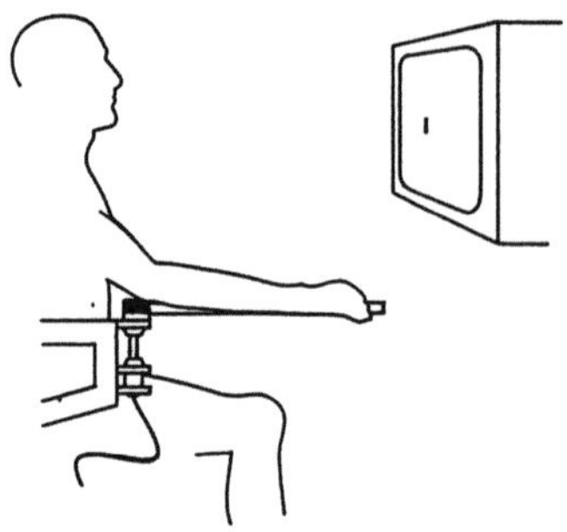

Fig. 5. Experimental arrangement of the studies of movement imitation reported (from Vogt [18]).

that has been processed by the cognitive system. For these reasons superimposed sinusoidal oscillations might be good candidates for collective variables or order parameters in the sense of synergetics [17].

The model of frequency coding in perceptual-motor systems may be proved or falsified by specifying the conditions under which it might work. One of our investigations, a motor learning experiment, tried to answer the question whether the human motor system is able to utilize information about the frequency structure of a movement pattern irrespective of the phase angle information of the different sinusoids [18]. As the experimental setup an imitation task was used where subjects learned to reproduce the movement trajectory of a visually presented moving line with the right arm by means of a lever (figure 5). In the first experiment, the main prediction was a positive transfer effect between two cyclical movements (pattern A and B in figure 6) with the same frequency composition but with different phase relations. On the other hand, a smaller transfer should be found between a learned and a transfer pattern (B and C in figure 6) different in frequency composition, even if similarities in the topological structure are obvious.

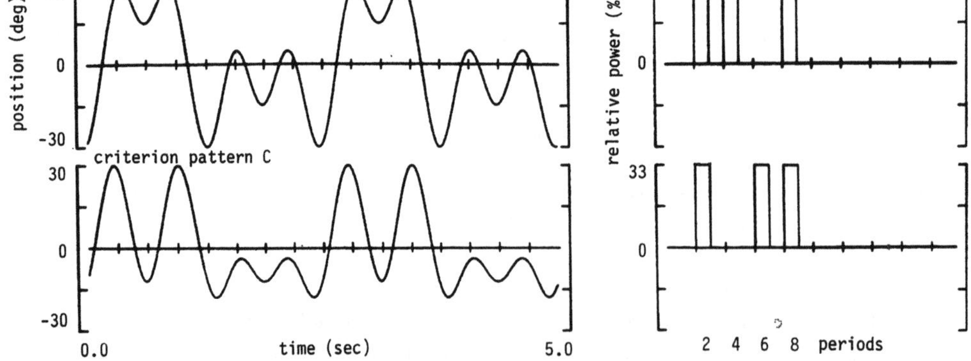

Fig. 6. Criterion patterns A, B, C (left) and the related periodograms (right) (from Vogt [18]).

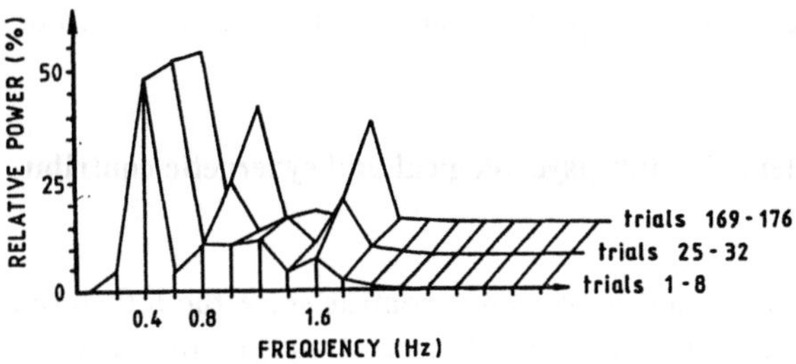

Fig. 7. Power spectra of the reproductions at 3 learning stages (from Vogt, Stadler & Kruse [5]).

But while an analysis of the learning process (see figure 7) revealed a good spectral correspondence between the criterion pattern and the reproductions (with a delayed stabilization of the high frequency components [19]); the results of the transfer experiment, however, did not confirm this hypothesis and showed a small positive transfer effect only between the two patterns (B and C) with a similar topological structure (see figure 8). Similar results were obtained in a recent study by Stanley and Franks [20]. Thus, a relatively safe conclusion from these experiments is that frequency-specific processing in the motor system - if it exists at all - cannot be dissociated from the phase angles.

Summarizing, we assume that the attractiveness of the Fourier coding model lies mainly in the integrative character of its components, each of which contributes to a movement pattern over the entire time span. The model thereby offers an alternative to serial chaining models of motor output and, moreover, allows the couplings of an organism with its environment to be described in terms of resonances [21]. The Fourier coding model is limited, however, in its strong determinateness between input/output variables on the one hand and internal cognitive processes on the other hand, as well as in its linearity, that is, its inability to explain nonlinear perceptual and behavioral phenomena

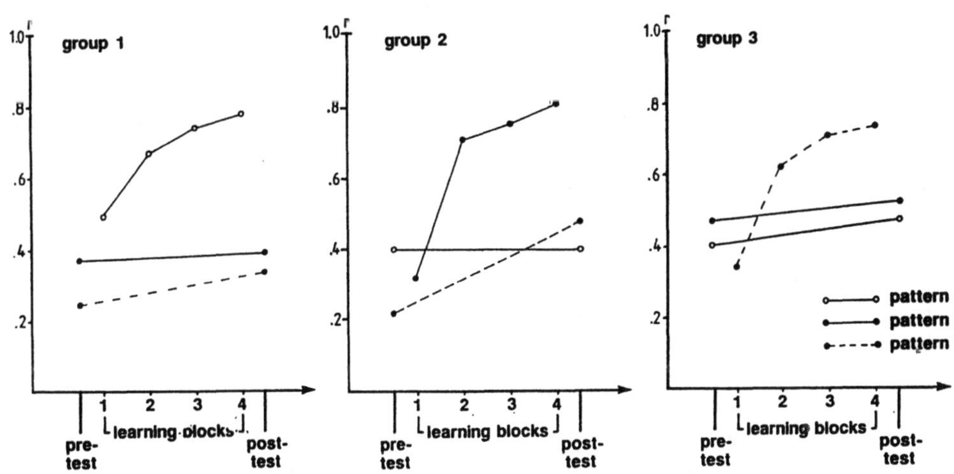

Fig. 8.Results of the transfer experiment; correlations between the criterion patterns and the reproductions (from Vogt [18]).

[22]. For this reason we shall discuss some principles of self-organization in the motor system in the next section.

4. Self-organization in the motor system: Gestalt psychological and synergetic contributions

Two properties of motor systems have puzzled movement scientists since the influential research of Nicolai Bernstein: The nature of movement representations and the degrees-of-freedom problem.

Bernstein was strongly convinced of the existence of a **central representation** (an 'engram') of a movement as a gestalt-like image that guides the performance of complex movements. He argued: "Therefore it is necessary that there should exist in the supreme nervous organ an exact representation of what will later occur at the periphery" [23]. The precise nature of such central representations, however, is still a matter of discussion [24]. While in the last decade the abstract, qualitative nature of a movement representation has been emphasized, further reconceptualizations might follow the lead of synergetics' view of non-prescriptive macroscopic order states [25]. This means that the macroscopic behavior as well as its anticipated representation should function as an order parameter for the microscopic events in the motor units and muscles that actually perform the complex movement and that this order parameter organizes the microscopic elements [26].

The **problem of the degrees of freedom** of the motor system emphasizes the fact that one and the same action-goal can be achieved - considering the domain of neuromuscular control - in many different ways, depending on the environmental field forces, on the initial conditions of the movement and on perceptual factors. Bernstein postulated that in motor control there exist many 'lines of equal simplicity' (for instance writing a sentence on paper or on the blackboard). Degrees of freedom in such number again require a self-organizing principle. In this context Turvey [27], Kelso et al. [28] and other 'action' theorists developed the concept of **coordinative structures** as a "temporary marshalling of many degrees of freedom into a task-specific functional unit". Also in our research group we attempted to understand the conspicuous ability of living systems to generate organized, goal-directed patterns of activity under changing environmental and organismical states as principally free from the necessity to fixate microscopic details of execution [29]. One important task of modelling in the field of motor behavior is to find the particular system forces and their modes of interaction in the emergence of movement patterns and to identify the related attractor states.

Our starting point was the idea to apply the gestalt principle of **'Prägnanz'** (economy) to the motor domain, but not in the sense that the overall energy level of a movement pattern is minimized but that the spatio-temporal movement trajectory evolves in a way that supplies an **optimal conduction** of the existing kinetic energy in the sense of the gestalt law of good continuation in perception. The gestalt psychologist Wolfgang Metzger [30] formulated this hypothetical principle as follows:

> "A new principle ... which is of critical importance for the smoothness, regularity, and homogeneity of movements and which is still not sufficiently studied, is the 'swing', i.e., the exploitation of kinetic energy in the creation of movement gestalts. An unskilled movement is apparently, from its beginning to its end, kept going by inner-

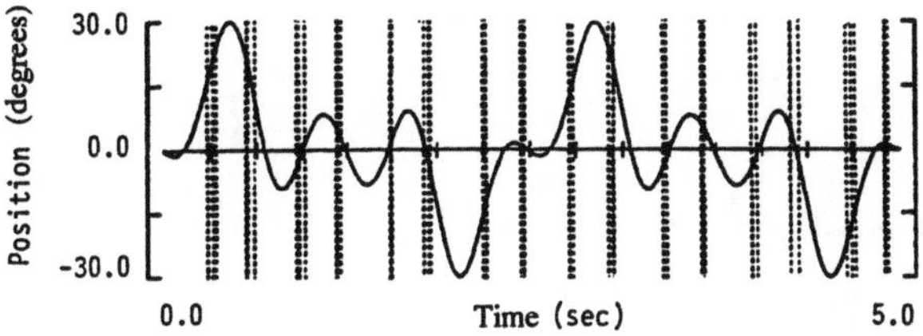

Fig. 9. Criterion pattern A with overlaid tapping rhythms (dotted lines) of several subjects (from Vogt [18]).

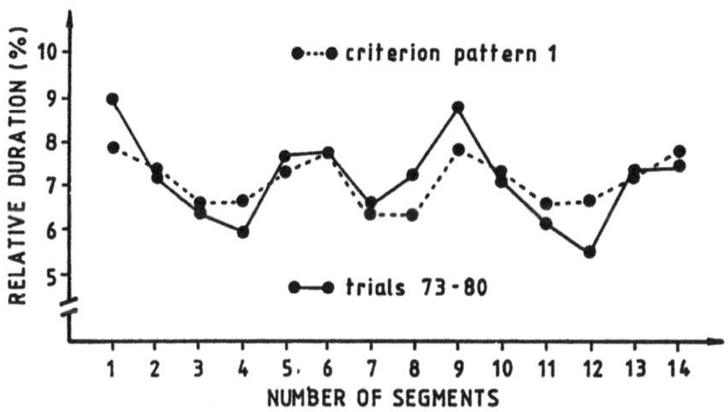

Fig. 10. Temporal structure of the rhythm of a criterion pattern (dotted line) and of a set of reproductions at the end of the learning process (continuous line) (from Vogt, Stadler & Kruse [5]).

vations and hindered from deviations from the intended way. The further the mastering of the movement has advanced, the larger becomes the portion of kinetic energy of the performance and the smaller the remaining portion of innervation. A perfectly mastered movement is characterized by a maximum of wholeness and uninterruptedness and by a minimum of innervational effort. Thereby, the movement is to a high degree 'left to itself'".

The rationale of our related empirical studies was to analyze the spatio-temporal structure of complex movement patterns in a movement imitation situation, while a number of temporal and sequential variables of the criterion pattern were manipulated. In order to find a psychologically adequate temporal segmentation, subjects were asked in a pilot study to synchronize a repetitive tapping rhythm with the displayed criterion pattern A as shown in figure 9, and in a further condition, with a regular sine wave of 0.8 Hz. In contrast to the expectations of most subjects, synchronization to the points of peak velocity (the points of inflexion in the position-time curves) revealed a more stable and precise phase locking than the attempt to synchronize with the spatial extrema.

In accordance with these results, skilled reproductions of pattern A were segmented into single acceleration phases and the interval durations of 14 phases were plotted as (relative) time-per-segment diagrams. Figure 10 shows the results of 8 trials for 10 subjects and the related temporal structure of the criterion pattern. In contrast to predictions from linear oscillator models [31] and from the isochrony principle [32], the temporal variations in the criterion pattern (which were correlated with its spatial structure) were not reduced toward a periodic timing but were even accentuated to an interval ratio

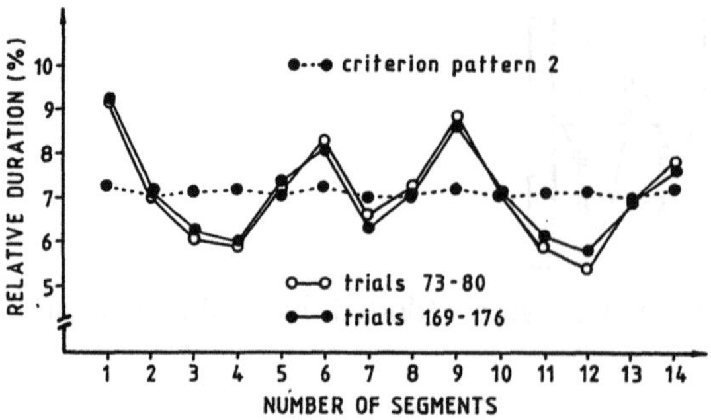

Fig. 11. Temporal structure of the rhythm of a modified (strictly periodic) criterion pattern (dotted line) and two sets of reproductions at the end of the learning sessions (continuous lines) (from Vogt, Stadler & Kruse [5]).

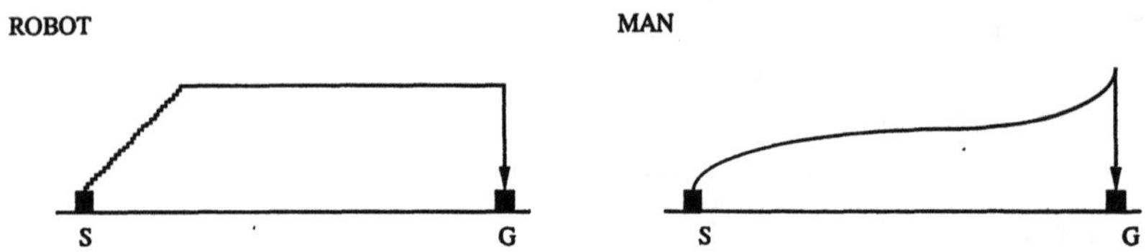

Fig. 12. Schematic representations of the trajectories of goal-directed movements in robot and man [34]).

of 1.49 for long vs. short cycles. This result was confirmed in a second experiment, where, as a first context manipulation, the temporal structure of the criterion pattern was modified toward strict periodicity. Here, subjects produced the same variations in the temporal structure as in the former condition (see figure 11; interval ratios 1.56 on day 1 and 1.51 on day 2 (for the results of further context variations see [33]). In general an autonomous self-amplifying rhythmic temporal structure seems to be produced even when the stimulus has a strictly linear periodicity. In the language of music this would be the transition from metronomic measure to rhythm and swing. In the domain of behavioral phenomenology the rhythmic structure is to be observed (and sensed) as an experienced stressing or amplification of the higher amplitudes and a suppression of the lower ones. And on the level of kinetic energy a conduction from one high amplitude to the next is facilitated. This marks the influence of the supposed principle of **exploitation of kinetic energy** in movement control.

The principle of exploitation of kinetic energy seems to be even more important in the organization of goal-directed action. This becomes obvious in a comparison between the movement control of industrial robots and human beings (figure 12). While the robot's movement towards the goal is programmed in every particular aspect from the beginning to the end, man produces an approximate amount of kinetic energy and controls his arm's swing towards the goal at the point of instability by a minimal correction.

The next experiments give a tentative answer to the question of what happens if very complex movement patterns (from the point of view of frequency composition) are "left to themselves": Which are the attractors of cyclic horizontal arm movements? For this purpose we performed serial reproduction experiments (the so-called Bartlett scenario [35]) as a means of demonstrating different stages of a self-organizing process [36]. Serial reproduction experiments consistently use the principle of recurrence to find eigen-structures in perception and performance. In contrast to the imitation experiments as described before, each pattern was presented only once to a subject, and her/his reproduction was used as a criterion pattern in the same manner as for the first subject (with 10 subjects participating in each Bartlett series). Under these conditions, an unexpected homogeneity of the final reproductions of a number of highly different starting patterns could be observed (2 examples are shown in figures 13 and 14): Whatever our initial stimulus conditions were and whatever the frequency composition of the initial criterion pattern was, the process converged to an attractor state of 1 and 3 periods per 5 seconds, that is a superimposition of frequencies in the relation of 1 : 3 (periodograms at the right side of the figures 13 and 14). The topological structure of the attractors was very different in the individual experiments (left side of the figures) but after Fourier analysis the movements were in most cases sufficiently described by the 1st and 3rd harmonic (figure 15) with highly varying phase angles. The 1 : 3 relation of the periods resembles what was found already by von Holst [37] in his medulla fish experiments as stable equilibrium of coupled oscillators. Without implying an unflattering comparison, we might say that brainless fish seem to love waltz rhythms as much as men do.

It can be summarized that there seems to exist a general tendency in the learning and self-organization of movement control from complex frequency composition to the prevailing of a few harmonic oscillations. This obviously resembles the so-called 'Prägnanz' tendency found in serial reproduction experiments in the perceptual domain [38].

5. Conclusions: Are actions driven or pulled?

Let us finally come back to the central problem of movement science. How are movements organized? On the microscopic or the macroscopic level? How is the infinite number of degrees of freedom to be handled? Is the organization of motor actions - for instance the goal-directed arm movement of grasping a glass of water - to be explained in terms of an innervation pattern of a great number of motor units working together in the exact calibration of tension and length of a chain of at least 5 pairs of antagonistic muscles resulting in well-defined variations of joint angles directing the hand finally to the glass? Are actions driven like that or is the hand simply attracted by the goal, as is suggested by phenomenological experience. We are surely not too daring in saying that both views are right. No goal-directed action will be possible without any muscular innervation and, on the other hand, without any visual guidance. But even a visually guided action may be described and programmed as a continuing process of adjustments of the goal approach elaborated successively by the visuo-motor control system. If, on the other hand, the goal and the body-scheme are part of the same functional unit in the cognitive domain, goals might be **attractors** or order parameters of action synchronizing all the neural oscillations on the microscopic level in a way that is necessary for aimed movements. At present, we do not know much about such processes; yet, there are enough hints to assume a shift or even a **phase transition** in the mode of organization during an evolutionary process of learning that may be interpreted as a transition from the control by means of goal adjustments to a strategy of goal attraction. In the latter strategy the kinetic forces - once established - are exploited and only minimal interventions of voluntary control are necessary at points of unstable equilibrium in the movement trajectory to

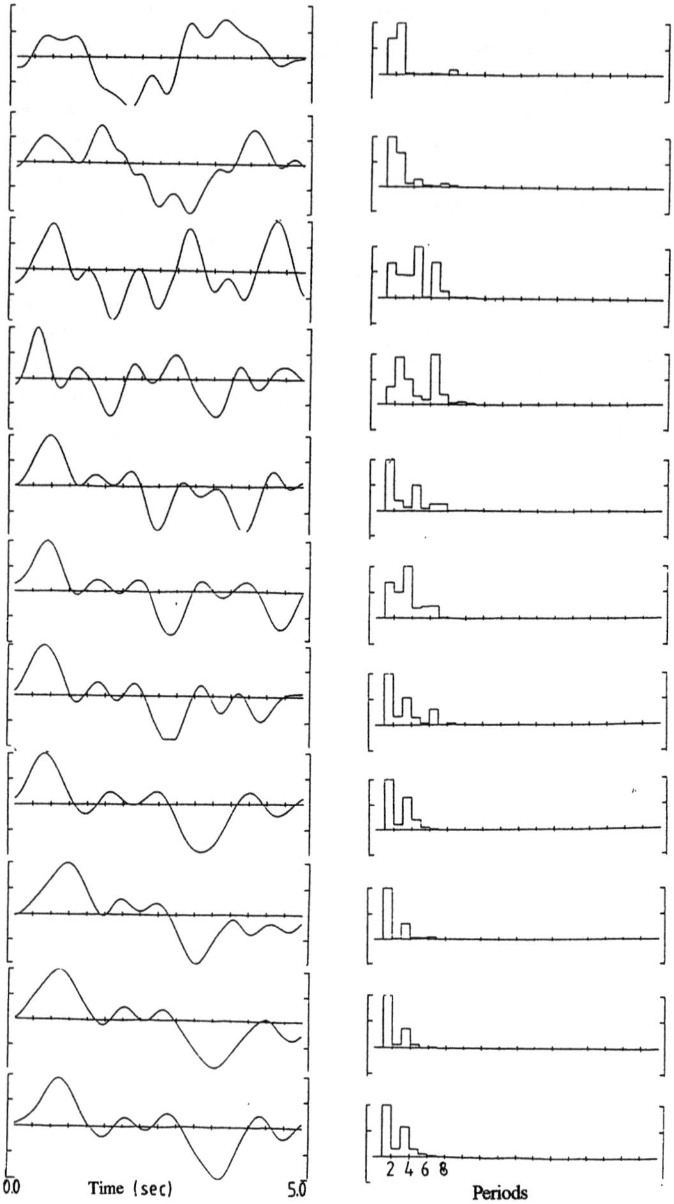

Fig. 13

Fig. 14

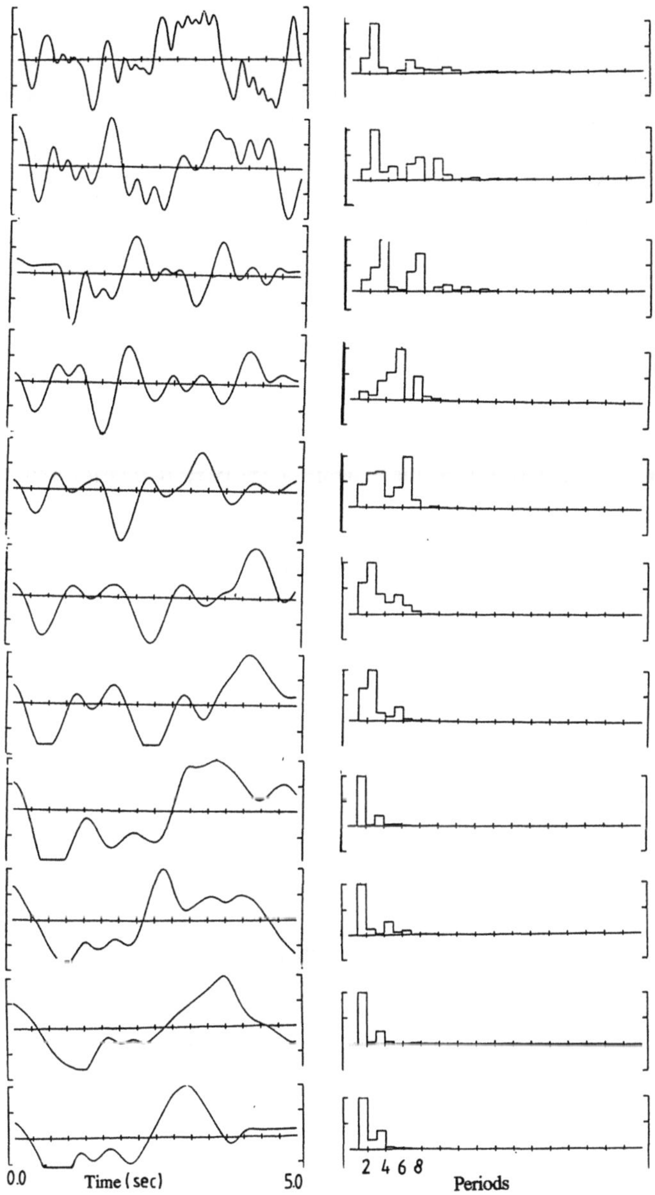

Fig. 13 and 14. Serial reproductions of complex motion trajectories (left) and the related periodograms (right); Criterion patterns upper left [36]).

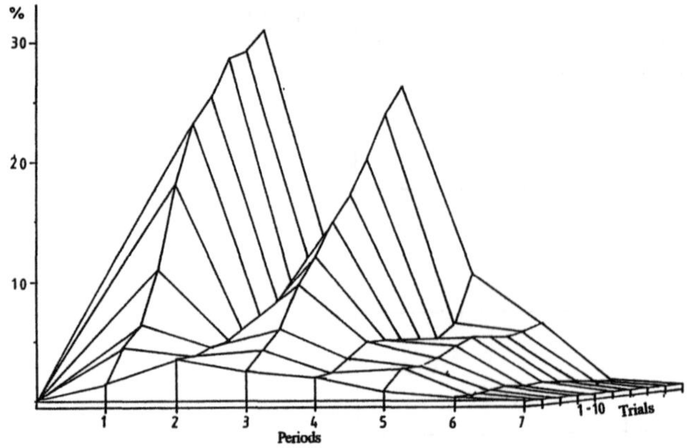

Fig. 15. Cumulated periodograms of 10 serial reproductions of a criterion pattern composed of periods in the relation of 2:3:4:6; the result is again a 1:3 period relation [36].

reach the goal. While such recourses to phenomenological experience have become rare in movement science during the last decades, they might nevertheless stimulate a more comprehensive view of motor organization, embracing both cognitive and physical aspects of human motor action.

Reference notes

[1] K. Koffka: Principles of Gestalt Psychology, (London, Routledge & Kegan Paul 1935).
At the same time Lewin used the terms 'psychological field' or 'life space'. K. Lewin: Dynamic Theory of Personality, (McGraw-Hill, New York 1935)

[2] J.A.S. Kelso, G. Schöner: Toward a physical (synergetic) theory of biological coordination, in Lasers and Synergetics, ed. by R. Graham, (Springer, Berlin 1987), pp. 224-237

[3] E. v. Holst: Naturwissenschaften 25, 625-631; 641-647 (1937)
E. v. Holst: Erg. Physiol. 42, 228-306 (1939)
see also C.R. Gallistel: The Organization of Action: a New Synthesis, (Lawrence Erlbaum, Hillsdale, N.J. 1980)

[4] N. Bernstein: The Co-ordination and Regulation of Movement (Pergamon Press, Oxford 1967)
see also K.H. Pribram: Languages of the Brain, Prentice-Hall, Englewood Cliffs, N.J. 1971) and
H.T.A. Whiting: Human Motor Actions: Bernstein Reassessed, (North-Holland, Amsterdam 1984)

[5] W. Köhler: Die physischen Gestalten in Ruhe und im stationären Zustand, (Vieweg, Braunschweig 1920)
D. Katz: Gestaltpsychologie (Schwabe, Stuttgart 1969),4th ed.
W. Metzger: Gestalt-Psychologie, ed by M. Stadler, H. Crabus, (Kramer, Frankfurt/M. 1986)
see also S. Vogt, M. Stadler, P. Kruse: Human Movement Sci. 7, 365-406 (1988)

[6] H. Haken: Synergetics, an Introduction, (Springer, Berlin 1977)
H. Haken: Advanced Synergetics, (Springer, Berlin 1983)

[7] H. Haken, J.A.S. Kelso, H. Bunz: Biol Cybernetics 51, 347-356 (1985)
J.A.S. Kelso, G. Schöner: Human Movement Sci. 7, 27-46 (1988)
J.J. Jeka, J.A.S. Kelso: The dynamic pattern approach to coordinated behavior: a tutorial review, in Perspectives on the Coordination of Movement, ed. by S.A. Wallace, (North-Holland, Amsterdam 1989)
see Kelso in this volume

[8] E. v. Holst (1939), see note 3

[9] T. Wehner, M. Stadler: Z. Psychol. 190, 183-201 (1982)

[10] see note 3

[11] see note 4

[12] see notes 3 and 4

[13] K.H. Pribram, A.Sharafat, G.J. Beekmann: Frequency encoding in motor systems, in Human Motor Actions: Bernstein Reassessed, ed. by H.T.A. Whiting (North-Holland, Amsterdam 1984), pp. 121-156

[14] G. Johansson: Perception and Psychophysics 14, 202-211 (1973)

[15] P. Kruse, M. Stadler, S. Vogt, T. Wehner: Gestalt Theory 5, 83-113 (1983)
P. Kruse, M. Stadler, T. Wehner: Vision Res. 26, 327-335 (1986)

[16] W. Metzger (1986), see note 5
K.H. Pribram, Towards a holonomic theory of perception, in Gestalttheorie in der modernen Psychologie, ed. by S. Ertel, L. Kemmler, M. Stadler (Steinkopff, Darmstadt 1975), pp. 161-184

[17] H. Haken: Information and Self-organization, (Springer, Berlin 1988), pp. 151f.

[18] S. Vogt: Einige gestaltpsychologische Aspekte der zeitlichen Organisation zyklischer Bewegungsabläufe (Doct. Thesis, Bremen 1988), ch. 4

[19] R.G. Marteniuk, S.K.E. Romanow: Human movement organization and learning as revealed by variability of movement, use of kinematic information, and Fourier analysis, in Memory and Control of Action, ed. by R.A. Magill, (North-Holland, Amsterdam 1983)

[20] M.L. Stanley, I.M. Franks: Learning to organize the frequency components of a perceptual motor skill (in press)
see also H.T.A. Whiting, S. Vogt, B. Vereijken: Human skill and motor control: some aspects of the motor control - motor learning relation, in Approaches to the Study of Motor Control and Learning, ed. by. J.J. Summers (North-Holland, Amsterdam 1991, in press)

[21] W. Metzger, see note 5
R.N. Shepard: Psychological Review 91, 417-447 (1984)

[22] K.H. Pribram: Prolegomenon for a holonomic brain theory, in Synergetics of Cognition, ed. by H. Haken, M. Stadler, (Springer, Berlin 1990), addendum p. 178

[23] N. Bernstein, in H.T.A. Whiting (1984), p. 101, see note 4

[24] H.T.A. Whiting, B. den Brinker: Image of the act, in Theory and Research in Learning Disabilities, ed. by J.P. Das, R.F. Mulcahy, A.E. Wall (Plenum, New York 1982)
see e.g. Whiting & den Brinker's integration of Bernstein's view and Pribram's (1971), note 4, concept of an 'image of achievement', and our discussion of the Fourier coding model in section 3

[25] see discussion in H.T.A. Whiting, S.Vogt, B. Vereijken, note 20

[26] H. Haken (1988), see note 17

[27] M.T. Turvey: Preliminaries of a theory of action with reference to vision, in Perceiving, Acting, and Knowing, ed. by R. Shaw, J. Bransford, (Lawrence Erlbaum, Hillsdale, N.J. 1977)

[28] J.A.S. Kelso, J.P. Scholz: Cooperative phenomena in biological motion, in Complex Systems - Operational Approaches, ed. by H. Haken, (Springer, Berlin 1985), pp. 124-149

[29] S. Vogt, M. Stadler, P. Kruse (1988), see note 5

[30] W. Metzger, p. 185, see note 5, English translation by the authors

[31] H. Heuer, R.A. Schmidt: J. Exp. Psychol.: Human Perception and Performance 14, 241-252 (1988)

[32] P. Viviani: Do units of motor action really exist?, in Generation and Modulation of Action Patterns, ed by H. Heuer, C. Fromm, (Springer, Berlin 1986)

[33] S. Vogt, M. Stadler, P. Kruse (1988), see note 5

[34] The experiments were performed by Z. Franko

[35] M. Stadler, P. Richter, S. Pfaff, P. Kruse: Psychological Research (in press)

[36] The experiments were performed by B. Pavlekovic

[37] E. v. Holst (1939), see note 3

[38] M. Stadler, P. Kruse: The selforganization perspective in cognition research: Historical remarks and new experimental approaches, in Synergetics of Cognition, ed. by H. Haken, M. Stadler (Springer, Berlin 1990), pp. 32-52

Part V

Basis of Circadian Rhythmicities

Interactions Between Human Circadian and (About 90 min) Sleep Rhythms: Problems in the Simulation and the Analysis

R.A. Wever

Max-Planck-Institut für Psychiatrie,
W-8138 Andechs, Fed. Rep. of Germany

1. Introduction

Biological rhythms which are manifested in a wide range of frequencies, are generated endogenously but commonly modified exogenously. In most rhythms this means that all parameters of the autonomous rhythms depend on the internal and external conditions constituting the relevant environment of the rhythmic system under consideration. It is obvious that this environment fluctuates regularly (possibly due to superimposed rhythms of lower frequencies) and irregularly (randomly). In that rhythmic system the properties of which are best known, i.e., in the circadian rhythmicity, moreover, an external modification is expressed in a synchronization of the endogenously generated rhythms by periodicities in the external conditions (i.e., by "zeitgebers"), at least within limited "ranges of entrainment" [1]. In most mammals including humans, apart from circadian rhythms only seasonal (yearly) rhythms correspond to external periodicities; in other organisms, external tidal and lunar rhythms can be relevant.

The properties of rhythmic systems are governed by relatively simple regularities. Originally only outgoing from some few generalizations of animal circadian experiments, these regularities have been expressed in a mathematical model [2,3,4]; it had been shown later that this model is applicable also to rhythms of greatly deviating frequencies as rhythms of nervous activities [5] and sleep rhythms [6]. Details of the mathematical model equation (van der Pol equation, modified by two non-linearities describing the apparently typical biological peculiarities) can be taken from the places mentioned. In nearly three decades a great multiplicity of divergent predictions had been confirmed in relevant experiments (in the last 25 years particularly in human experiments which had been started only after the establishment of the model), without to have been confronted with the necessity to change even one term of the original equation. Not one model prediction could be found to be in contradiction to biological facts. It seems to be reasonable, therefore, to include model postulations supplementing the biological results, also in the considerations to be discussed in this paper.

2. Model Considerations

It is a fundamental result to be deduced from solutions of the mathematical model that the self-sustaining capacity of an autonomous rhythm is manifested only within a limited range of environmental conditions (or "external forces"), i.e., within the "oscillatory range". Inside of this range, all features of the "free-running" rhythms (period, amplitude, mean value, waveshape, duration of transient states and many more) depend on the external force; outside of this range, in fact, the feed-back mechanism is still operating, but it is no longer able to overcome (to use an obvious mechanical picture:) the increasing friction [7]. When operating outside of the oscillatory range the system seems to show, with a rough inspection, a damped oscillation after any deflection; a closer inspection, however, indicates a special behavior which is due to the steadily varying (non-linear) coefficient of the damping (which depends on y, i.e., on the deflection of the system from a fixed value). This behavior is of particular interest when

Springer Series in Synergetics, Vol. 55 **Rhythms in Physiological Systems**
Eds.: H. Haken and H.P. Koepchen

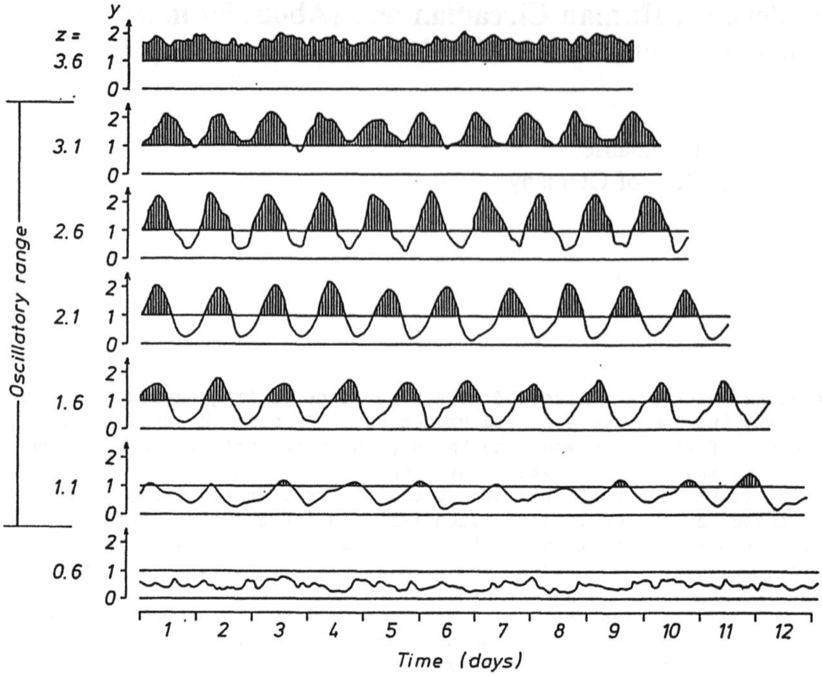

Figure 1: Solutions of the model equation in the circadian domaine, with 7 different values of the external force z (each superimposed by similar random fluctuations). A fixed threshold devides the course of the solutions into "wake" (above the threshold; hatched areas) and "sleep" (below the threshold). Outside of the "oscillatory range" the solutions do not describe rhythms.

the system is under the influence of a superimposed random noise; the answer of the system, though released by random fluctuations, is not a truely random course but a course which is determined, in addition, by the regularly varying coefficient of damping.

Figure 1 illustrates solutions of the model equation in the circadian domaine, with 7 different values of the external force z, and superimposed by comparable random noise each; the presentations include 10 successive cycles each. In the lowermost and the uppermost solutions is the system outside of the oscillatory range (which is positioned between z = 0.847 and z = 3.189) and, hence, does not show rhythms but almost (see above) random fluctuations; period analyses of any type do not result in any rhythmicity. As common in modelling circadian sleep-wake rhythms, the course of the dependent variable y is devided by a fixed threshold (at y = 1) into "activity" (above the threshold; hatched areas) and "rest" (below the threshold). As can be seen, with increasing z the total period shortens, but the activity-time lengthens, eventually leading from "continuous rest" to "continuous activity" (with remaining rhythmicity as long as the system is inside of the oscillatory range).

Figure 2 shows solutions of the model equation of the same type but for rhythms with a much higher frequency (with increasing frequency the model solutions shift from the pendulum type to the relaxation type of oscillation). Again, solutions with 7 different z-values are shown (with z-values always close to the lower limit of the oscillatory range). And again, the solutions inside of the oscillatory range show a dependency of the period on the z-value (to a much higher extent than with the low-frequency solutions); moreover, the superimposed random fluctuations affect less the amplitude (as they do in the low-frequency

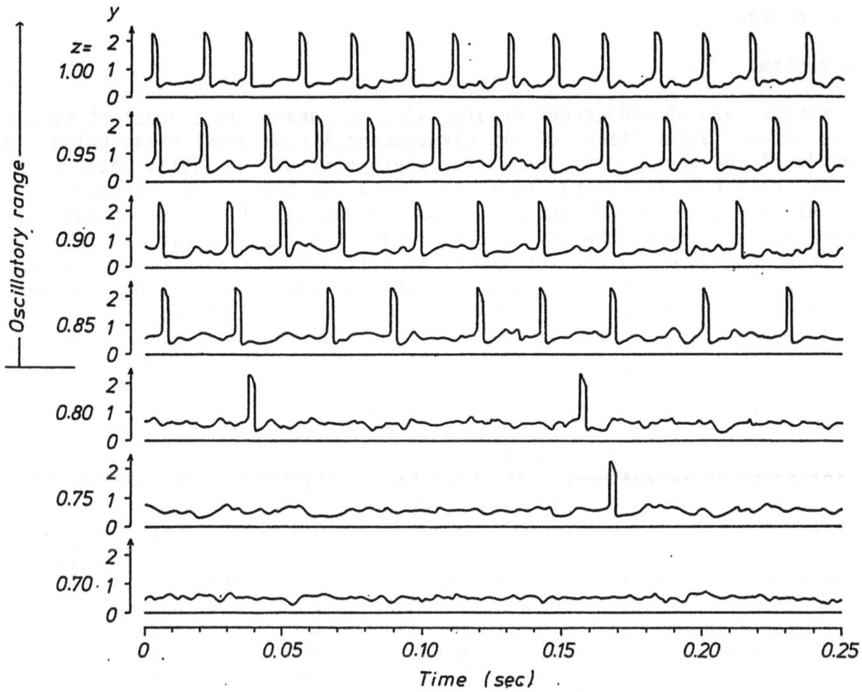

Figure 2: Solutions of the model equation in the domaine of nervous activities, with 7 different values of the external force z (each superimposed by similar random fluctuations). Outside of the "oscillatory range" the solutions do not describe rhythmic courses.

solutions) but more the frequency of the generated rhythm. Outside of the oscillatory range, any deflection of the external force transgressing the limit of the oscillatory range, either due to an external stimulus ("reactive action of a nerve cell") [5] or a random fluctuation (Fig. 2), lead to a "spike" the features of which are nearly independent of the releasing agent. This means, with the high-frequency solutions the shape of a single "spike" is nearly independent of the z-value, and it is nearly independent of the mode of the generation of this spike, either (inside of the oscillatory range) by an internal rhythmicity or (outside of the oscillatory range) by a single stimulus or a random fluctuation. Figure 2 shows that the probability for "spontaneous actions" is the lower the farther away the (mean) external force is from the oscillatory limit. But again, even under the influence of random fluctuations the reactions do not succeed one another in random intervals but according to some regularity (for instance, the interval between two spikes cannot remain under a - z-dependent - threshold, because of a "refractory phase" following every reaction).

In the context of this paper, the interaction between two rhythmic systems is of special interest; it is described by two (or more) model equations of the same type which are mutually coupled [2,3,4]. In the special case to be discussed here, the output of the rhythmic system generating circadian rhythms constitutes relevantly the environment (external force) of the other system which generates the (higher frequency) sleep rhythms. The second system, therefore, is controlled by the first system as well as by environmental stimuli. This case will be discussed later, after the regularities of the separated rhythms have been reviewed briefly.

3. Circadian Rhythms

Autonomous rhythms

Circadian rhythms are of endogenous origin, also in humans. In a constant condition (i.e., after elimination of all environmental time cues) which holds the system inside of the oscillatory range, the rhythms of all variables persist but with periods slightly deviating from 24 hours (in humans, the "freerunning" period is always close to 25 hours). In most animal experiments the transition from inside to outside of the oscillatory range, i.e., from a clear rhythmical state to arrhythmia (or vice versa) can be released by slight changes in the environmental conditions (e.g., the intensity of illumination), or even spontaneously. Figure 3 shows, as an example, the behavior of a Green Finch under an illumination the intensity of which had been increased and (several weeks later) decreased steadily and very slowly. Under the originally constant dim illumination the bird showed a clear circadian rhythmicity in locomotor activity. A few days after starting the increase in light intensity, it became arrhythmic from one day to the next; although the course of locomotor activity was not truely random (see above), various period analyses did not show any rhythmicity. When the light intensity decreased again, the bird became rhythmic again, also in this case mostly from one day to the next [8]. Typically, the rhythmicity re-occurred at a higher light intensity (here at 7 lux) than it disappeared with increasing light intensity (here at 3 lux). This difference is opposite to that what had to be expected if it would be due to hysteresis; rather, it is due to the fact that the system (and hence, also the oscillatory limit) is controlled, not only by the absolute value of the external stimulus but also by its time derivative [1]. In summary, the picture shows transitions between a highly ordered rhythmic state and a completely irregular state which are due to only slight changes in the environmental conditions; similar pictures have been obtained in 7 more birds under the same experimental condition. And in addition, there are a lot of examples where such transitions occur spontaneously, i.e., without any (intended, or recognized) change in the constant experimental conditions.

In healthy human subjects, a state of complete arrhythmia (i.e., a state outside of the oscillatory range) had never been observed. In the actually observed states inside of the oscillatory range, mostly the rhythms of all variables run in mutual synchrony ("internal synchronization"). In a minority of subjects, however, the rhythms of different variables run, in the steady state, with different periods ("internal desynchronization"). Figure 4 presents a comparison between the two rhythm states, in the left diagrams the more frequent state of internal synchronization and in the right diagrams the more rare state of internal desynchronization. The upper diagrams show the courses of the rhythms of sleep-wake and body temperature, at left in the common "double-plot" manner (cf. Fig.3), and at right in the - in case of internal desynchronization more obvious - "triple-plot" manner. The lower diagrams show results of periodogram analyses of the two time series; in case of internal synchronization the analyses of all rhythms (including the rhythms which had been measured but not presented in Fig.4) result in identical prominent periods; and in case of internal desynchronization the analyses of different rhythms (here represented by only two rhythms) result in different prominent periods [1].

The state of internal desynchronization shows obviously that the assumption of only one internal clock is insufficient; we deal, rather, with a multi-oscillator system (where mostly the acceptance of two different, mutually coupled oscillators is sufficient - represented by the rhythms of sleep-wake and body temperature). In most cases, all the different oscillators run in mutual synchrony so that they appear outwardly as only one oscillator; the herewith realized state of internal synchronization allows the evaluation of the properties of single oscillators. In other cases, the intrinsic periods (and also other rhythm parameters) of the involved oscillators are so different (i.e., outside of the mutual range of entrainment) that a mutual synchronization is no longer possible, in spite of the remaining mutual coupling. In these cases, the different oscillators run separated from each other with (at least on average) their own intrinsic periods;

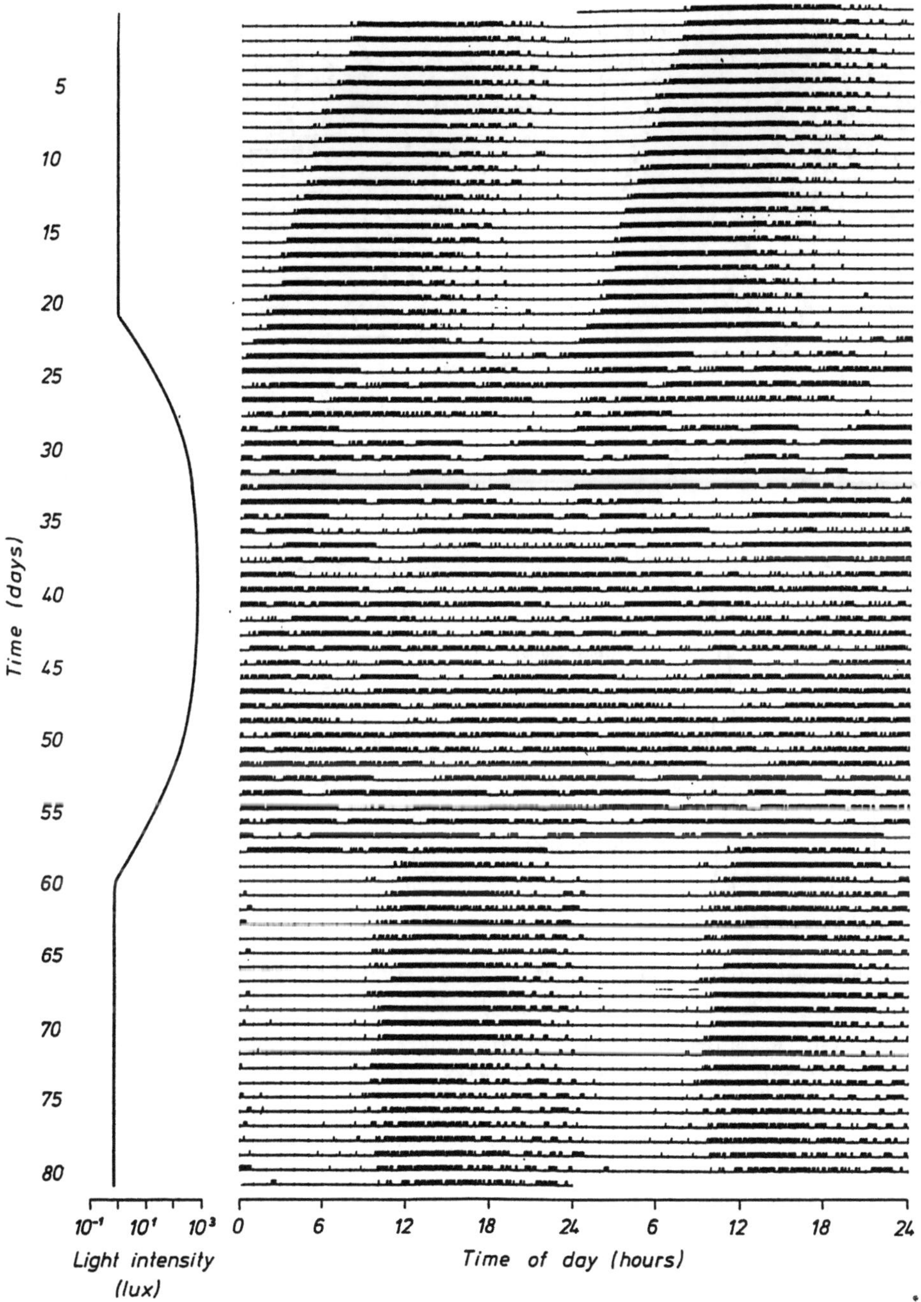

Figure 3: Locomotor activity of a Green Finch (perch hoppings, measured with microswitches under the perches) living under constant conditions. The data are presented in the common "double-plot" manner (the activity records from successive days are plotted one beneath the other; for clarity, in addition, data from two successive days each are plotted side by side). The intensity of illumination (left border) was 0.7 lux initially and finally; in between, it increased slowly and steadily up to 560 lux, and it decreased slowly and steadily later.

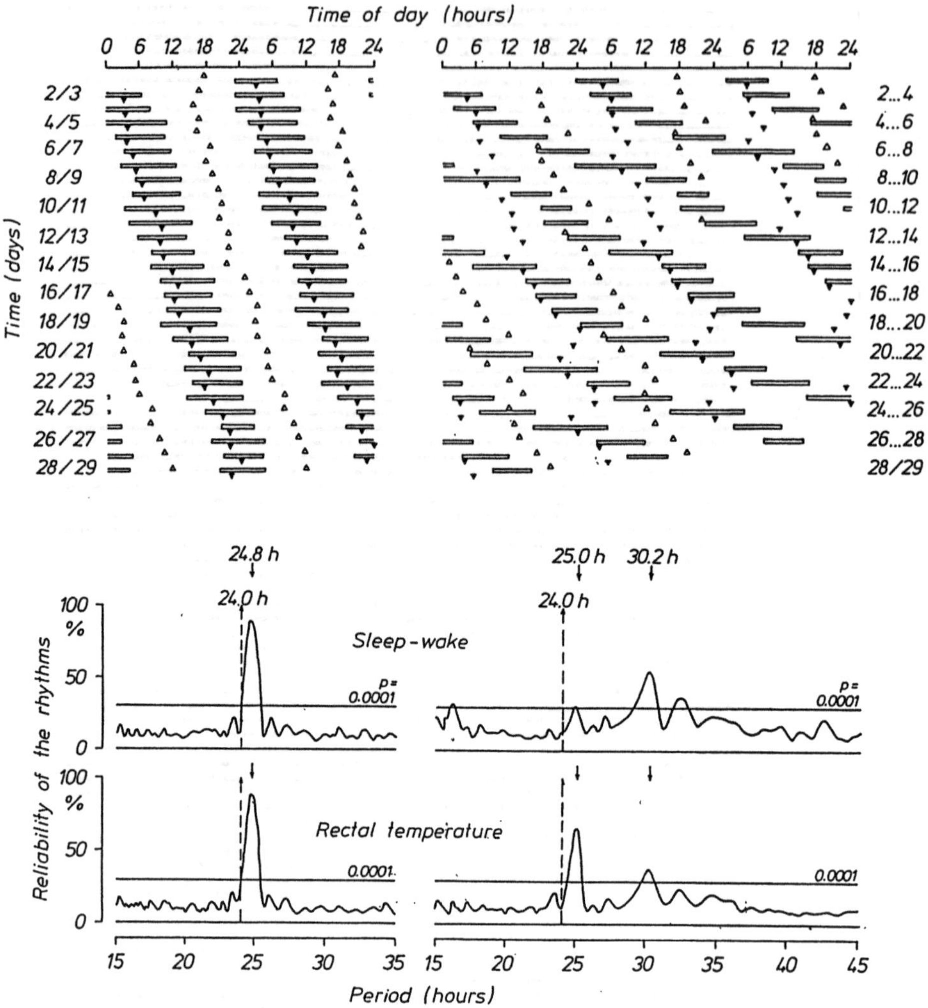

Figure 4: Two examples of human circadian rhythm which freerun under constant conditions, at the left hand side with internally synchronized rhythms and at the right hand side with internally desynchronized rhythms. Upper diagrams: courses of the rhythms of sleep-wake (bars: sleep episodes) and body temperature (triangles: temporal positions of the extremum values of body temperature), presented in a double-plot manner (left; cf. Fig.3), or a triple-plot manner (right), respectively. Lower diagrams: periodogram analyses of the two time series each as presented in the upper diagrams; "reliability" means that fraction of the total variability of the data which is due to a rhythmicity with the indicated period (supplementing that fraction which is due to irregular fluctuations); a normalized reliability of 30% means an error probability for the occasional occurrence of a rhythm with the indicated period of p = 0.001.

the still existing mutual coupling can be recognized by the presence of "relative coordination". The herewith realized state of internal desynchronization allows the evaluation of the interactions between the oscillators. A complete understanding of the circadian system demands the knowledge of the properties of the single oscillators as well as the properties of the mutual interaction. In the following, therefore, some few basic regularities of internally synchronized and desynchronized rhythms will be discussed separately.

In the state of internal synchronization the rhythms show a remarkable long-term stability which is due to an inherent mechanism: there is a highly significant negative serial correlation among successive cycles; i.e., every deviation of a cycle duration from the individual mean is corrected, with high probability, by opposite deviations of the following and the following but one cycles [9]. In the model, an even quantitatively identical serial correlation can be shown to be due to a non-linearity in the restoring force (the sign of this non-linearity is determined by mathematical stability conditions). Another negative serial correlation (again in accordance with model predictions) have been observed between adjacent wake- and sleep-episodes [9].

In addition to the serial correlations within every rhythm, there are highly significant inter- (and intra-) individual correlations between different rhythm parameters: for instance, the larger in a special subject the amplitude of the body temperature rhythm is, the shorter is, with high probability, the mean free-running period, the larger is the fraction of sleep within the sleep-wake cycle, the higher is the mean value of the body temperature rhythm, the steaper is the slope of the rising section of the temperature cycle relative to the declining section, the less is the temperature rhythm phase-advanced relative to the sleep-wake rhythm, and the larger are the variabilities of all these rhythm parameters, and, hence, the higher is the overall reliability of the rhythms. Of course, the relation of all other rhythm parameters to the rhythm's amplitude is arbitrary; also with reference to any other rhythm parameter corresponding significant intercorrelations can be found. And also these correlations are in agreement with corresponding model predictions [10].

Finally, there exist several dependences of rhythm parameters on the age and the sex of the subjects. All parameters of the sleep-wake rhythm (e.g., period and fraction of sleep) show significant sex differences but no dependence on the age (in females, the length of the total sleep-wake cycle is shorter and the length of sleep within this cycle is longer than in males [9]). On the other hand, all parameters of the rhythm of body temperature show significant differences between data from male subjects and younger female subjects (up to the age of the menopause) but no difference between data from male subjects (of any age) and older female subjects. Consequently, there is no age dependence in males but a clear age dependence im females; to be sure, this age dependence is not steady, but there seems to be a threshold age (around the age of the menopause) above and below of which the data differ significantly from each other, while there are no age dependences in the data from females above the threshold and in data below the threshold when considered separately [11]. Finally, there are dependences of other types in other data; for instance, the effectiveness of the internal mechanism mentioned above, stabilizing the period of freerunning rhythms, shows a steady dependence on the age but no dependence on the sex of the subjects.

In internally desynchronized rhythms, in fact, the steady state periods of rhythms of different variables differ from each other in the long run; the actual periods, however, show a more or less marked dependence on the internal phase relationship between the different rhythms ("relative coordination"). The difference between the periods of the different rhythms, therefore, varies systematically with the internal phase relationship (it is the smaller the closer the internal phase relationship is to its common value which is accepted in internally synchronized rhythms); and the amplitude of this variation is a measure of the strength of the mutual interaction between the different rhythms (or better: between the underlying basic oscillators). To be sure, not only the period but many more rhythm parameters vary systematically with the internal phase relationship. As an experimental result (and, simultaneously, a model consequence), the length of a wake or a sleep episode is commonly shorter when it is positioned around a temperature maximum than when positioned around a temperature minimum (the actual angular velocity of the sleep-wake rhythm is positively correlated to the actual value of body temperature); or the probability for sleep onset is commonly higher during the descending than the ascending section of the temperature cycle, and the probability for wake onset is commonly higher during the ascending than the descending section of the temperature cycle. Furthermore, the actual amplitude of a cycle of body temperature is larger when this cycle is in phase with a sleep-wake cycle (as common in internally synchronized rhythms) than when it is counterphased to a sleep-wake cycle. And again, the degree of the fluctuations of all these rhythm parameters is a measure of the strength of the interaction between the involved rhythms. In the border-line case (observed in about 10% of all experiments showing internal desynchronization) the interaction between the different rhythms is so weak that the relations to the internal phase relationship as mentioned, cannot be observed [12].

Apart from this type of "oscillatory interaction" there is the "masking effect". This is a (at least apparently) direct action from changes in one variable (or an environmental stimulus) on another variable. In humans, the masking effect is frequently reduced to the influence of changes between wake and sleep on body temperature: body temperature is higher during wake than sleep (on average, by about 0.3 °C), nearly independent of circadian variations [13]. The most obvious way to establish masking effects is to calculate educed, or average cycles from a consistent time series, as it is common in circadian analyses. However, here the educing process had be done not only, as common, by using all temperature data of the time series, but to calculate it, additionally, by using temperature date obtained exclusively during wake episodes and exclusively during sleep episodes. The result are separated mean "wake temperature cycles" and "sleep temperature cycles"; they run, in all tested cases, remarkably parallel up to individual details but are shifted against each other, firstly in temperature (the mean wake cycle is 0.28 ± 0.06 °C higher than the sleep cycle), and secondly in time (the sleep cycle is phase advanced relative to the wake cycle by 0.87 ± 0.09 hours). Another way to establish masking effects is to interpose in the experimental series cycles with a "constant routine". Here, however, the problem arises that the experimental manipulation itself may influence the rhythm and, hence, confuse the desired result (for instance, it is well known that sleep deprivation as a type of a constant routine delays the rhythm by about one hour per day, corresponding to the phase delay of the mean wake cycle relative to the mean sleep cycle by about one hour per cycle).

The masking effect had been considered, in the foregoing paragraph, as a variation which is independent of circadian variations, and which superimposes more or less additively the circadian variations (i.e., "by-passing" the clock mechanism). Model considerations, however, have suggested, rather, the masking effect to be an inherent part of the clock mechanism. Alternative models had been compared, and that is, on the one hand, the (more simple) original model which includes the masking effect inseparably, and, on the other hand, a modified (more complicated) model which does not include a masking effect inherently but had to be supplemented subsequently (i.e., superimposed) by a second model describing exclusively the masking effect [3]. Of course, with regard to the masking effect both models lead to identical predictions (otherwise a calculation error would be present), but they differ in predictions concerning a great multiplicity of various other effects: only solutions of the original model lead, in all these cases, to agreement with biological results but not solutions of the modified model. Hence, only the original model which includes the masking effect as an inseparable and inherent part of the clock mechanism, constitutes a useful model.

The transition between internally synchronized and desynchronized rhythms can be induced by a corresponding change in the environmental conditions (e.g., by the transition between normal and bright constant illumination, the transition between constant illumination and "self-controlled" illumination, or the transition between the presence of a constant electric 10-Hz field of low intensity and field-free conditions); in other cases, this transition occurred spontaneously, i.e., without any intended and recognized change in the experimental conditions (always from internal synchronization to desynchronization, but never in the opposite way). The tendency towards the occurrence of internal desynchronization does not only depend on the experimental conditions but also on personality data. The most relevant is the age of the subjects: in older subjects (from about 45 years on) the probability for the occurrence of internal desynchronization is about four to five times greater than in younger subjects. Moreover, in subjects

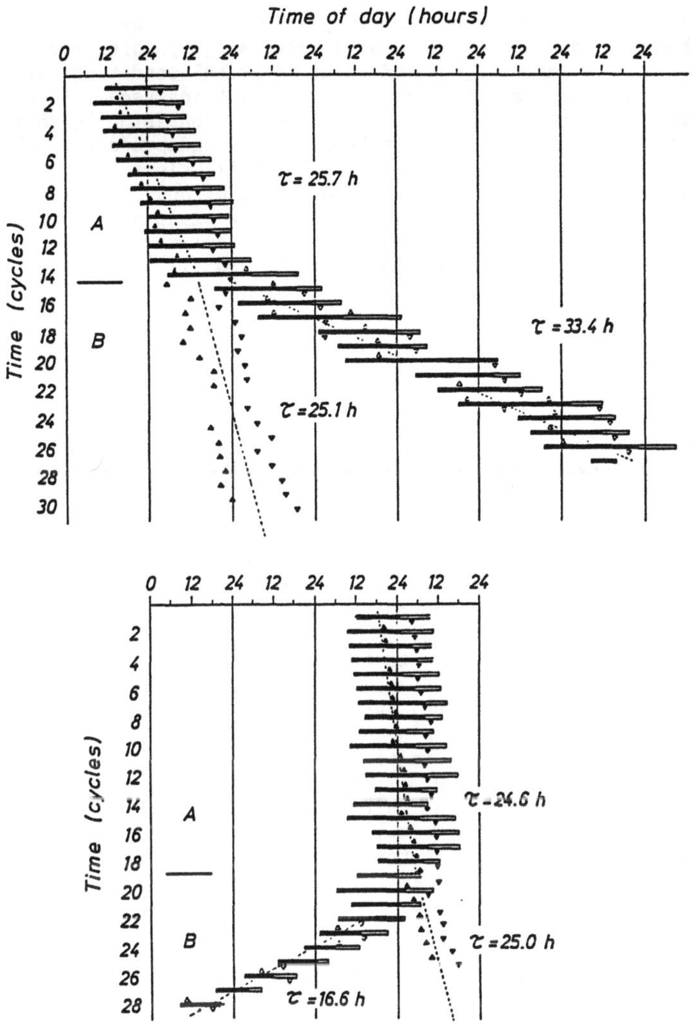

Figure 5: Two examples of freerunning rhythms in which internal desynchronization occurred spontaneously during the running experiment, in the upper diagram by a considerable lengthening and in the lower diagram by a considerable shortening of the sleep-wake cycle (with the period of the body temperature rhythms showing only slight changes in the opposite direction each). Presented are the courses of the rhythms of sleep-wake (bars; white: wake; black: sleep) and body temperature (triangles: temporal positions of the extremum values; open triangles: temporally correct repetitions of already drawn triangles); successive cycles of every rhythm are plotted one beneath the other.

with increased neurotic scores the probability for the occurrence of internal desynchronization is significantly greater than in the average of all subjects [14].

Frequently (i.e., in about 35 experiments), a spontaneous transition occurred during the running experiment (i.e., this state was not present from the beginning of the experiment as in Fig.4). This means, a possibly very slight change in one of the controlling parameters (which is too small to be recognizable in itself), i.e., either in the difference between the periods of the involved rhythms, or in the strength of the mutual coupling, leads to a transgression of the limit of mutual entrainment resulting in a very well recognizable change in the overt periods of, at least, one of the involved rhythms. The slight change may also be due to a particularly large random fluctuation (a spontaneous internal re-entrainment by another random fluctuation is much more unlikely, because then the clearly smaller limit of catching a rhythm must be transgressed, and not the clearly larger limit of holding a rhythm [7]). Figure 5 shows two cases of a spontaneous occurrence of internal desynchronization, and that in the two opposite directions (i.e., by an - each instantaneous - considerable lengthening and a considerable shortening of the sleep-wake cycle). These transitions give us the chance to compare sections with internally synchronized and desynchronized rhythms in the same subject and under identical conditions (the sections A and B each are indicated arbitrarily for subsequent analyses). Such a comparison, for instance, shows that the fraction of sleep is consistently smaller under internal desynchronization than synchronization, independent of the direction of the desynchronization, i.e., whether the total sleep-wake cycle lengthens or shortens considerably. The period of the rhythm of body temperature changes always in a direction opposite to the change of the sleep-wake period, and to a much smaller extent (on the average, and with only small variations from subject to subject, by a 1:12-ratio) than the sleep-wake period [1].

Heteronomous rhythms

Under natural conditions, circadian rhythms of all organisms are always under the influence of the 24-hour day-night cycle; fortunately, also artificial zeitgebers are effective in synchronizing circadian rhythms within limited "ranges of entrainment" under laboratory conditions. In corresponding experiments two questions can be answered, firstly for the mechanism of the synchronization and, secondly, for the stimuli which are effective as zeitgebers (and if so, in what ranges they are). To anticipate the answer to the first question: all zeitgebers which are effective, operate via phase-control and not via frequency-control. The acrophases of synchronized rhythms are concentrated within special ranges of phase-angle differences to the zeitgeber which, in addition, depend on the inherent period of the synchronized rhythm relative to the zeitgeber period [1].

In nearly all animal circadian experiments, a light-dark alternation is the most effective zeitgeber modality. In humans, surprisingly enough, a light-dark alternation had been shown originally to be nearly ineffective as a zeitgeber; this is true, at least, if the intensities of illumination do not exceed the range of normal artificial illumination (i.e., with intensities $<$ 1000 lux), and if the subjects have the option to elude the zeitgeber (by having the chance to switch on some dim light during the dark phase allowing them to be active; "relative light-dark zeitgeber"). A light-dark alternation, in fact, gaines zeitgeber effectiveness after the option of eluding the zeitgeber is removed, by an illumination during the dark phase which is too low to allow the performance of any activities (and without the availability of any auxiliary lamps; "absolute light-dark zeitgeber"). It had been shown, however, that such a zeitgeber is not effective by a direct action of light, but by the requests ("go to bed" and "get up") which are inevitably linked to the transitions between light and darkness; i.e., it is effective due to its behavioral component and not its physiological component. When the same requests are given acoustically under constant illumination (or even constant total darkness), the zeitgeber effectiveness of this pure "information zeitgeber" is the same as that of the absolute light-dark alternation; this means, without any participation of a direct effect of light (i.e., with the information zeitgeber), the full zeitgeber effectiveness is present, and there is no more any space for an additional direct effect of light with the absolute light-dark alternation [15].

These results change dramatically when the intensity of illumination during the light phase of a light-dark alternation increases to values which exceed a threshold in the range of about 2500 lux; light intensities above this threshold are rarely to be present indoors. With an absolute light-dark alternation (i.e., with darkness during the dark phase and no auxiliary lamps available) as with a relative light-dark alternation (i.e., even with an intensity of 300 lux during the "dark phase" corresponding to normal room illumination and allowing every kind of activity), such a "bright-light zeitgeber" shows a remarkable strength [16]. To be sure, a subject must be exposed to bright light for, at least, six hours per cycle, to get a sufficient effectiveness of the bright-light zeitgeber. In our industrial society, however, only a small minority of people fulfils these conditions; the ubiquitous behavioral, or social zeitgeber, therefore, is the most relevant zeitgeber in humans in spite of the fact that it is not the strongest [17].

The strength of a zeitgeber can be measured by the size of its range of entrainment. It turned out, for instance, that a pure (relative) light-dark zeitgeber (with normal light during the light phase) results in a range of entrainment of less than half an hour to either side of the freerunning period (it must be recalled to mind that a range of entrainment of, at least, one hour is necessary to overcome the average difference between the freerunning period and the period of the natural day-night cycle); an absolute light-dark zeitgeber (which is effective nearly only due to its behavioral component; see above) results in a range of entrainment of slightly more than two hours (i.e., it ranges from about 23 to 27 hours). Also the pure information zeitgeber results in a range of entrainment of about two hours. The bright-light zeitgeber, on the other hand, is considerably stronger; even in a relative mode (i.e., without any participation of a behavioral component) it results in a range of entrainment of about four hours to either side; and in an absolute mode (i.e., supplemented by a behavioral component) the range of entrainment is measured to be slightly more than six hours; (i.e., it ranges from about 19 to 31 hours). These numbers show that the strengths of the different zeitgeber modalities superimpose each other nearly additively [15]. The result of the additive superimposition implicates practically the chance to supplement or substitute a zeitgeber which is too weak for full entrainment, by another zeitgeber. Mostly this means to supplement a social zeitgeber which may be too weak in old or ill patients (social deprivation is a well known psychical problem in such patients which now gaines, in addition, a physiological dimension, and that is because of the insufficient synchronization of rhythms also of physiological variables). Other practical implications are the chance to reduce the "jet-lag syndrome" tremendously, and to help shift-workers to overcome their problems to a great extent [16]. The result of the nearly additive superimposition of different zeitgeber modalities implicates theoretically a new access to the evaluation of physiological pathways in the control of circadian rhythmicity. In fact, it is just the great diversity of different stimuli controlling human circadian rhythms, and that in the same way (apart from bright light and social contacts, only a weak electric - 2.5 V/m - 10 Hz square wave field is able to exert such a control), which leads to promising hypotheses [18].

The multi-oscillatory structure of the circadian system which had been demonstrated firstly in autonomous rhythms (see above), results, in heteronomous rhythms, in different ranges of entrainment of the rhythms of different variables. For instance, the weaker rhythm of sleep-wake should show a much wider range of entrainment than the stronger rhythm of body temperature. The phenomenon of only "partial external synchronization", in fact, had been observed in numerous experiments. For instance, this phenomenon occurs always under the influence of an absolute light-dark alternation where the sleep-wake rhythms are synchronized within wide ranges while the rhythms of physiological variables like body temperature can be synchronized only in ranges of about two hours to either side of the freerunning period (see above). Between the limits of the different ranges of entrainment, i.e., within the ranges of partial synchronization (in the case mentioned, this is with periods shorter than 23 hours and longer than 27 hours), internal desynchronization is present in every subject, not occurring spontaneously as under constant conditions (see above) but forced by the experimental conditions (without that this "forcing" became aware to the subjects); this gives us the chance to evaluate this particularly interesting state of the rhythms in every subject, and to evaluate properties of a special rhythm independent of the other rhythms [1].

The instability close to the entrainment limits can be utilized for an amplification of effects which are too small to become obvious in themselves. As an example, the experiment underlying Figure 6 should assist in answering the question whether or not behavioral stress affects circadian rhythms [12]. There were some suggestions that such a stimulus lengthens the period of autonomous rhythms; the change in period to be expected (by about 0.2 hours), however, was to small to become guaranteed unambiguously. In the relevant experiment a subject was exposed to a period of an absolute light-dark alternation of 22:40 hours which was

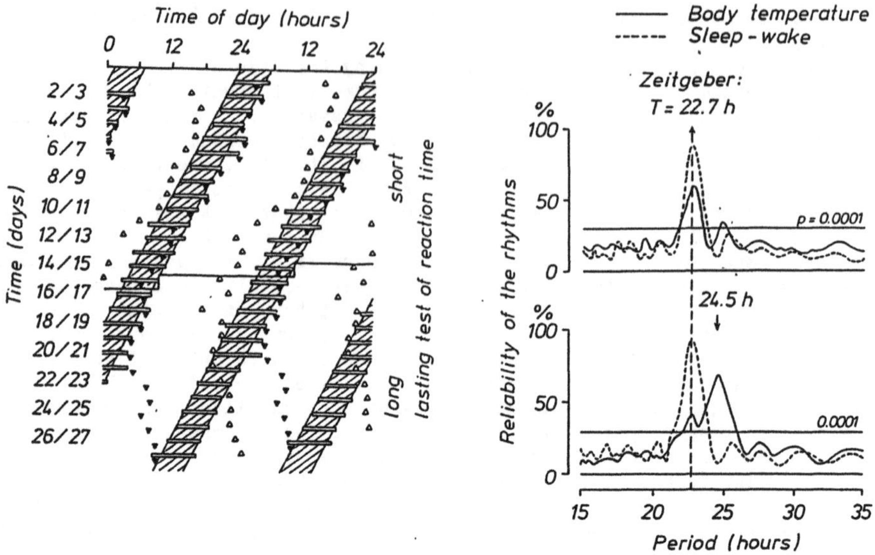

Figure 6: Example of a circadian rhythm under the influence of an absolute light-dark zeitgeber (300:0.03 lux) with a constant period of 22:40 hours. The subject had to perform (on regular requests, 7 times per cycle) a reaction time test which lasted, during the first two weeks of the experiment 5 min per test session, and during the second two weeks 20 min per test session (during the first half of the experiment the subject liked the test, according to his diary, but he hated it, and experienced it as a troublesome stress during the second half of the experiment, without to realize consciously the reason for this difference). The left hand diagram shows the course of the experiment, with the same designations as in Fig.4 (upper diagrams). The right hand diagram shows results of periodogram analyses from the two time series, computed separately from the two sections of the experiment, with the same designations as in Fig.4 (lower diagrams).

known to be close to the entrainment limit. During the first two weeks of the running experiment the rhythms of all variables ran in synchrony with the zeitgeber; then, a behavioral stress was introduced (by lengthening an initially rather stimulating test of reaction time, which was then experienced by the subject as a troublesome stress). As the result of the experiment, the initially externally and internally synchronized rhythms desynchronized immediately after the introduction of the stress: the sleep-wake rhythm remained still synchronized to the zeitgeber, but the rhythm of body temperature loosened from the zeitgeber and started to freerun. This can be seen obviously in the left hand diagram of Fig.6 which shows the courses of the rhythms of sleep-wake and body temperature. And this can be seen objectively in the right hand diagram which shows the results of periodogram analyses: in the first section the two rhythms show identical prominent periods (coinciding with the zeitgeber period), and in the second section they show considerably deviating prominent periods. The interpretation is that the introduction of the behavioral stress (by lengthening the sessions of the test), in fact, lengthened the inherent period of the temperature rhythms (but not that of the sleep-wake rhythm which stood synchronized to the zeitgeber); the lengthening by the expected small amount of about 0.2 h led to a transgression of the entrainment limit which resulted in an overt lengthening in the period of nearly 2 hours; and this lengthening is unambigously recognizable. Since also some additional experiments of the same type led to similar results, the question for an effect of behavioral stress can now be answered positively.

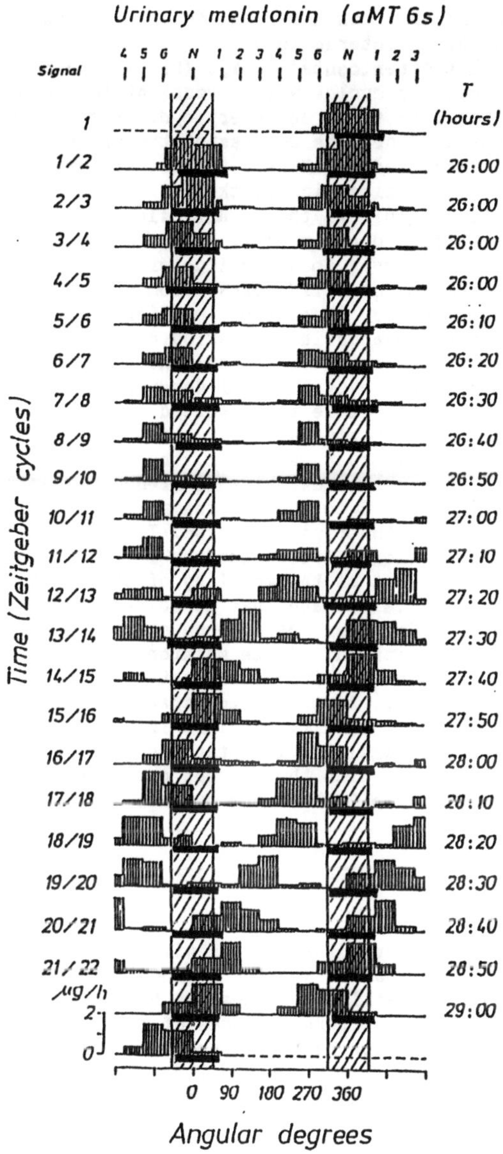

Figure 7: Example of an experiment in which a subject was exposed to a zeitgeber (abs. LD 300:0.03 lux, supplemented by regular requesting signals; cf. upper border) the period of which lengthened steadily (the actual zeitgeber periods are indicated at the right border). Presented are the rhythms of sleep-wake (black bars: sleep episodes) and the urinary melatonin excretion (vertically hatched areas); successive zeitgeber cycles are plotted one beneath the other, and for clarity, two successive zeitgeber cycles each are plotted side by side; the zeitgeber cycles are always normalized to 360°, independent of their actual lengths. While sleep-wake ran synchronously to the zeitgeber during the entire experiment, the rhythm of the melatonin excretion ran synchronously to the zeitgeber only during about half the experiment, and it freeran later (with a constant period of 24.9 hours, i.e., a period which was shorter than the respective zeitgeber period).

In a special experimental design the limits of entrainment can be evaluated precisely, and that is by the introduction of a zeitgeber the period of which changes slowly but steadily (e.g., consistently by 5 or 10 min per cycle). Under such a "fractional desynchronization" the rhythms of nearly all variables start to freerun (i.e., they transgress the entrainment limit) not simultaneously but at different days, and that means, at different periods [19]. At least for a few days, therefore, internal desynchronization can be induced, not only between sleep-wake and body temperature but between the rhythms of nearly all variables (e.g., between body temperature and cortisol excretion, between sodium and potassium excretion, between body temperature and performance, between simple and complex types of performance, and between several more variables); and this means that there cannot be a functional interdependence between the variables mentioned. Hence, investigations of circadian rhythms can assist in the evaluation of basic properties of physiological and psychical interactions.

Again, an example should illustrate this experimental design (Fig.7). In this case, the experiment started with 4 successive cycles with a constant length of 26:00 hours (to get a steady state), and from there on, every cycle was 10 min longer than the previous cycle. Demonstrated are the courses of the sleep-wake rhythm and that of the urinary melatonin (measured as aMT6s); the latter variable is known to constitute the best marker of the circadian clock (for instance, it does not show a masking effect; see above) [10]. As the result, sleep-wake ran synchronously to the zeitgeber during the entire experiment; on the other hand, the rhythm of melatonin excretion ran synchronously to the zeitgeber and, hence, sleep-wake only up to cycle # 12, or a period of 27:10 hours; from here on, it freeran with its inherent period of 24.9 hours, independent of light-dark and independent of sleep-wake. The considerable change in the period occurred from one day to the next, as instantaneous as the change of the sleep-wake period commonly occurs in case of a spontaneous occurrence of internal desynchronization (cf. Fig.5). With melatonin as the considered variable the reported course is so obvious that there cannot be any doubt about the alteration in the state of the rhythm during the experiment (in the rhythms of other variables, though behaving basically in the same way, the same picture can be established frequently only after sophisticated computer analyses). With regard to the melatonin rhythm, this experiment is not an exception but represents the typical course. It belongs to this typical picture that the melatonin excretion reduces when the zeitgeber period approaches the entrainment limit (independent of the modality of the zeitgeber), down to about a third of its original value, and that it raises again to its original value instantaneously at the day when the freerun starts. And it belongs to the typical picture that the instant of the melatonin excretion shifts, with changing zeitgeber period, temporarily into the light phase of the zeitgeber and/or the wake time of the subject. According to established rules which, to be sure, had been deduced only from results obtained in the 24-hour day, the hormone melatonin should be secreted exclusively during the dark phase and during the sleep; these rules have been shown with the present experiments to be only borderline cases which may be valid in a 24-hour day but not in general [20].

4. Ultradian Sleep Rhythms

The biological rhythms which are closest to circadian rhythms within the hierarchy of biological rhythms in the direction to shorter periods (ultradian rhythms), are the about 90-min sleep rhythms. At least in healthy humans, sleep is a highly ordered process during which the different stages of sleep follow each other in a rhythmic fassion [21]. To illustrate such sleep rhythms, Figure 8 presents a polygraphic sleep record (cutout) from a subject who lived for 29 days under constant conditions (for clearness, an experiment with a particularly long period had been selected). In addition, the course of body temperature during the sleep episodes is presented, to make possible a consideration of intercorrelations between the actual pattern of the sleep stages and body temperature. It is striking that this (typical) picture of a freerunning rhythm deviates from the usual pattern to be observed in the 24-hour day, in two respects: firstly, the minimum of body temperature is always at the beginning and not at the end of every sleep episode; and secondly, in every sleep episode the first REM phase is always the longest, and the subsequent REM phases rather shorten, whereas in the 24-hour day the lengths of the successive REM phases increase in every sleep episode, so that the first REM phase in every sleep episode is the shortest. This parallelism between changes in the REM pattern and the course of body temperature with the transition between freerunning and 24-hour rhythms suggests a correlation: the actual REM phases are the longer the lower the body temperature at the same time is. This suggestion is supported by the inspection of sleep records from internally desynchronized rhythms where (see above) the course of body temperature alters from sleep episode to sleep episode, with the minimum in several

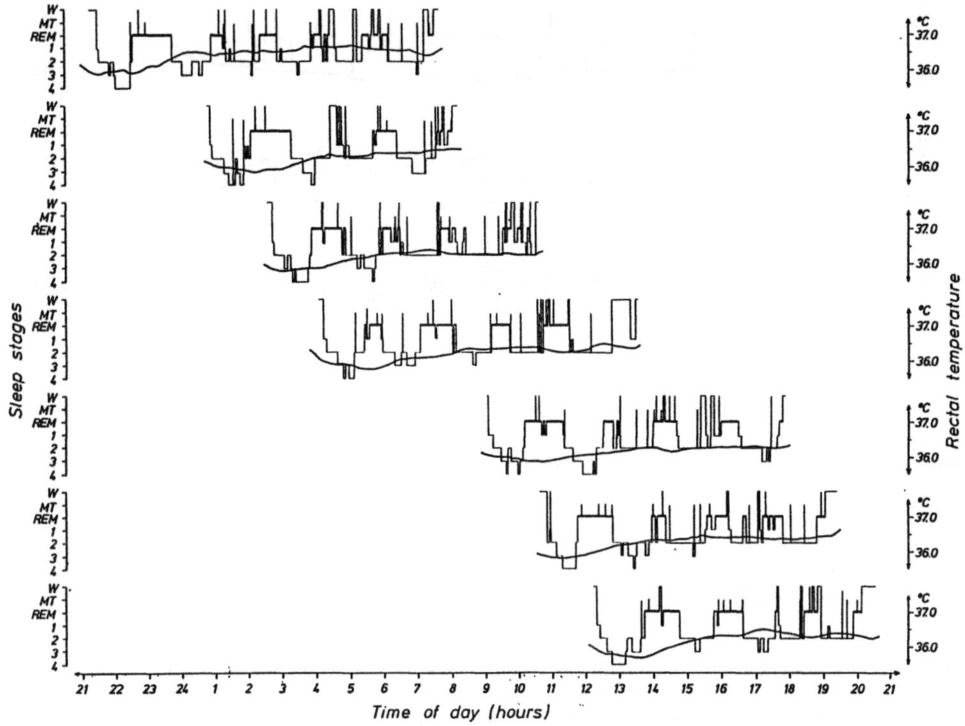

Figure 8: Example of a polygraphic EEG sleep record from a subject living under constant conditions, with a freerunning period of 26.5 hours (cutout of a 29-day experiment). Presented are the succession of the different stages of sleep during 7 successive sleep episodes, and the courses of body temperature during these sleep episodes; data from successive days are plotted one beneath the other.

cycles at the beginning and in other cycles at the end of a sleep episode. As the result of such an inspection, the lengths of the different REM phases during a sleep episode reflect the course of body temperature in the sense mentioned. In opposition to the REM sleep pattern which obviously is controlled by the rhythm of body temperature, the pattern of the slow wave (deep) sleep (stages 3 and 4), rather is controlled by the sleep-wake rhythm: slow wave sleep is found always at the beginning of a sleep episode, as in 24-hour rhythms also in freerunning rhythms, and that in internally synchronized and desynchronized rhythms, and independent of the actual course of body temperature.

To illustrate another type of interaction between circadian and ultradian rhythms, Figure 9 shows another example of a polygraphic sleep record obtained under constant conditions. This record originates from a subject the freerunning rhythms of whom were desynchronized in a special way (the total experiment again lasted 29 days so that only a cutout is shown): while the body temperature rhythm showed a common period of 24.6 hours, the period of the sleep-wake rhythm was just twice as long (49.2 hours; "circa-bi-dian sleep-wake rhythm"), so that every sleep-wake cycle covered precisely two cycles of body temperature (hence, the internal phase relationship between the two rhythms did not alter from cycle to cycle as in case of real internal desynchronization; it was, rather, temporally constant but in an ambiguous way; "apparent internal desynchronization") [1]. It is worth mentioning that also in such an extreme case the subject had not realized consciously the length of his day; as all other subjects living under constant

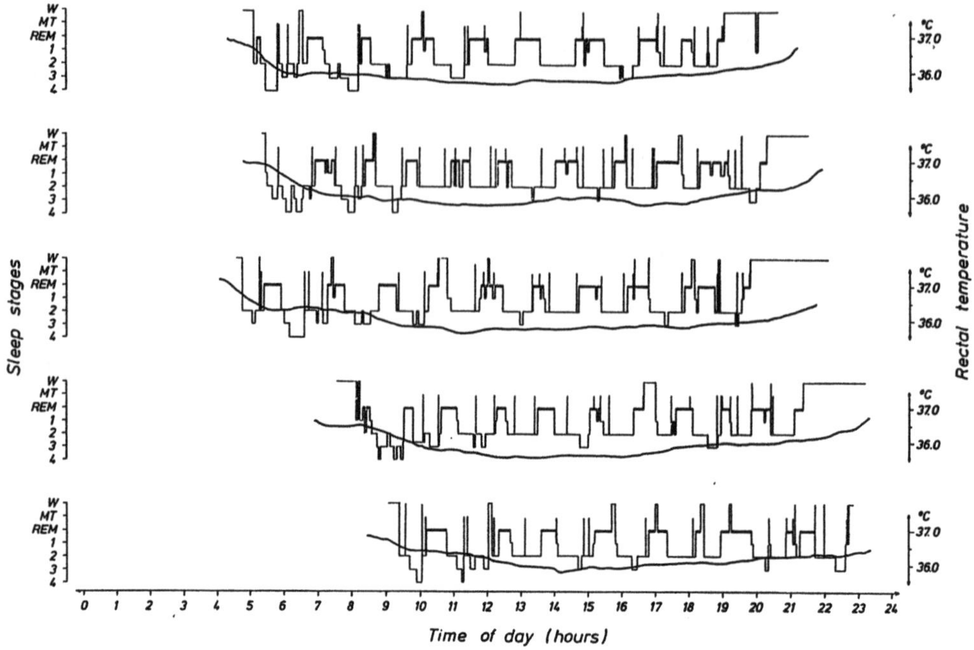

Figure 9: Example of a polygraphic EEG sleep record from a subject living under constant conditions, with a circa-bi-dian sleep-wake rhythms (period: 49.2 hours) and a circadian rhythm of body temperature (period: 24.6 hours) (cutout of a 29-day experiment). Presentations from 5 successive sleep episodes according to Fig.8, exept that the sleep data originate only from every second day.

conditions including those showing internal desynchronization in either direction, he was rather convinced to live according to a 24-hour schedule. Since the sleep/wake ratio remained nearly unchanged, every sleep (and also wake) episode was just twice as long as usual. As Figure 9 shows, the sleep rhythm shows its common period of about 90 min, so that every sleep episode does not cover, as usual, about 5 REM-nonREM cycles but about 10 cycles. This result shows that the circadian rhythmicity, in fact, modifies the ultradian sleep rhythmicity (cf. Fig.8), but that this modification concerns mainly the shape of the rhythms but nearly not the period of the rhythms. Simultaneously, this result demonstrates the autonomous nature of the sleep rhythms; if the sleep cycles would be subintervals of circadian cycles, their lengths would change proportionally to the length of the circadian cycle (as it is the case, for instance, with regularly repeated meals [17]).

Particularly the record of Figure 8 demonstrates a dilemma in the analysis of sleep rhythms: commonly, not more than about 4 to 5 successive cycles (within every sleep episode) are available for analyses, and even during these cycles the rhythm is not in a steady state (records analogous to that shown in Fig.9, which include considerable more cycles per sleep episode, are available only in very rare cases). A meaningful analysis would be possible if the records from several successive sleep episodes could be combined; such a combination, however, presupposes a rhythmicity which continues also during the wake episodes with an unchanged period. In fact, several types of ultradian rhythms during wake have been established (e.g. stomach movements); but they all are - by definition - different from real sleep rhythms, and they all show periods which are clearly longer than those of the sleep rhythms (mostly in the range of 3 to 4 hours). Hypotheses utilizing the model equation are possible in many different versions; but they all meet in the fact of a frequency modulation of the ultradian rhythms by the circadian variations, i.e., in all different versions of ultradian models the period is different during sleep and (also when hypothetically continuing) during wake. Hence, an improvement of the analyses from experimentally measured sleep rhythms by combining successive sleep episodes is not possible at the present state of knowledge.

5. Model: ultradian (sleep) rhythms as influenced by circadian rhythms

Systematic changes in the influence of circadian on ultradian variations can be investigated easiest in model considerations. The established model equation for two-oscillator systems [2,3,4] had been standardized in that way, that the threshold separating wake from sleep (cf. Fig.1) constitutes simultaneously the (upper) limit of the oscillatory range of the ultradian rhythm, i.e., it is simultaneously the threshold only below of which the ultradian system is able to generate autonomous rhythms [6]. Since the period of the sleep rhythm is considerably shorter than the circadian period, the resulting rhythm tends considerably more into the direction of a relaxation oscillation (see above), though not so drastically as with the solutions modelling nervous activities (cf. Fig.2). Of special interest is the influence of changes in the circadian amplitude (but with unchanged superimposed random fluctuations) on the ultradian system; with reduc-

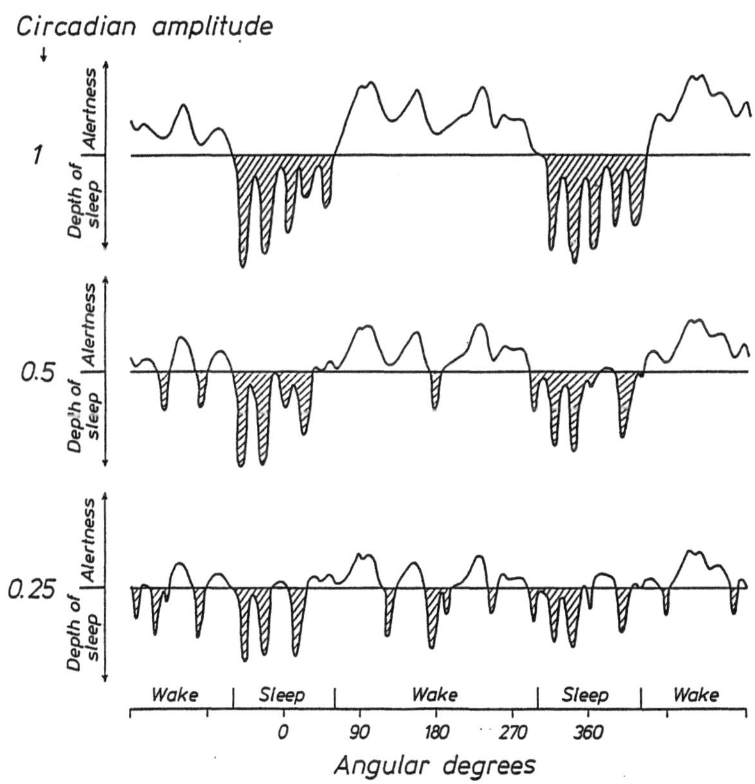

Figure 10: Solutions of the model equation of a two-oscillator system: circadian sleep-wake rhythms with 3 different amplitudes (with the same superimposed random fluctuations each) controlling an ultradian (90-min sleep) rhythm. The threshold separating wake and sleep (cf. Fig.1) is simultaneously the lower limit of the oscillatory range (cf. Figs. 1 and 2) of the ultradian system.

ing amplitude, the relative weight of the random fluctuations increases. Figure 10 presents solutions of such a model with 3 different circadian amplitudes. With the largest amplitude (and, hence, the smallest relative amount of the random fluctuations), the result is a highly ordered sleep pattern, with regularly 5 sleep cycles per sleep episode. With reducing circadian amplitude the sleep pattern becomes more and more irregular. With the middle amplitude the sleep episodes are still recognizable, though also during the wake episodes several "naps" are interspersed. And with the smallest circadian amplitude, any regularity in the sleep pattern is hard to discover; the pattern is rather "chaotic". The fraction of sleep within the circadian sleep-wake cycle (hatched areas below the threshold) remains nearly unchanged with the changes in the circadian amplitude; it is only the pattern of sleep which changes dramatically (with shiftings of the threshold, on the contrary, the total amount of sleep per cycle changes considerably but nearly not the pattern of sleep). Possibly, the model solutions shown in Fig.10 can serve as a model for depression. Also in depressives, the circadian amplitude is reduced, and the sleep is irregular (and insufficient) though rather normal in its total duration; the model shows now that both these phenomena are linked together necessarily; possibly, this is another example in what way model considerations can facilitate the insight into medical problems [12,17].

References

1. Wever, R.A.: The Circadian System of Man (XI and 276 p). (Springer-Verlag, New York - Heidelberg - Berlin, 1979).
2. Wever, R.: Zum Mechanismus der biologischen 24-Stunden-Periodik. III. Mitteilung: Anwendung der Modell-Gleichung. Kybernetik 2: 127-144 (1964).
3. Wever, R.A.: Toward a mathematical model of circadian rhythmicity. In: Mathematical models of the circadian sleep-wake cycles (eds.: M.C. Moore-Ede and C.A. Czeisler), pp.17-79 (Raven Press, New York 1984).
4. Wever, R.A.: Mathematical models of circadian one- and multi-oscillator systems. In: Some Mathematical Questions in Biology: Circadian Rhythms (ed.: G.A. Carpenter), pp. 205-265 (The Amer. Math. Soc., Providence RI 1987).
5. Wever, R.: Ein mathematisches Modell für biologische Schwingungen. Z. Tierpsychol. 21: 359-372 (1964).
6. Wever, R.A.: Modes of interaction between ultradian and circadian rhythms: toward a mathematical model of sleep. Exper. Brain Res., Suppl. 12: 309-317 (1985).
7. Wever, R.: Zum Mechanismus der biologischen 24-Stunden-Periodik. II. Mitteilung. Der Einfluß des Gleichwertes auf die Eigenschaften selbsterregter Schwingungen. Kybernetik 1: 213-231 (1963).
8. Wever, R.A.: Circadian rhythms of finches under steadily changing light intensity: are self-sustaining circadian rhythms self-excitatory? J. Comp. Physiol. A 140: 113-119 (1980).
9. Wever, R.A.: Properties of human sleep-wake cycles: parameters of internally synchronized freerunning rhythms. Sleep 7: 27-51 (1984).
10. Wever, R.A.: Characteristics of circadian rhythms in human functions. J. Neural Transm., Suppl. 21: 323-373 (1986).
11. Wever, R.A.: Geschlechtsunterschiede von Schlafparametern. In: Biologische Psychiatrie (ed.: B. Saletu), pp. 375-379 (Thieme-Verlag, Stuttgart 1989).
12. Wever, R.A.: Order and disorder in human circadian rhythmicity: possible relations to mental disorders. In: Biological Rhythms and Mental Disorders (eds.: D.J. Kupfer, T.H. Monk and J.D. Barchas), pp. 253-346 (The Guilford Press, New York 1988).
13. Wever, R.A.: Internal interactions within the human circadian system: the masking effect. Experientia 41: 332-342 (1985).
14. Wever, R.A.: Behavioral aspects of circadian rhythmicity. In: Rhythmic Aspects of Behavior (eds.: F.M. Brown and R.C. Graeber), pp. 105-171 (L. Erlbaum Ass. Inc., Hillsdale NJ 1982).

15. Wever, R.A.: Light effects on human circadian rhythms: a review of recent Andechs experiments. J. Biol. Rhythms 4: 161-185 (1989).
16. Wever, R.A.: Use of light to treat jet-lag: differential effects of normal and bright artificial light on human circadian rhythms. In: The Medical and Biological Effects of Light. Ann. New York Acad. Sci. 453: 282-304 (1985).
17. Wever, R.A.: Possible relations between disorders in circadian rhythmicity and mental disorders. In: Integrative Biologic Psychiatry (eds.: H.M. Emrich and M. Wiegand), in press (Springer-Verlag, Heidelberg 1991).
18. Wever, R.A.: The electromagnetic environment and the circadian rhythms of human subjects. In: Biological Effects and Dosimetry of Static and ELF Electric Fields (eds.: M. Grandolfo, S.M. Michaelson and A. Rindi), pp. 477-523 (Plenum Press, New York 1985).
19. Wever, R.A.: Fractional desynchronization of human circadian rhythms: a method for evaluating entrainment limits and functional interdependences. Pflügers Arch. 396: 128-137 (1983).
20. Wever, R.A.: Schlaf und Melatonin. In: Biologische Psychiatrie (ed.: B. Saletu), pp.397-401 (Thieme-Verlag, Stuttgart 1989).
21. Wever, R.A.: Circadian aspects of sleep. In: Methods of Sleep Research (eds.: S. Kubicki and W.M. Herrmann), pp. 119-151 (Gustav Fischer Verlag, Stuttgart 1985).

Part VI

Rhythms in Electrical Activity of the Brain

The Information Content of the Human EEG

H. Petsche

Institut für Neurophysiologie der Universität Wien,
Währinger Str. 17, A-1090 Wien, Austria

Abstract. The field potentials recorded by the EEG emerge from cooperative processes and are commonly explained by "synchronization", a still poorley understood inherent feature of the nervous system. Several properties of the potential fields underlying this phenomenon are demonstrated in rabbits and in humans. Such cooperative processes have functional significance. This was demonstrated for the theta rhythm in the rabbit's hippocampus and in the characteristic local changes of power and coherence accompanying various mental processes in humans.

1. Introduction

In his "Synopsis and Introduction" to the 1983 Elmau Symposium [6] Prof.Haken mentioned a central concept of synergetics, namely "order parameters", in analogy to the "macroscopic observables" utilized in thermodynamics (e.g. temperature and pressure). Both parameters change in time more slowly than the microscopic processes underlying them. It seems useful to apply this concept also to the EEG. If this is done, one of the principal features of brain electric activities is the phenomenon of synchronization, without which no oscillating potential fields would arise, the effects of which are recorded by an EEG machine. The generation of spatially confined potential fields and their properties seems to be the kernel of electroencephalography. Moreover, as "structure is frozen function" [36], the problem of synchronization should not be approached by electrophysiological methods alone, but must also find support from neuroanatomical studies. Because the principal purpose of my occupation with the EEG has been to understand how electrical events in the microregions of the nervous system lump together to form potential fields, my search was equally distributed among neurophysiological and neuroanatomical strategies. However, before entering in more detail into the properties and possible meaning of synchronization, a few words should be said about the generation of the EEG and its present field of application.

Springer Series in Synergetics, Vol. 55 **Rhythms in Physiological Systems**
Eds.: H. Haken and H.P. Koepchen

For most people the EEG is a more or less familiar diagnostic tool for the detection of localized or generalized irregularities of brain metabolism; it is known to be of some use for ascertaining the diagnosis "epilepsy", sometimes also a brain tumour and to allow gross statements as to the severeness of vascular diseases of the brain. Its significance as a localizing method for diseases of the brain, however, has declined with the increasing use of computerized methods such as CT, PET and SPECT. Moreover, the EEG is known to represent merely electrical mass activities of the nervous system and, thus, hardly any statements about the more refined functions of the brain were thought to be expected from EEG recordings.

This attitude of resignation towards the EEG is further boosted if its generation is considered. There is general agreement today that the essential events underlying the EEG are currents flowing through the membranes of nerve and glial cells of the cortex. The various impedances and capacities of the tissue layers through which the ionic currents flow distort these currents in so many ways that other than fairly gross statements with respect to extended pathological processes (let alone normal functions of the brain) could not be expected from the EEG. This is the proper reason why neurophysiologists have mainly being studying intracellular events by using microelectrodes and seldom rely on the "field potentials" which are always and everywhere present in the brain and which form the basis of the EEG.

Another reason why the human EEG, recorded from the scalp, does not seem to be of great use as a research tool for the study of brain functions is that its main generators, the layers of the perpendicularly arranged cortical neurons, are curved in several planes, due to the brain's convolutions. This layering with the vertical arrangement of the major part of cellular elements in the cortex is the structural base for the innumerable vertical intracortical dipoles oscillating at the main frequency of the EEG. Due to the complex geometry of the cortex and the attenuating effect of CSF and the various tissue layers above the cortex with their low impedances it is in fact surprising that an EEG may be recorded at all through the skull. Already 25 years ago this question was studied by Cooper et al. [1]: they found that cortical areas of at least 6 cm have to be fairly synchronously active in order to produce EEG patterns detectable by scalp electrodes.

The word "synchronous" is not only among the most used but also least understood concepts in clinical electroencephalography. The term "synchronization" or "synchrony" gave rise to much dissent as soon as it became evident that even the brain waves seen in the seemingly most "synchronous" event, the so-called spike-and-wave pattern, as found in children during petit mal seizures, do not appear exactly simultaneously over the entire skull but display phase shifts, as can be shown when recorded at higher than the usual speed of EEG recording. As we demonstrated in 1954 [14], the "spikes" as well as the "waves" of this phenomenon behave as travelling waves, sweeping at velocities in the m/s range in a longitudinal direction over the scalp, the "spikes" faster than the "waves". This not only holds true for the spike-and-wave phenomenon but for all sorts of synchronous activities, even if the phase shifts are not always so uniform as in the spike-and-wave pattern.

Based on these findings in cases of epilepsy and other pathological processes [3,23] we became more and more intrigued by the question of whether this apparent spreading of regular EEG phenomena could also have any physiological significance. For this reason we studied this problem in both animals and man. The answer may be anticipated: by studying at first such cooperative processes in rabbits we were able not only to elucidate some mechanisms of the rising and maintenance of seizure patterns but also to learn more about the generation of the physiologically occurring theta rhythm in the hippocampus; finally, our work on humans gave us at least a partial answer to Prof.Haken's question put in the already mentioned synopsis of the 1983 symposium [6]: how can we measure thinking? For these reasons, this paper is to be focussed on two subjects: the first relates to experimental work and concerns mainly the events taking place in epileptic seizures in rabbits. The second part concerns the background EEG of man and the spectrum of consequences its thorough analysis may bring about.

2. Strategy of Research

Before presenting a few results from our work on rabbits and on man, it may be worth mentioning our concept of the electric activity of the brain as a 5-dimensional continuum and the methods we devised to explore this continuum along several dimensions by means of interpolation.

The macro-domain of electric activities, the realm of clinical electroencephalography, has been explored fairly thoroughly; the same holds true for the exploration of membrane events by micro-electrodes. In between these two orders of magnitude, however, there is a zone of terra incognita: hardly any attempts have been made to explore this range by simultaneously recording from a larger number of semi-microelectrodes in order to explore how the electrical properties can be related to the structural peculiarities of the cortex. For this reason a method called "Micro-EEG" was developed [16], a combination of multiple surface electrodes at small distances for the study of the behaviour of the equipotential fields and probes with multiple contacts for intracortical microrecordings [29] in order to find out how the structural peculiarities of the different cortical layers are conditional for the electrical phenomena observed in the different cortical layers. The analysis of power spectra, coherence and phase estimates contributed to the understanding of spontaneous activities [34] and seizure patterns [15]. In addition, the recordings from the 16 equidistant contacts of the intracortical probe were used for the computation of current source and sink densities (CSD) that offered clues as to the spatio-temporal distribution of accumulations of EPSPs in the cortex [31]. From the side of neuroanatomy, these studies were supplemented by various staining procedures for the myelin sheath and cell bodies as well as by HRP staining for establishing long-distance fibre connections [9]. Because of its unfolded cortex the rabbit was chosen as the experimental animal. In this way one could get rid of geometrical complications.

3. Animal Studies

Before dealing with the neocortex, however, a few words ought to be said about a very characteristic activity, the rabbit's theta rhythm generated by a much simpler cortical structure than the neocortex, the hippocampus, which consists of only three layers. In this region by far the most frequent cell type are pyramidal cells whose bodies form a densely packed stratum with concise boundaries; their apical dendrites are arranged palisadelike and constitute a fairly voluminous layer. The stratum oriens, on the other side of the layer of the pyramidal cell bodies, contains their axons that leave the hippocampus via fimbria. In this highly ordered structure one of the most regular brain activities, the theta rhythm, is generated. It oscillates usually between 4 and 7 Hz and both its frequency and amplitude correlate with the strength of sensory stimuli, whereby proprioceptive and olfactory stimulation are most effective.

Our toposcopic studies of this rhythm [24] hinted at a point of origin in the center of the septum from where it spreads symmetrically backwards over the hippocampus. In the search for a possible anatomical substrate of its pacemaker we explored this region more closely and found that electrical stimulation in the center of the septum was able to suppress the theta, whereas stimulation of more lateral regions instead increased both its amplitude and frequency [25]. Thus, we investigated this region by microelectrodes and found that the neurons of Broca's nucleus of the diagonal band in the median portion of the septum discharge in bursts synchronously to the theta rhythm [5]. In a neuroanatomical study, fibre connections from septum to hippocampus could be established [12]. Bursts of action potentials in the medial septum nucleus thus evidently impinge upon hippocampus neurons to give rise there to oscillations of membrane potentials, the theta rhythm that propagates along the hippocampus. This nucleus of the septum, thus, acts as a pacemaker for the theta rhythm.

The purpose of this rhythm is not yet completely understood. According to O'Keefe and Nadel [10] the theta is connected to learning of behaviour related to place ("local memory"); the cells in the hippocampus could be shown to be place coded. The theta rhythm evidently has some function that involves scanning the hippocampus, which is considered to be a cognitive map, and selecting the respective cells to issue signals to the motor systems of the brainstem so that the animal moves in the right general direction.

These observations on the hippocampus-septum system were a first clue that the spontaneous EEG, though being a neuronal mass activity, may be of some significance for brain functions.

A further step in our attempts to understand the processes underlying the poorly defined phenomenon of synchronization in EEG was to study seizure patterns in the rabbit's cortex. Seizures were chosen since these phenomena were thought to be more easily understood than the apparently stochastic spontaneous activities. For these studies, the Micro-EEG proved particularly effective for a better comprehension of the spatio-temporal electrical continuum of the cortex and for obtaining correlations with respect to neuroanatomical data. From the many findings with this method I report on just a few obtained with surface and a few others with intracortical electrodes to illustrate the manifoldness of the electric mass activity.

In our first studies on seizures we used rows of equidistant electrodes put on the rabbit's cortex [11]. To make visible the time relationships, the voltage oscillations of seizure activities were transformed into light modulations. In these studies several properties of the potential fields of generalized seizures (elicited by electrical stimulation or local application of penicillin) could be established: (1) potential fields underlying so-called brain waves are confined potential structures propagating in different directions over the cortex at speeds in the cm/s range, (2) no preferred zone of origin was found, (3) there is no preferred direction of the trajectories of the potential fields, the direction of spreading may even reverse from one seizure to the next, (4) at architectonic boundaries these travelling waves either change shape or disappear, (5) their propagation may be prevented by vertical incisures into the cortex provided that layer VI is also lesioned, (6) the spreading pattern becomes very regular and is maintained over long periods of time after lesioning the capsula interna [19]. These findings clearly demonstrate that the properties of the potential fields underlying the EEG as well as their propagation depend on structural peculiarities of the cortex.

Because the unidimensional electrode array used in these experiments did not prove sufficient to reveal the obviously intricate pathways of propagation of seizure potential fields, let alone the generation of a seizure, in the following experiments we used two-dimensional arrays and studied seizures emerging from a single, well-defined focus elicited by the minimum possible dosis of penicillin. In doing so, a number of substantial properties of the emerging epileptic seizure were detected [18,26].

One of these is the tendency of the trajectories of the potential fields to circle around the penicillin focus at its beginning (its tonic stage) [21]. Immediately after the epicortical application of penicillin interictal spikes arise. They are caused by sinks in layers II/III, accompanied by sources above and in the deep pyramidal layer [18]. The latter could be shown to be due to large IPSPs found in the early stages of the interictal spikes [2]. With the increasing invasion of the cortex by penicillin the intensity of the sinks increases, the corresponding spikes gain in voltage and the IPSPs fade away so that, at this stage, and if penicillin is applied to the motor cortex, the discharges of layer V pyramidal cells impinge now upon the anterior horn of the spinal cord and give rise to muscle jerks [2]. Then, gradually, and if the amount of penicillin is larger than about 1000 I.U., the spikes tend to come in groups until finally a large spike is followed by the tonic stage of a self-sustained seizure. With current source density analysis [31] this stage presents oscillating sinks and sources in different layers of the cortex. With the transition into the clonic stage, the time course of these sinks and sources becomes more irregular and complex interactions with deep cerebral structures are found. A remarkable property of both sinks and sources is their tendency to spread, in the cm/s range and in a vertical direction of the cortex, mainly from the deeper to the more superficial parts of the cortex. These are aspects of the epileptic seizure found with the multiple intracortical electrode probes.

If the potential fields of the developing seizure are studied with a net of cortical electrodes, the regular tonic oscillation is initially confined to a small region (roughly 3 mm in diameter); with time this region enlarges, whereby the amplitude of the field potential increases and the frequency in the EEG decreases. By plotting the trajectories of the potential fields they turned out to circle around the penicillin focus; the radius of these circles gradually enlarges while the velocity of the revolving potential fields decreases [21]. Therefore, the frequency of the seizure pattern observed in the EEG is a function of the velocity at which potential fields are passing underneath the electrodes.

The synchronization processes leading to the potential fields of a seizure evidently are due to cooperative processes. These could be interpreted by the following model: let us assume the existence of elementary generators in the cortex. These may be either single nerve cells or groups of nerve cells which may be either activated or inhibited. By the exciting action of penicillin the number of elementary generators increases and forms groups of elementary generators ("nests" of synchronization). Such "nests" successively activate adjacent regions and inhibit others. At the beginning of the penicillin's action, the amount of inhibition produced by the firing elementary generators is strong enough to produce a broad area of inhibition around the focus and to prevent the discharges from becoming generalized [28]. However, as soon as a critical number of those elementary generators is sufficiently stimulated by penicillin, they have the tendency to propagate, to encompass adjacent ones, while others are inhibited. But the extensive zone of inhibition produced around every epileptic focus by the wall of IPSPs allows the potential fields only to propagate along the inner wall of this basin of inhibition. Moreover, the bottom of this basin is formed by the large IPSPs in the deep layers of the cortex as mentioned above [2]. Thus the focal excitation remains spatially confined. Only within this basin of inhibition may the excitatory processes of a seizure develop; since the zone of excitation within this basin has the tendency to involve adjacent regions - this means, to spread - it can do this only within the wall of the surrounding inhibition which, however, is gradually overcome by the increasing excitation and impelled into an outside direction until finally both the wall and the bottom of the basin of inhibition break down and the seizure becomes generalized.

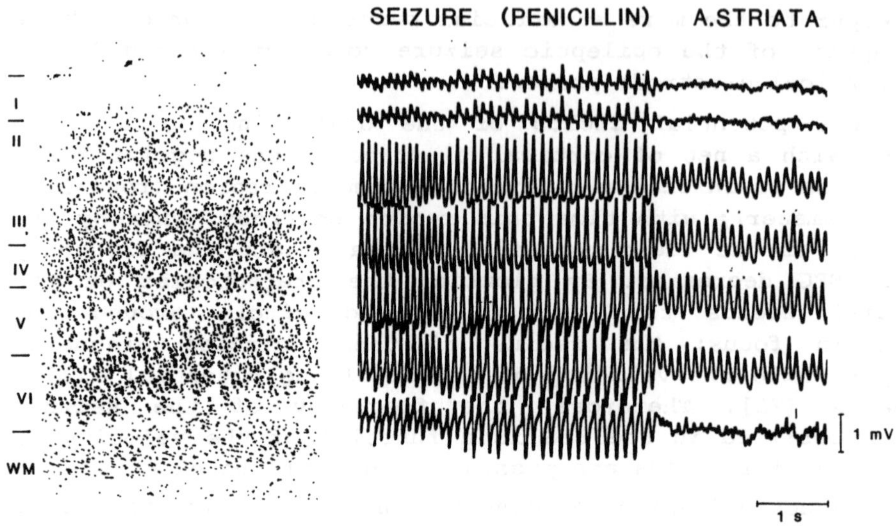

Fig.1 Section through the rabbit's visual cortex with the layers indicated by Roman numerals. Simultaneous recording from the 8-fold intracortical probe (8 semi-microelectrodes at 300 µm separation)

When the pattern gradually changes into the so-called "clonic stage", the propagation becomes saccadic and more irregular, the wave shapes in the EEG also become less consistent. At this stage, the trajectories of the potential fields become more irregular and the spreading less continuous: the potential fields now seem to jump to different places without any recognizable rule. With increasing dwelling times between the jumps, the conditions for the emergence of cooperative processes decrease until finally the seizure stops. According to these observations, synchronization in seizures seems to be due to a discontinuous activation of elementary neuronal generators; the continuity of the phenomena in the tonic stage is feigned by the initial overweight of excitatory processes with respect to inhibitory ones in the area of focus. Only with decreasing excitation towards the end of a seizure does the discontinuity of the phenomenon become visible.

Before leaving this topic, however, one phenomenon is to be presented which frequently occurs, specifically when larger doses of penicillin are used. The record of Fig.1 was made with 8 semi-microelectrodes at 300 µm distances. The position of the traces corresponds to the recording sites within the visual cortex of the rabbit. The record shows a typical tonic pattern recorded from different layers as shown on the left side. This seizure pattern may suddenly change both in freqency and in amplitude throughout the cortex. As could be shown by a model based on data from phase and coherence analyses, such changes may be induced by the superposition of synchronized and uncorrelated activities (e.g. noise being injected into layer IV [33], the layer where the main sensory input is found).

The observations presented up to this point are able to demonstrate the manifoldness of synergetic processes taking place even at microlevels of the EEG. They also prove that their knowledge seems to be of some importance for a better understanding of the brain processes involved and, therefore, also emphasize the urgent need for an algorithm to make them scientifically accessible rather than to just describe them by simple words of the colloquial language. They also demonstrate that the usual objection against EEG recordings because of the blurring by volume conduction is no longer valid, as current source density analyses have shown. The need for appropriate algorithms for the evaluation of EEG data is even more pressing when the human EEG is considered as described below.

4.Studies in Humans

There were several reasons that prompted us to also make use of the human EEG to look for the significance of cooperative processes outside pathology. One of these was our experience with the hippocampus theta rhythm in rabbits; another the enormous literature on ERP recordings, which shows that psychic phenomena are reflected by the EEG; a third was Freeman's work on the significance of the EEG of the olfactory brain for olfactory perception [35]; and a final reason were Livanov's reports on the functional significance of the similarity of brain patterns in different areas of the cortex [8]: when stimulated by flickering light, rabbits sometimes reacted with movements of their legs and sometimes not. Motor reactions proved to occur only if the electrical activities of the visual and the motor cortex were synchronized. Livanov therefore computed the cross-correlation between the activities of these two locations and found that the light stimulus was more readily transmitted to the motor area when the correlation coefficient was high than when it was low. The final incentive to enter this field of research was a request of the Herbert von Karajan-Foundation of the Wiener Gesellschaft der Musikfreunde to look for possible effects of music on brain electric activities. This latter was also one reason why at first questions in connection with listening to music were studied before additional mental activities were investigated with respect to their reflections in the EEG.

The principles of the method devised for this purpose are as follows [32]: For the study of possible reflections of thinking in the background EEG it was thought to be more convenient to use a top-down rather than a bottom-up strategy, this means to give a subject a mental task to deal with for at least one minute and to compare the EEG during this period of time with the EEG at rest, recorded before and after this task or with the average of all periods of EEG at rest. In order to include the entire skull, at least 19 equally distributed electrodes should be used as in clinical EEG. 4 parameters were chosen: the recording site, 5 frequency bands between 4 and 32 Hz, absolute power and coherence, the latter between adjacent electrodes and between electrodes on homologous recording points of the two hemispheres. The significant changes of the parameters "absolute power" and "coherence" were computed, whereby either changes caused by a certain task were measured in single subjects or in groups of subjects, or differences between individuals or groups of individuals at rest or at a specific task were established. For statistics the paired Wilcoxon test was used. As a matter of fact, this test is descriptive and only yields clues to which of the various comparisons may exhibit differences. The last step was the representation of the results in schematic brain maps displaying the probabilities of power and coherence changes. Since group studies have more weight in showing certain tendencies of EEG changes most emphasis was put on these.

In the following a few samples are shown. The first concerns the difference of the EEG changes in a group of 78 volunteers caused by opening the eyes and by listening to a Mozart quartet [13]. As far as the changes of power caused by these two tasks are concerned: opening of the eyes significantly and generally reduces power in all frequency bands, whereas listening to the Mozart quartet does this almost only in the alpha range, and in the left temporal region more so than on the right side. There are also remarkable differences between opening the eyes and listening to Mozart in the parameter coherence: if coherence changes are estimated between adjacent electrodes ("local coherence"), the main findings while opening the eyes is a decrease in the frontal regions, probably caused by the involvement of the frontal regions in the processing of visual stimuli from the surrounding. When listening to Mozart, on the other hand, only local zones of increased coherence are found in the theta range and in the frequency range between 18 and 32 Hz. If coherence is estimated between corresponding electrodes of the two hemispheres ("interhemispheric coherence"), opening of the eyes induces coherence to increase between the two occipital regions in all frequency bands (probably caused by the increasing occipital interaction of information when the eyes are opened), and a decrease of coherence in the frontal regions (most likely due to a functional decoupling of the two frontal lobes during the act of viewing). By listening to music, on the other hand, the most remarkable finding is an increase of coherence in a frequency band between 18 and 24 Hz between the temporo-occipital and the parietal regions, possibly due to an activation of musical memory contents.

It cannot be the purpose of this paper to enter into possible interpretations of these phenomena but rather to present data that may give some hints towards possible future prospects of electroencephalography. In our desire to get to know the scope of information hidden in the EEG, as a next step we inquired whether or not the EEG might also, in some way, reflect the extent of training. Therefore the subjects of this study were divided into two groups according to their musical training. At least five years of training in any instrument or in singing was chosen as the criterion [13].

Among the 75 right-handed subjects of the above group 52 complied with this criterion, 23 did not. Considerable group differences were seen in their EEG when listening to Mozart: particularly coherence increases were much more pronounced in the trained than in the total group; especially the increase of coherence in the frequency range of 18 - 24 Hz extended in the trained group all over the posterior half of the hand, whereas, in the group without musical training, listening to Mozart even lowered the interhemispherical coherence in the occipital region. In parenthesis it should also be mentioned that considerable sex differences were found in this and all the other studies concerning mental task and EEG.

The next question was whether or not even the spontaneous EEG alone, the on-going EEG at rest, without any task being performed, might display differences between musically trained and untrained subjects. This too has turned out to be the case: with males, in almost all frequency bands the local coherence of the on-going EEG is larger in the right temporooccipital area; with females, larger and more areas are involved. There are also considerable differences in interhemispheric coherence: musically trained females evidently display more interplay between the two hemispheres than untrained; with males the opposite happens.

At first sight it may seem strange to claim that the spontaneous EEG of persons trained in a music instrument should differ from that of an untrained person. Musical education as such cannot be so important for the brain as to produce a different EEG pattern at rest. However, if one bears in mind the number of abilities that are trained by musical education, this finding loses something of its oddness: musical training on whatsoever instrument not only improves manual skill of both hands, but also increases the ability to discriminate between sounds and to follow and to reconstruct acoustic architectonic structures. Moreover, the sense for both perceiving and producing rhythms is improved. That musical training not only fosters the ability to play an instrument but also contributes to the general training of intellectual abilities was shown recently by Hassler and Nieschlag [7].

There is not enough space to report in detail on our further studies on the reflection of mental activities by the EEG, such as during silent reading [22], watching pictures [17], playing chess, doing mental arithmetic [27], mental cube rotation [30], perception of rhythm [20], creative thinking and others. In all these activities, characteristic changes of the EEG with respect to the EEG at rest were found. In several cases also clear relations were found with regard to intellectual functions, e.g. in mental cube rotation where, on calculating the rank correlations between the coherence changes and the scores of performance, positive correlations were found in men and negative ones in females. There are even further hints indicating that some intellectual traits are reflected by EEG characteristics, a claim, which, incidentally, was also raised by other authors using other methods [4].

Several reasons induced me to mention all this: first, I wanted to illustrate the ubiquity of synergetic events in the EEG, secondly - and this mostly in animal experiments - to present the great variety, both in size and in shape, of cooperative processes, and thirdly, to point out their possible significance for thinking processes. What I wanted to show in addition was that basic thinking strategies manifest themselves in the ongoing EEG by changes of the parameters coherence and power. It is to be hoped that this knowledge may give rise in the not too distant future to mathematical tools for the quantification of EEG data with respect to thinking that might not only be of significance for a better grasp of the diffuse concept of intelligence but also give rise to useful means for the examination of psychiatric patients.

References

[1] R.COOPER, A.L. WINTER, H.J. CROW, W.G. WALTER: Electroenceph clin Neurophysiol, 18, 217-228 (1965)

[2] C.E.ELGER, E.J.SPECKMANN: Electroenceph clin Neurophysiol, 56, 604-622 (1983)

[3] G.FOITL, H.PETSCHE: Arch Psychiat 200, 36-51 (1959)

[4] D.GIANNITRAPANI: The Electrophysiology of Intellectual Functions. Karger, Basel, (1985)

[5] G.GOGOLAK, CH.STUMPF, H.PETSCHE, J.STERC: Brain Research, 7, 201-207 (1968)

[6] H.HAKEN: In: E.BASAR, H.FLOHR, H.HAKEN, A.J.MANDELL (eds): Synergetics of the Brain. Springer, Berlin, pp. 3-27, (1983)

[7] M.HASSLER, E.NIESCHLAG: In: H.PETSCHE (Hsg.): Musik - Gehirn - Spiel, Birkhäuser, Basel, pp. 191-202, (1989)

[8] M.N.LIVANOV: Spatial Organization of Cerebral Processes. Wiley and Sons, New York (1977)

[9] I.B.MÜLLER-PASCHINGER, T.TÖMBÖL: Anat Embryol, 180, 81-88 (1989)

[10] J.O'KEEFE, L.NADEL: The Hippocampus as a Cognitive Map. Clarendon Press, Oxford (1978)

[11] H.PETSCHE: Wiener Nervenheilkunde, 25, 373-387 (1967)

[12] H.PETSCHE, G.GOGOLAK, CH.STUMPF: Hirnforschung, 8, 129-136 (1966)

[13] H.PETSCHE, K.LINDNER, P.RAPPELSBERGER, G.GRUBER: Music Perception, 6, 133-159 (1988)

[14] H.PETSCHE, A.MARKO, H.KUGLER: Wiener Nervenheilkunde, 8, 294-323 (1954)

[15] H.PETSCHE, H.POCKBERGER, P.RAPPELSBERGER: In: M.R.KLEE, H.D.LUX, E.J.SPECKMANN (eds). Physiology and Pharmacology of Epileptogenic Phenomena. Raven Press, New York, pp. 53-63, (1982)

[16] PETSCHE, H.POCKBERGER, P.RAPPELSBERGER: Neuroscience, 11, 1-27 (1984)

[17] H.PETSCHE, H.POCKBERGER, P.RAPPELSBERGER: In: F.H.DUFFY (ed): Topographic Mapping of the Brain. Butterworth, Stoneham, pp. 63-98, (1986)

[18] H.PETSCHE, H.POCKBERGER, P.RAPPELSBERGER: In: H.G.WIESER, E.H.SPECKMANN, J.ENGEL, (eds): The Epileptic Focus, John Libbey, London, pp. 59-81, (1987)

[19] H.PETSCHE ,P. RAPPELSBERGER: Electroenceph clin Neurophysiol, 28, 592-600 (1970)

[20] H.PETSCHE, P.RAPPELSBERGER, O.FILZ, G.H.GRUBER: Wenner-Gren-Symposium "Music - Language - Speech and Brain" (in press)

[21] H.PETSCHE, P.RAPPELSBERGER, R.LAPINS, R.VOLLMER: In: E.J.SPECKMANN, H.CASPER, (eds): Origin of Cerebral Field Potentials. Thieme, Stuttgart, 60-79, (1979)

[22] H.PETSCHE, P.RAPPELSBERGER und H.POCKBERGER: In: H.M.WEINMANN (Hsg.): Zugang zum Verständnis höherer Hirnfunktionen durch das EEG. Zuckschwerdt, München, pp.59-74, (1987)

[23] PETSCHE, H.SCHINKO, F.SEITELBERGER: In: J.VAN BOGAERT (ed): Elsevier, Amsterdam, pp 363-385 (1961)

[24] H.PETSCHE, CH.STUMPF: EEG clin Neurophysiol, 12, 589-600 (1960)

[25] H.PETSCHE, CH.STUMPF, G.GOGOLAK: EEG clin Neurophysiol, 14: 202-211 (1962)

[26] H.PETSCHE, P.RAPPELSBERGER, R.TRAPPL: Electroenceph clin Neurophysiol, 29: 567-578 (1970)

[27] POCKBERGER, P.RAPPELSBERGER, H.PETSCHE, S.KRIEGLSTEINER: In: E.BASAR (ed): Dynamics of Sensory and Cognitive Processing of the Brain. Springer, Berlin, pp. 266-274 (1988)

[28] D.A.PRINCE, D.J.WILDER: Arch Neurol Psychiat Chicago, 16, 194-202 (1967)

[29] O.PROHASKA, F.PACHA, P.PFUNDNER, H.PETSCHE: EEG clin Neurophysiol, 47, 629-631 (1979)

[30] P.RAPPELSBERGER, S.KRIEGLSTEINER, M.MAYERWEG, H.PETSCHE, H.POCKBERGER: J.clinical Monitoring, 3: 320-322 (1987)

[31] P.RAPPELSBERGER, H.POCKBERGER, H PETSCHE: Pflüger's Arch, 389, 59-170 (1981)

[32] P.RAPPELSBERGER, H.PETSCHE: Brain Mapping, 1, 46-54 (1989)

[33] P.RAPPELSBERGER, H.PETSCHE, R.VOLLMER, P. LAPINS: In: E.J.SPECKMANN, H.CASPERS, (eds): Origin of Cerebral Field Potentials. Thieme, Stuttgart, pp, 80-97 (1979)

[34] P.RAPPELSBERGER, H.POCKBERGER, H.PETSCHE: Electroenceph clin Neurophysiol, 54, 252-269 (1982)

[35] A.SKARDA, W.J.FREEMANN: Brain Beh Res, 10, 161-195 (1987)

[36] D.W.THOMPSON: On Growth and Form. Cambridge, University Press (1942)

Synergetics of Evoked Alpha and Theta Rhythms in the Brain: Topographic and Modality-Dependent Aspects

E. Başar, C. Başar-Eroglu, E. Rahn, and M. Schürmann

Institut für Physiologie, Medizinische Universität zu Lübeck, Ratzeburger Allee 160, W-2400 Lübeck 1, Fed. Rep. of Germany

Evoked electrical activity of the brain and brain resonance phenomena are discussed in the context of synergetics /1,2/ as an extension to a previous article /3/. The analysis methods given in the previous work /3/ are applied for analysis of topographic and modality-dependent properties of evoked potentials.

Experimental data were obtained (1) from auditory and visual cortex in 5 cats and (2) from vertex and occipital derivations in 12 human subjects. The evoked potentials were studied in the frequency domain by means of Fourier transforms. Furthermore, digital filtering was applied to confirm and extend the analysis. Preliminary results indicated that resonances were observed mainly in 5, 10, 20 and 40 Hz frequency bands. The frequency of resonance maxima depended on the sites of measurement electrodes and the stimulus modality.

We conclude that the analysis of resonance phenomena in the brain is a useful approach to understand the relation between evoked potentials and single-cell recordings.

1. Introduction

The **formation of temporal and spatial patterns of neural activities** is one of the problems treated by the discipline called **"synergetics"** /1,2/. Spontaneous electrical activity of the brain also gives rise to such patterns. Such records are called **electroencephalograms (EEG)**. This activity can be measured by using either intracranial macroelectrodes (50-100 μm) implanted in nervous masses or by using scalp electrodes. The term **"evoked potential" (EP)** is used for responses elicited by means of an external stimulus.

1.1. Relation between EEG and evoked potentials

There are several interpretations concerning the genesis of EPs and their relation to the spontaneous EEG (for an overview, see e.g. /4/).

According to our working hypothesis, the brain is a dynamic system having the property of continuously and spontaneously changing state. "At any instant, each state variable has a rate of change which depends on the current state of the system. Therefore if we stimulate or excite such a dynamic system (brain), it is to be expected that for every state a different **excitability** or different **response susceptibility** should exist. Accordingly, we use single records of brain responses together with the brain electrical activity prior to stimulation" /3/. At present, these "single sweep records" would be called records of **"induced rhythmicities"** /5/. "Our approach to the understanding of neural populations is derived from a combined analysis of EEG and EPs. This combined analysis provided us with results which could be interpreted by using analogies from laser physics, magnetism theory, concepts of statistical physics and dynamic systems theory and on cooperative phenomena in populations of neurons /6,7,8,9/. Accordingly, the new discipline 'synergetics', which established links between dynamic system theory and statistical physics, provided us with useful integrative concepts and methods for the understanding of evoked potentials and the brain waves" /3/.

1.2 Evoked potentials: resonance phenomena in the brain - transitions to coherent states - reduction in entropy

According to the working hypothesis outlined above, EPs can be considered as manifestations of **resonance properties** of the EEG. Furthermore, we stated that general resonance phenomena do reflect most dominating transfer functions of the brain. We defined the resonance phenomenon as the responsiveness of brain structures where the EP contains frequency components already present in spontaneous EEG activity; responses are time-locked, magnified in amplitude and frequency-stabilized. In other words: "resonance" and/or "selectivity" is defined as the ability of brain networks to facilitate (or activate) electrical transmission within determined frequency bands, when a stimulation is applied to the brain /9/. This working hypothesis was based on our own measurements of the cat and human brain, and took into consideration experimental findings of the groups of Spekreijse and Van der Tweel /10/ and of Lopes da Silva and Storm van Leeuwen /11,12/.

In terms of synergetics, **coherent states** in the electrical activity of the brain can be obtained by the application of an external stimulus. This is an analogy both to a simple harmonic oscillator and to the classical examples of synergetics in the field of **laser physics** /1,2/: the transition from an incoherent to a coherent state is also a feature of the laser at the emission threshold. In parallel to this analogy, the external stimulus giving rise to the EP is regarded as the **control parameter** or **order parameter**. Fig. 1 demonstrates this analogy in a plot of the relative phase angle between the EEG-EP-waveform and an "appropriate" pure sine function (see caption). The large fluctuations of the relative phase angle show a decrease (being time-locked to stimulus) for a "coherence time" after which

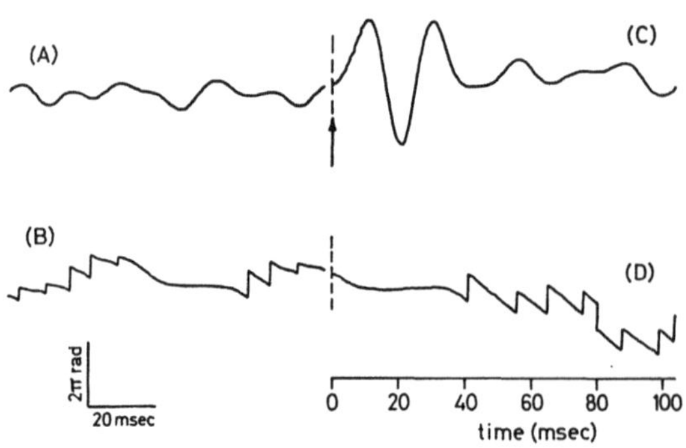

Fig. 1; upper part: EEG-EP epoch filtered in the 30-80 Hz range; (A) spontaneous EEG just before stimulation; (C) EP. Time of stimulation (0 ms) indicated by an upward arrow. **Lower part** (B, D): phase angle ϕ of EEG-EP epoch (A,C) relative to the phase of a pure sine wave having a frequency of 55 Hz, i.e.: the center of the 30-80 Hz filter range

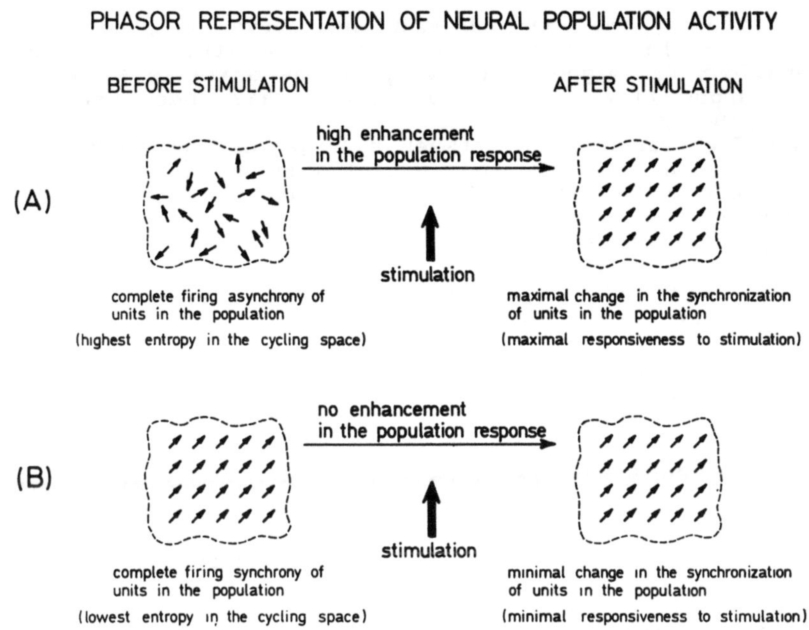

Fig. 2: A phasor representation of neural poulation activity. The phasor represents the state of individual neural elements in the physico-chemical cycling space, in analogy with the moment vectors of the elementary magnets in a magnetic field

fluctuations set in again /3/. However, epochs of constant phase angle can be observed in the "spontaneous" EEG as well.

This transition can also be interpreted in terms of **entropy**. According to our working hypothesis and as far as its electrical activity is concerned, the neural system can be compared to a system of **coupled oscillators**. The excitability of such a system is high when the oscillators are in a state of **asynchrony**, i.e. in a stage of high entropy. During **regular firing** or low entropy the neural system is not responsive. An analogy between the genesis of EPs and the **ordering of elementary magnets** in a paramagnetic substance exposed to a magnetic field has been put forward previously /9/. Fig. 2 is a schematic presentation of this analogy.

1.3 Evoked potentials and single cell recordings

At present, resonance phenomena of the brain merit increasing consideration. This is due to several recent neurophysiological investigations both at the cellular level and in EEG and magnetoencephalographic (MEG) measurements.

Some examples may demonstrate the essential progress made at the cellular level. The concept of Llinás /13/, who demonstrated in single cells 6 and 10 Hz intrinsic resonances, opened a new window to our knowledge at the level of field potentials and single neuronal recordings. Studies by Gray and Singer /14,15/ showed 40 Hz coherent oscillations in the visual cortex to light stimulation. Their measurements - confirmed and extended by Eckhorn et al. /16/ - provided another important step forward to indicate the relation between EEG and single unit recording. These advances have been accompanied by a number of measurements at the EEG and MEG level. In some of these studies, the approach mentioned above (1.2) has been applied: For example, Narici et al. have used the concept of resonance behaviour for magnetic activity of the brain in 6 and 10 Hz frequency ranges /17/. Similar approaches have been used in EEG-EP studies /18,19/.

In the past few years, links between synergetics and electrical activity of the brain have also been established in the context of nonlinear and chaotic dynamics. This topic, however, is beyond the scope of this paper. The reader is referred to the literature (for a collection of papers, see e.g. /20/).

1.4 Topography- and modality-dependent properties of evoked potentials: extending earlier studies

The results mentioned above and our EEG measurements as well as our investigations in cooperation with Saermark's MEG group /21/ raise the following questions extending our previous studies:

(a) What can be the physiological properties of the so-called "alpha and theta resonance phenomena" in various parts of the brain? (b) Do resonant responses of the brain elicited by acoustical and visual stimulation have different codes? (c) What do we learn from a comparison of the effects a visual stimulus elicits in the visual cortex with the effects elicited by the same stimulus in the auditory cortex ("cross-modality" experiments)? (d) Are cognitive processes also reflected in resonant behaviour?

A final interpretation is not possible for the time being. However, it will be shown that comparative studies of resonance phenomena in various parts of the cat and human brain lead to a working hypothesis on global functional correlates of theta and alpha resonances.

2. Methods

The methodology to evaluate EPs, amplitude frequency characteristics and digitally filtered data was described in several previous publications (e.g. /9/). Therefore we describe the methods shortly as follows:

1. **Recording of EEG-EP epochs:** With every stimulus presented segments of 1 s EEG activity preceding and 1 s of EP following the stimulus were digitized, labelled and stored on computer disc memory. This operation was repeated about 100 times.
2. **Selective averaging of EPs:** The stored raw single EEG-EP epochs were selected with specified criteria after the recording session: Movement artifacts were eliminated with double check; EEG segments showing sleep spindles or slow waves were also eliminated. (Such an a posteriori approach permits more efficient artifact rejection than various direct online methods /9/).
3. **Computation of amplitude frequency characteristics:** The selectively averaged EP is transformed to the frequency domain with the Fourier transform (FFT) in order to obtain the amplitude frequency characteristic [$G(j\omega)$] of the studied brain structure:

$$G\,(j\,\omega) = \int_0^\infty \frac{d\,\{c(t)\}}{dt}\,\exp\,(-j\,\omega\,t)dt.$$

   c(t) is the step response of the system, here, the sensory EP. Details of this method, which we called the transient response frequency characteristics method (TRFC), are given in references /9,22/.
4. **Filtering:** Digital filters (without phase shift /9/) were applied to compute EP frequency components.

For the description of surgical procedures and exact location of implanted electrodes in the cat brain we also refer to earlier reports /9/.

3. Results

3.1 Amplitude frequency characteristics in the cat cortex

Figs. 3A and 3B show auditory and visual EPs in the time domain. Comparison of the curves recorded in the primary auditory cortex of the cat indicated that the amplitude of the EP elicited by tone was significantly higher than the amplitude of the EP elicited by light (Fig. 3A). Visual cortex responses showed the opposite behaviour: The amplitude of visual EPs in the visual cortex is much higher than the amplitude of auditory EP (Fig. 3B). This behaviour was expected: It is already known that in the primary sensory areas of the cortex an adequate stimulation (e.g. auditory cortex/auditory stimulation) elicits much higher responses than an inadequate stimulation. In addition to the differences in amplitude the evoked potentials depicted completely different waveforms.

Fig. 4 represents amplitude frequency characteristics computed from the transient EPs of Fig. 3 by means of Fourier transforms. The solid curves present responses to acoustical stimulation, whereas the dashed curves present the visual evoked responses. In the auditory cortex (Fig. 4A) the amplitude frequency characteristics triggered with an adequate stimulation (auditory EP) had a dominant maximum at a frequency of 10 Hz. A second smaller peaking was observed at around 20 Hz and a third one between 40 and 60 Hz. The amplitude frequency characteristics obtained upon inadequate stimulation (auditory cortex/visual EP) had a completely different frequency behaviour. There are at least 3 dominant maxima centered at about 4 Hz, 20 Hz and 40 Hz. The first peak around 4 Hz was a

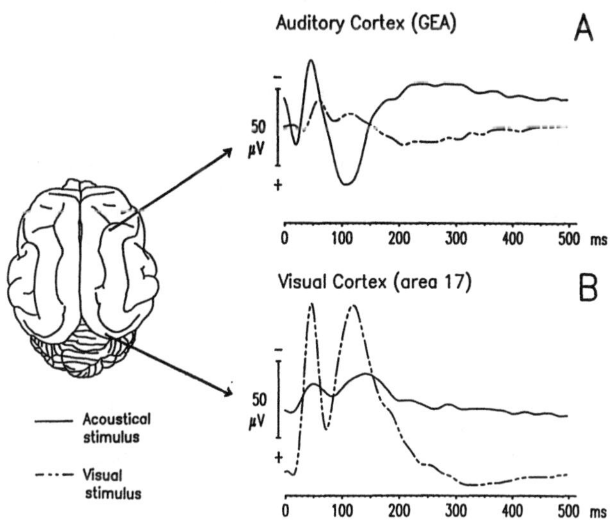

Fig. 3: Averaged EPs recorded from auditory and visual cortex of the cat brain. (A) Auditory cortex; (B) visual cortex. Solid curves: auditory stimulation; dashed curves: visual stimulation. Averages of N=12 experimental sessions with 4 cats (GEA, Gyrus ectosylvianus anterior)

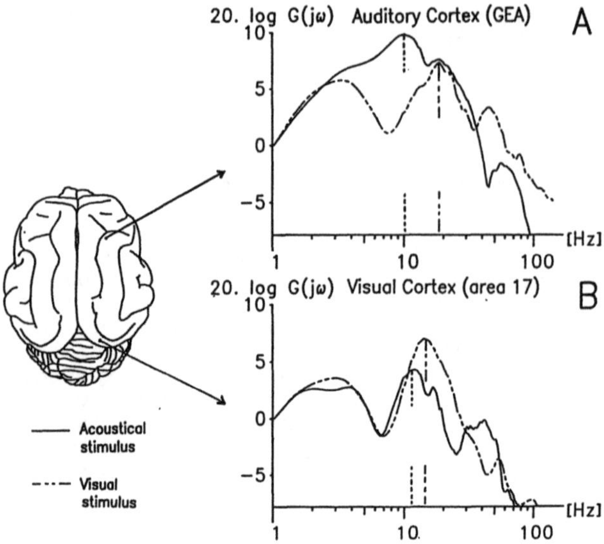

Fig. 4: Amplitude frequency characteristics computed from the cat EPs of Fig. 3A and 3B. Abscissa: frequency in Hz; ordinate: amplitude in relative units (dB, decibels). (A) Auditory cortex; (B) visual cortex

distinct one, whereas in the amplitude frequency characteristics upon acoustical stimulation the peak around 4 to 5 Hz was a type of shoulder. The beta resonance maxima were similar in the two amplitude frequency characteristics.

In the visual cortex the response to visual stimulation was dominated by a 12 to 15 Hz maximum with an amplitude of about 8 dB (Fig. 4B). A second peak was in the delta/theta frequency range with a maximum around 4 Hz. The presence of a delta/theta selectivity in visual evoked potentials recorded from the cat visual cortex was in good agreement with our earlier results /9/. The amplitude frequency characteristics of auditory EPs in the visual cortex (inadequate stimulation) revealed resonance maxima in the delta/theta frequency range similar to those of the visual EP/visual cortex recordings. However, the amplitude maximum in the alpha frequency range is shifted to lower frequencies almost around 10 or 11 Hz. There is also a marked response in the 40 Hz frequency range (Fig. 4B; dashed curve).

3.2 Digital filtering of transient EPs recorded from cat cortex

Fig. 5A shows filtered EPs recorded in the auditory cortex. We first consider the responses to auditory stimulation (solid lines). The filtered response in the frequency range of 10 Hz (filter 8-13 Hz) had an amplitude of around 30 μV. The theta component (3-7 Hz) had comparable amplitudes. On the other hand, the visual EP recorded in the auditory cortex can be decomposed into alpha and theta responses which showed very low amplitude. However, the theta response was higher in comparison to the alpha response (10 and 5 μV, respectively, Fig. 5A). In other words, the theta response oscillation dominates the responses to inadequate (visual) stimulation and is less reduced in comparison to the alpha component. Fig. 5B shows the respective data for the visual cortex.

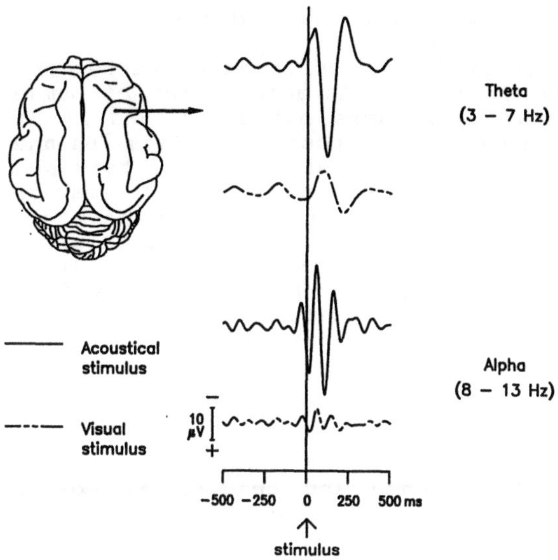

Fig. 5A: Filtered components of the averaged EPs of Fig. 3A. Auditory cortex recordings. Filter limits are chosen adequately according to the amplitude frequency characteristics of Fig. 4A

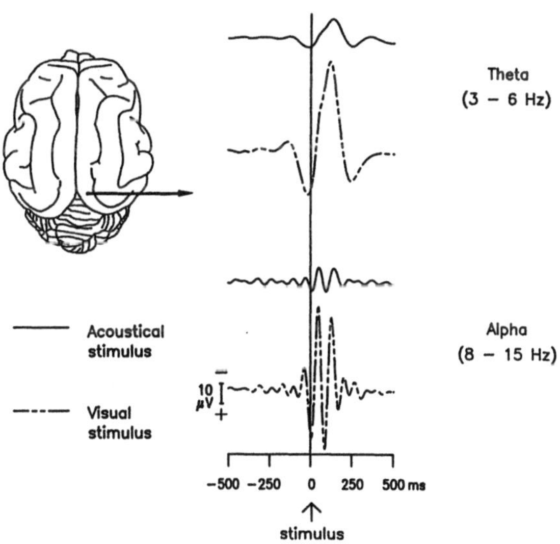

Fig. 5B: Filtered components of the averaged EPs of Fig. 3B. Visual cortex recordings. Filter limits are chosen adequately according to the amplitude frequency characteristics of Fig. 4B

3.3 Amplitude frequency characteristics of human subjects

In this paper the comparative analysis was confined to vertex and occipital recordings. Amplitude frequency characteristics of human evoked potentials showed maximal amplitudes between 5 and 15 Hz depending on stimulus modality and electrode position. Figs. 6A and 6B illustrate amplitude frequency characteristics of auditory and visual evoked potentials recorded from vertex and occipital areas of 12 subjects. The vertex frequency response of an auditory EP was characterized by a theta-alpha compound response covering a broad frequency range. The maximum is at about 7 Hz. However, in the 10 Hz range usually a small shoulder or minor peaking was observed in single subjects. The evoked response to auditory stimulation had small maxima at around 3 and 9 Hz and also a maximum at 20 Hz (Fig. 4A). The 9 Hz response had small amplitudes in comparison to the vertex response.

In the human visual evoked responses the maximum of frequencies was centered at 10 Hz but it is mostly a broader maximum covering also the 20 Hz range (in both vertex and occiput, Fig. 6B). A comparison of human occipital response and cat visual cortex response indicated common features: There is also a responsiveness to visual stimuli in the theta frequency range, as was the case in the response of cat evoked potentials in the frequency domain (Fig. 4B). However, the alpha maximum is dominant in the frequency response.

Fig. 7 presents also a comparison between the frequency characteristics of vertex auditory and visual amplitude response as mean values of 12 subjects. Contrary to the response to auditory stimulation, the resonance maxima elicited by visual stimulation were shifted to the higher alpha frequency range. The center frequencies of the peaks were about 2 Hz higher.

Fig. 8 shows the amplitude frequency characteristics for one typical subject. These results are in good agreement with results of the mean value response of 12 subjects presented in Fig. 7. Here again the differences of visual vs. auditory responses were clearly seen in the frequency domain.

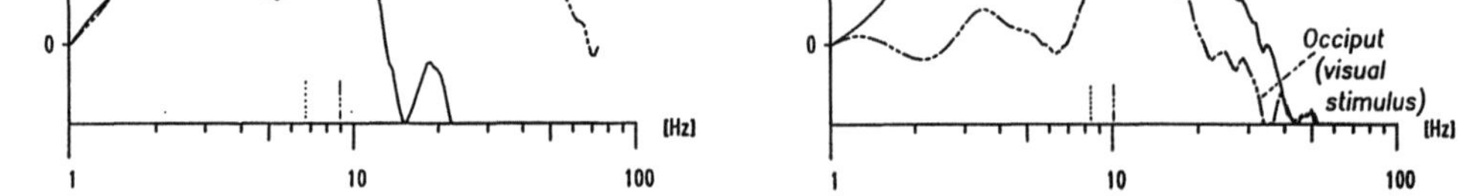

Fig. 6: Amplitude frequency characteristics of human EPs (computed from the grand averages of N=12 subjects). (A) Acoustical stimulus; (B) visual stimulus. Solid curves: vertex; dashed curves: occiput

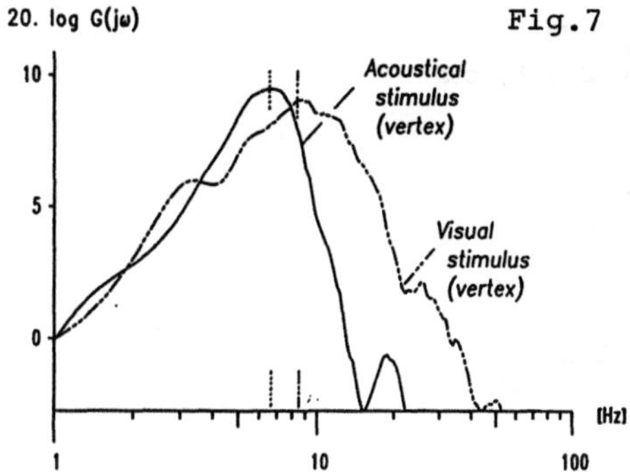

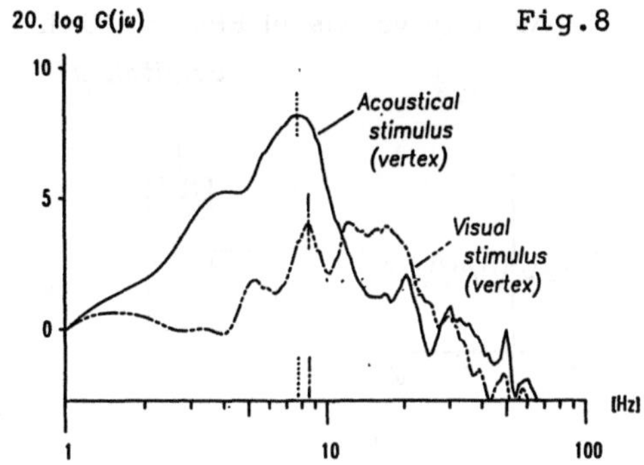

Fig. 7: Comparative presentation of amplitude frequency characteristics of vertex responses to auditory and visual stimulation (joint presentation of the curves of Figs. 6A and 6B; computed from the grand averages of N=12 subjects)

Fig. 8: Same presentation as Fig. 7; only one typical subject

3.4 Adaptive filtering of transient EPs recorded from human subjects

The results of digital filtering (with zero phase shift) are presented in Figs. 9A and 9B. A comparison of frequency ranges of 5-8 Hz and 8-13 Hz shows that in vertex the theta auditory component was significantly higher than the alpha component, thus confirming the results of amplitude frequency characteristics in Fig. 6A. In the occipital position (inadequate stimulation) theta and even slower frequency responses were dominant.

Visual stimulation elicits a large occipital alpha response, whereas the slower frequencies were not increased significantly. In

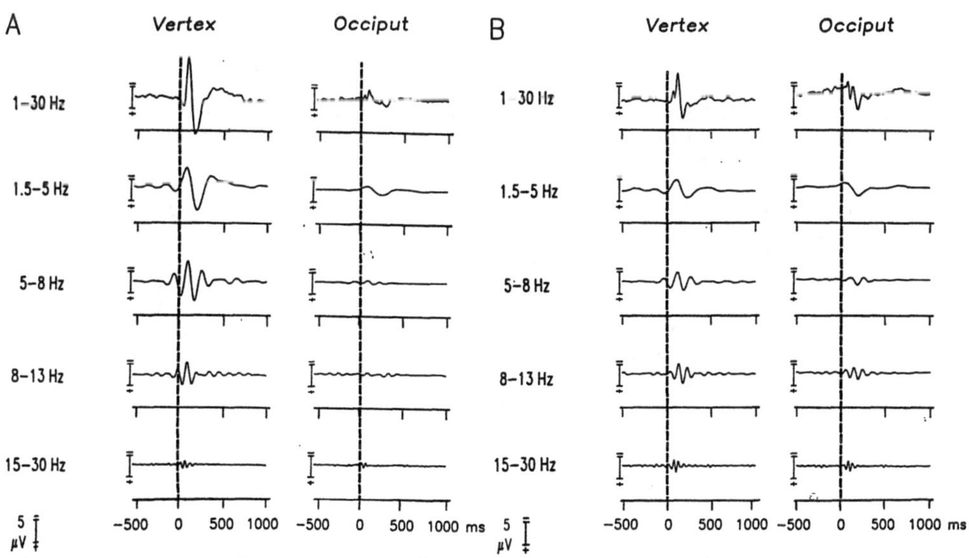

Fig. 9: Comparative presentation of human averaged auditory (A) and visual (B) EPs (grand averages; N=12), filtered in various filter bands. Vertex and left occipital recordings

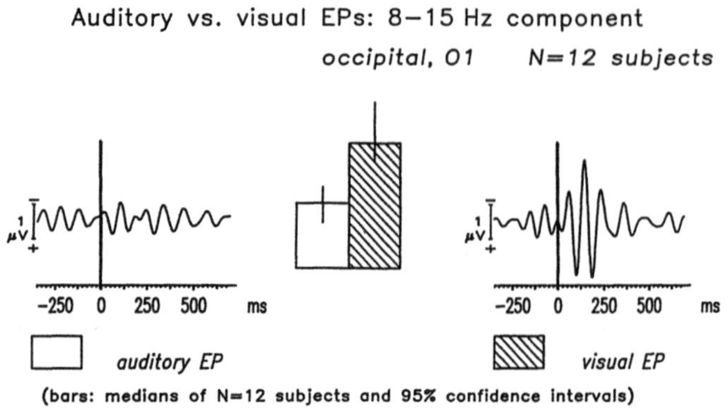

Fig. 10: Alpha components of left occipital auditory and visual EPs. Human scalp recordings. Time series are grand averages of N=12 subjects filtered in the alpha range. Bars represent the medians of the maximal post-stimulus alpha amplitudes in 12 single subjects

the vertex frequency response theta, alpha and beta components of the visual EP had similar amplitudes.

This is also illustrated by Fig. 10 showing alpha components of left occipital auditory and visual EPs. A marked alpha response is only elicited by visual stimulation.

4. Discussion

4.1 Resonance phenomena and the "excitability rule"

Based on various experiments using the methodology of EEG-EP epochs, we have tentatively formulated the following working hypothesis, which we called "the excitability rule":

"Various brain structures depict spontaneous rhythmic activity in a wide frequency range between 1 Hz and 1000 Hz. Without application of external sensory stimulation, the spontaneous activity of a given brain structure can often show frequency-stable and high-magnitude electrical activity. If regular spontaneous oscillations can be detected in the electrical activity of a defined brain structure during a determined period, it is to be expected that upon external sensory stimulation, this structure will have a response susceptibility in the same frequency channel (for example, 40 Hz activity of the hippocampus and cortex, 250-300 Hz activity of the cerebellum, 10 Hz activity of all brain structures, and 10 Hz activity of human scalp responses). The response susceptibility of a brain structure depends mostly on its susceptibility to its own **'intrinsic rhythmic activity'**. Frequency stabilization, time-locking, and amplification of the spontaneous activity upon stimulation contribute greatly to the genesis of large potential changes which are called evoked potentials" /9/.

In earlier studies we demonstrated that the existence of maxima in amplitude frequency characteristics was a manifestation of the enhancement (amplitude increase, frequency stabilization) of the EEG activity following sensory stimulation. This fact can also be clearly shown by using the filtered EPs (or **evoked rhythmicities** as they are called by Başar /5/) of Figs. 9A and 9B. In the 8-13 Hz frequency range the acoustical stimulation elicits no **"alpha enhancement"** in the occipital location (Figs. 9B, 10), whereas light stimulation evokes high alpha enhancement. If we compare amplitude

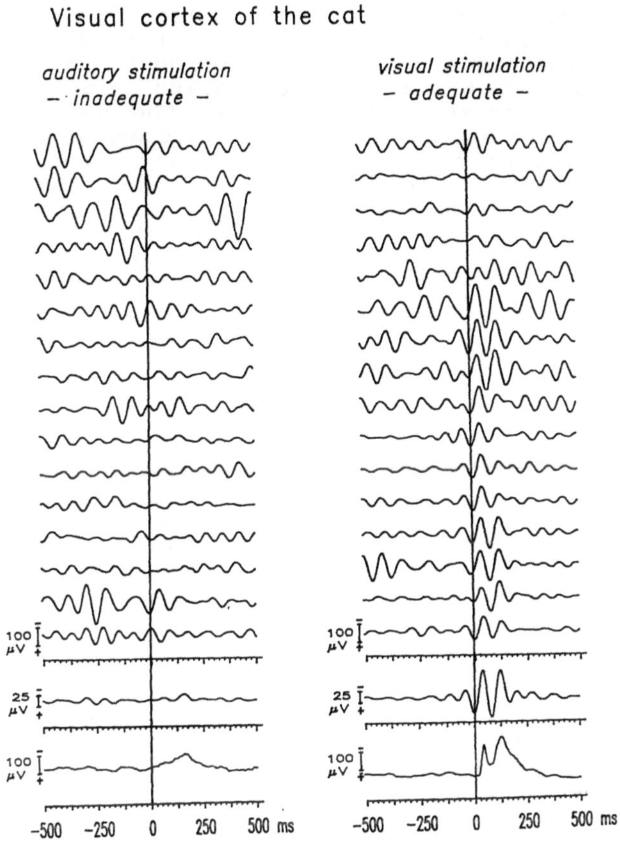

Fig. 11: EEG-EP-epochs (EEG-EPograms, "single sweep EPs") recorded from the visual cortex of a cat, filtered in the 8-15 Hz range. Last but one row: averaged EP, filtered in the 8-15 Hz range. Lowermost row: unfiltered averaged EP

characteristics of Fig. 6A (vertex) it is easy to see that there is no maximum in the 10 Hz frequency range upon auditory stimulation. On the other hand, a minimum is recorded at around 11 Hz. However, the amplitude characteristics which describe the visual response in the occipital location are dominated by a prominent maximum in the 10 Hz frequency range (Fig. 6B).

Single epoch analysis of evoked rhythmicities in the 10 Hz range for typical subjects confirmed the global description of mean values reflected in Figs. 6A and 6B (vertex). A detailed account of this analysis will be given in a future report. Some examples of single sweeps (cat visual cortex recordings) are given in Fig. 11 (see below, section 4.2).

4.2 Possible functional relations of different resonance components

In the present study we confined our attention to the frequency range between 1 and 15 Hz and did not analyze in detail resonant responses at the beta and 40 Hz frequency component.

One of the most important findings of this study concerns the "alpha response". We define this term as a responsiveness in a frequency channel between 8 and 15 Hz. In the cat evoked potentials the maximal alpha amplitudes in the primary auditory area are recorded when the applied stimulation is an auditory one. On the other hand, a high alpha resonance maximum was recorded in the primary visual area of the cat brain if visual stimulation was applied. These findings can be interpreted as follows: The adequate stimulation of the brain elicits maximal alpha responsiveness in the appropriate or adequate primary sensory areas. In the case of inadequate stimulation the alpha response is greatly reduced. Two sets of single sweeps (Fig. 11) recorded from the cat visual cortex support this interpretation: In the case of visual stimulation there are a large number of single sweeps resembling the averaged EP filtered in the alpha range. This does not apply to the case of auditory stimulation.

The differences between "adequate" and "inadequate" stimulation are less striking for the delta/theta responses (Figs. 6A, 6B, 7). This means: the alpha responses appear to be more dependent on the adequacy of the stimulus. In other words, the "adequate/inadequate" ratio is higher for alpha than for theta responses. However, we did not record pure alpha or pure theta responses (compare Figs. 4, 6, 7, 9).

A large number of neurophysiological studies showed that primary sensory stimuli elicit impulses or volleys converging over thalamic centers to primary sensory areas. In contrast the "sensory stimulation of second order" usually reaches the cortex over association areas (see e.g. Shepherd /23/). Due to this consideration it is conceivable that the responses in lower frequency ranges (theta, delta) might reflect the responsiveness of various brain areas in cases of association processes involved in global associative cognitive performance.

Moreover, the term **"alpha reponsiveness"** is not sufficient to interpret the modality-dependent differences of frequency coding within the alpha band: In the cat brain the auditory response occurs at around 10 Hz, whereas the visual stimulation gives rise to alpha responsiveness in higher frequency ranges between 12 and 15 Hz. The differences recorded in responses of the human brain are similar: The acoustical stimulation elicits a resonant response in the vicinity of 7 to 9 Hz, whereas the responses to light stimulation contain slightly higher frequencies (around 10 Hz).

There is a conceptual similarity between our results showing the sensory modality-dependent differences in resonant responses within the alpha band and the observations of Pfurtscheller's group, who mentioned two types of alpha functional relationship for activities between 6 and 10 Hz, and also activities above 10 Hz. These authors stated that the lower alpha frequency band has a prevailing cognitive significance /24/. We also emphasize the cognitive significance of lower alpha or theta resonances.

Do these above-mentioned results mean that there are several different alpha generators, which are triggered by various kinds of sensory and cognitive processing? This question can also not be answered easily. However, there is a distinct possibility that in the frequency range covering the alpha frequency range several functional rhythmicities may exist. These rhythmicities might cover sensory, cognitive and motor function /13, 24, 25/.

4.3 Compound theta/alpha responses

In many cases (especially in the case of human vertex responses to auditory stimulation) it is very difficult to describe a pure frequency response. In frequency characteristics of mean value curves it is usually not possible to detect clearly separated alpha or theta peaks. Therefore we use the term "compound alpha/theta response" - a fusion of resonances in the fast theta and slow alpha range.

Single sweep analysis revealed that the evoked potentials of the brain usually consist of two subsets of oscillatory waveforms or oscillatory rhythmicities in theta and alpha frequency ranges /26/. Accordingly we suggest here that the most appropriate analysis should take into consideration single EPs and not the averaged evoked potentials.

Our yet unpublished observations showed that the amplitude frequency responses of frontal lobes (association areas) in the cat and human brain were dominated by theta resonances /27/.

4.4 Resonance phenomena - common features of evoked potentials, evoked fields and intracellular oscillatory activity

The first tentative interpretation of our results led us to formulate that the alpha response component mostly contains the primary sensory processing of signals, whereas the theta response and/or slower responses are mostly involved in procedures of association and cognition. We also emphasize that it is not our trend to strictly separate cognitive or sensory mechanisms from each other. It may become possible to describe in the brain global interacting areas or structures which would be defined as a diffuse "alpha response system" and a diffuse "theta response system". Exact locations and exact functions cannot be properly defined as yet /26/.

We have to note here that a direct comparison between the results from cat and human brain is only limited. The electrodes used in cat evoked potentials provide a a better space resolution than the surface electrodes applied for human evoked potential measurements.

In the interpretation of these results the recent advances in measurements at the single cell level, at the level of small neuronal populations and at the level of human EEG field potentials mentioned above should be taken into account.

There are links between intracellular recordings and the EEG-EP level both in the 40 Hz range and in the 10 Hz range:

Our studies starting in 1972 had already predicted the existence of a 40 Hz response in EPs upon auditory and visual stimuli in various cortical and subcortical structures of the brain /22,28/. Important experimental work by Freeman showed that the 40 Hz responsiveness was observed in the single cell recordings /29/. Recent studies by Gray and Singer /14, 15/ - confirmed and extended by Eckhorn et al. /16/ - showed that visual complex stimulation can trigger 40 Hz oscillatory responses at the cellular level.

Furthermore, there is a conceptual similarity between 10 Hz resonance responses in cat and human EPs and cellular recordings revealing 10 Hz resonances in the primary auditory cortex of the cat, as recorded by Schreiner /30/.

Accordingly for the time being we tentatively formulate the existence of a significant relation between field potential resonances and resonances at the neuronal level.

In this report we confined our attention to resonant behaviour in the human and cat cortical responses, and analyzed only the amplitude frequency characteristics. However, resonant phenomena and synergetic processes in the cortex can be better interpreted by additional use of phase relationships and the concept of phase coherence as introduced by Hoke et al. /31/, especially in studies of magnetic fields.

MEG measurements with a number of SQUID detectors may be a supplementary approach to enhance the study of "heterosensory" communication in the brain. The term heterosensory has been described by Fessard /32/: "The brain (...) must not be considered simply a juxtaposition of private lines, leading to a mosaic of independent cortical territories, one for each sense modality (...). Unfortunately, we still lack principles that would help us to describe and master such operations in which heterosensory communications are involved. These principles may gradually emerge in the future from an extensive use of multiple microelectrode recordings, together with a systematic treatment of data by modern electronic computers, so that pattern-to-pattern transformation matrices can be established and possibly generalized. For the time being, it seems that we should do better to try to clear up such principles as seem to govern the most general transformations - or transfer functions - of multiunit homogeneous messages during their progression through neuronal networks". We also suggest that it would be useful to take into consideration the results of resonant responses of the cat cortex for an appropriate analysis of dipole locations (or multiple heterosensory dipoles) in the study of magnetic field responses. We use the expression "heterosensory" dipoles since the cortex of the brain is dominated by association areas. An appropriate analysis should take account of the fact that every evoked response contains relevant cognitive response components besides the primary sensory response.

According to our working hypothesis outlined at the beginning, the EP is a manifestation of an EEG-transition from an incoherent to a coherent state, the external stimulus being the control parameter. This hypothesis points out an analogy to Haken's classical example of similar transitions in the laser /1,2/. Furthermore, our results presented above raise the question whether the genesis of EPs involves cooperative phenomena where coupling of several rhythms in the brain plays a part. There is a possible cooperation of structurally distinct subsystems within the brain. However, for the time being any conclusions from our limited set of data must remain preliminary. For future studies we propose conceptual strategies taking into account analogies between coupled brain rhythms and transition phenomena in multimode lasers.

Acknowledgements

We are grateful to F. Greitschus for expert software development, to B. Stier for excellent technical assistance and to J. Djordjevic for careful secretarial work.

References

/1/ Haken H. Synergetics: An introduction. Berlin Heidelberg New York: Springer, 1977

/2/ Haken H. Synopsis and introduction. In: Başar E, Flohr H, Haken H, Mandell AJ, ed. Synergetics of the brain. Berlin Heidelberg New York: Springer, 1983: 3-27

/3/ Başar E. Synergetics of neuronal populations. A survey on experiments. In: Başar E, Flohr H, Haken H, Mandell AJ, ed. Synergetics of the brain. Berlin Heidelberg New York: Springer, 1983: 183-200

/4/ Ruchkin DS. Measurement of event-related potentials.: signal extraction. In: Picton TW, ed. Human event-related potentials (EEG handbook, revised series, vol. 3). Amsterdam: Elsevier, 1988

/5/ Başar E. Induced rhythmicities of the brain (in preparation)

/6/ Başar E, Gönder A, Ungan P. Important relation between EEG and brain evoked potentials. I. Resonance phenomena in subdural structures of the cat brain. Biol Cybernetics 1976; 25: 27-40
/7/ Başar E, Gönder A, Ungan P. Important relation between EEG and brain evoked potentials. II. A systems analysis of electrical signals from the human brain. Biol Cybernetics 1976; 25: 41-48
/8/ Başar E, Demir N, Gönder A, Ungan P. Combined Dynamics of EEG and evoked potentials. I. Studies of simultaneously recorded EEG-EPograms in the auditory pathway, reticular formation and hippocampus of the cat brain during the waking stage. Biol Cybernetics 1979; 34: 1-19
/9/ Başar E. EEG-Brain dynamics. Relation between EEG and brain evoked potentials. Elsevier: Amsterdam, 1980
/10/ Spekreijse H, van der Tweel LH. System analysis of linear and nonlinear processes in electrophysiology of the visual system. Proc R Neth Acad Sci C 1972; 75: 77-105
/11/ Lopes da Silva FH, van Rotterdam A, Storm van Leeuwen W, Tielen AM. Dynamic characteristics of visual evoked potentials in the dog. I. Cortical and subcortical potentials evoked by sine wave modulated light. Electroencephalogr Clin Neurophysiol 1970; 29: 246-259
/12/ Lopes da Silva FH, van Rotterdam A, Storm van Leeuwen W, Tielen AM. Dynamic characteristics of visual evoked potentials in the dog. II. Beta frequency selectivity in evoked potentials and background activity. Electroencephalogr Clin Neurophysiol 1970; 29: 260-268
/13/ Llinás RR. The intrinsic electrophysiological properties of mammalian neurons: insights into central nervous system function. Science 1989; 242: 1654-1664
/14/ Gray CM, Singer W. Stimulus-specific neuronal oscillations in the cat visual cortex: a cortical function unit. Soc Neurosci Abstr 1987; 404: 3
/15/ Gray CM, Singer W. Stimulus-specific neuronal oscillations in orientation columns of cat visual cortex. Proc Natl Acad Sci USA 1989; 86: 1698-1702
/16/ Eckhorn R, Bauer R, Jordan W, Brosch M, Kruse W, Munk M, Reitboeck HJ. Coherent oscillations: A mechanism of feature linking in the visual cortex? Biol Cybern 1988; 60: 121-130
/17/ Narici L, Pizzella V, Romani GL, Torrioli G, Traversa R, Rossini PM. Evoked alpha and mu rhythm in humans: A neuro-magnetic study. Brain Res 1990; 520: 222-231
/18/ Dettmar P, Volke HJ. Time-varying spectral analysis of single evoked brain potentials. In: Klix F, Näätänen R, Zimmer K, ed. Psychophysiological approaches to human information processing. Amsterdam: Elsevier, 1985: 225-233
/19/ Klauck U, Heinrich H, Dickhaus H. Classification of EP's with respect to EEG activity. Proceedings of the North Sea Conference for Biomedical Engineering 1990 (in press)
/20/ Başar E, ed. Chaos in Brain Function. Berlin Heidelberg New York: Springer, 1990
/21/ Mikkelsen KB, Saermark K, Lebech J, Bak C, Başar E. Selective averaging in auditory magnetic field experiments. Proceedings of the VIIth International Congress of Biomagnetism, New York, August 1989
/22/ Başar E. A study of the time and frequency characteristics of the potentials evoked in the acoustical cortex. Kybernetik 1972; 10: 61-64
/23/ Shepherd GM. Neurobiology. Oxford: Oxford University Press, 1988
/24/ Pfurtscheller G, Steffan J, Maresch H. ERD mapping and functional topography: Temporal and spatial aspects. In:

Pfurtscheller G, Lopes da Silva FH, eds. Functional brain imaging. Toronto: Hans Huber, 1988: 117-130
/25/ Başar E, Başar-Eroglu C, Röschke J, Schütt A. The EEG is a quasi-deterministic signal anticipating sensory-cognitive tasks. In: Başar E, Bullock TH, eds. Brain Dynamics. Berlin Heidelberg New York: Springer 1989: 43-71
/26/ Başar E. EEG-Dynamics and evoked potentials in sensory and cognitive processing by the brain. In: Başar E, ed. Dynamics of sensory and cognitive processing by the brain. Berlin Heidelberg New York: Springer, 1988: 30-55
/27/ Demiralp T, Başar E. (in preparation)
/28/ Başar E, Rosen B, Başar-Eroglu C, Greitschus F. The associations between 40Hz-EEG and the middle latency response of the auditory evoked potential. Int J Neurosci 1987; 33: 103-117
/29/ Freeman W. Mass action in the nervous system. New York: Academic Press, 1975
/30/ Schreiner CE. Functional topography in the primary auditory cortex of the cat. Acta Otolaryngol (in press)
/31/ Hoke M, Lehnertz K, Pantev C, Lütkenhöner, B. Spatiotemporal aspects of synergetic processes in the auditory cortex as revealed by the magnetoencephalogram. In: Başar E, Bullock TH, eds. Brain Dynamics. Berlin Heidelberg New York: Springer, 1989: 84-105
/32/ Fessard A. The role of neuronal networks in communication within the brain. In: Rosenblith WA, ed. Sensory communication. Cambridge: MIT Press, 1961: 585-606

EEG Rhythms – Event-Related Desynchronization * and Synchronization

G. Pfurtscheller

Department of Medical Informatics, Institute of Biomedical Engineering,
Graz University of Technology,
and Ludwig Boltzmann Institute for Medical Informatics,
Brockmanngasse 41, A-8010 Graz, Austria

Abstract. Event-related desynchronization (ERD) designates a short-lasting and localized amplitude attenuation of rhythms within the alpha band; event-related synchronization (ERS) describes a short-lasting amplitude enhancement. The topographical pattern of the ERD may reflect activation or excitation of cortical areas. Localized patterns of ERS probably represent inhibition of cortical areas. During a visual task, ERD is found over occipital areas and ERS over central areas close to the electrodes C_3 and C_4. Intrinsic rhythms within the alpha band can therefore be used to study excitation and inhibition or idling of cortical areas.

1. Introduction

Cell assemblies have the ability to generate rhythmic potentials or oscillations of various frequencies, whereby the alpha (8-13 Hz) and gamma (35-85 Hz) bands play a dominant role in processing sensory information and in planning motor behavior. The main difference between the two rhythms is their localization on the scalp and extent. Alpha band activity is relatively widespread, meaning that it can be picked up over large cortical areas. Gamma rhythms are localized to specific neural structures and have been recorded from the olfactory bulbs [1], the motor cortex [2], the caudate nucleus and the globus pallidus [3] and visual cortex [4], and on man by Giannitrapani [5] and DeFrance and Sheer [6], as well as by other groups [7]. Beside the strict localization, the event-related occurrence or "event-related synchronization" is characteristic for the oscillations in the gamma band. These occur very often in the form of large-amplitude bursts (eg. from the olfactory bulb during sniffing or in the visual cortex in orientation-specific columns after light stimulation). From all these reports, we can speculate that high-frequency rhythms within the gamma range reflect circumscribed cortical excitability in limited subassemblies [8].

In contrast to the rhythms within the gamma range - also known as 40-Hz oscillations - alpha band rhythms can be found over wide areas of the scalp with a dominance over the posterior region. It is important to note that there is not only one, but rather a great variety of rhythms within the alpha band [9]. The existence of different types of alpha band rhythms may be documented by different reactivities. So, for example, the central alpha band rhythm or mu rhythm can be blocked or desynchronized with sensorimotor stimulation and voluntary movement and the occipital alpha rhythm with all types of visual input.

This reactivity of rhythms of the brain to externally or internally paced events can have a specific topographical pattern at one moment of time and a specific spatial pattern at one location on the scalp. The event-related EEG reactivity can be characterized either by an attenuation of alpha amplitudes or event-related desynchronization (ERD), or by an enhancement of amplitudes or event-related synchronization (ERS) [10, 11]. Also possible is a combination of alpha attenuation followed by an alpha enhancement [12]. This phenomenon was also described as transient alpha augmentation [13].

Excitability in limited cortical cell assemblies may be reflected, on the one hand, in oscillations or synchronized activity in the gamma range; on the other hand, they may be reflected by desynchronized alpha band rhythms. This relationship between desynchronized alpha band and synchronized gamma band rhythms is best documented at this time in the visual cortex following light stimulation: 40-Hz oscillations begin about 140 msec after stimulation [14]

* Supported by the Austrian "Fonds zur Förderung der wissenschaftlichen Forschung", Project S4902.

and the occipital alpha rhythm needs about the same time to desynchronize [12]. Therefore, it can be speculated that alpha desynchronization is a prerequisite for gamma synchronization, that both are characteristic of increased cortical excitability, and that the latter is directly linked with the encoding process.

In this paper, some types of spatial and temporal patterns of ERD and ERS within the alpha frequency range will be presented.

2. Quantification of ERD

ERD and ERS (actually a negative ERD) can be quantified in the time domain and frequency domain, whereby the latter method has a limited time resolution, but can give a good overview of reactive and non-reactive frequency components in an event-related situation. The quantification of ERD is based on a number (N) of event-related sampled EEG trials, whereby at least one second of time must elapse before the event and some seconds after the event. The event can be either internally-paced (e.g. voluntary movement) or externally-paced (e.g. stimulation) and has to be repeated N times in intervals of at least some seconds. In our experiments, we used N between 30 and 100 and as an event a visual stimulus or a voluntary finger movement. The length of event-related EEG trials varied between 4 and 7 seconds, depending on the experiment. The sampling rate was 64 Hz.

To obtain information on EEG reactivity, one-second power spectra calculated before the event (reference period) and after the event (activity period) can be compared. The difference between the two spectra indicate the most reactive alpha frequency and give a measure for the ERD (for examples, see Figs. 2 and 3). When the band power in the reference period is assumed to be 100%, the ERD can be expressed as a relative alpha band power decrease and ERS as a relative alpha band power increase.

When EEG data are band pass filtered, squared and averaged across trials, the average band power (ERD time course) is obtained for each channel. After normalization, the ERD can be expressed in percentage numbers. When multichannel EEG data - we use 29 and 30 channels - are analyzed, ERD maps can be computed.

The dynamics of the EEG on one scalp location can be documented by applying an autoregressive model to calculate non-stationary power spectra. Fig. 1 shows an example in a three-dimensional display during a 4-sec interval in a reading task. It can be seen that especially the rhythms within the alpha band demonstrate a dynamic behavior with a transient desynchronized state after the warning stimulus and again following the imperative stimulus. In

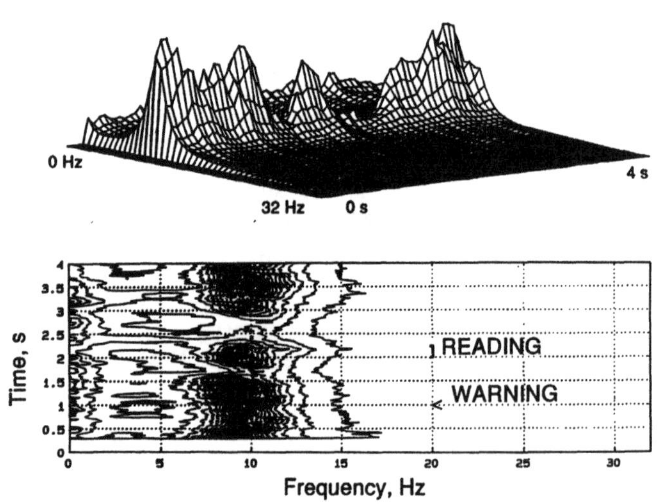

Fig. 1: Non-stationary power spectrum (upper panel) and 2-dimensional projection (lower panel) calculated from an occipital region during a reading experiment. The warning and imperative stimuli are indicated. For the calculation of the power spectrum, an autoregressive model of order 6 was used.

the two-dimensional display (Fig. 1, lower panel), the phasic amplitude changes of the alpha band rhythms are also evident and underscore the importance of studying rhythmic brain activities. From the two-dimensional display, it can be seen that, in addition to the alpha amplitudes, the frequency also changes with time. This interrelation of amplitude and frequency was studied by Kawabata during photic alpha blocking [15]. To describe the behavior of the electrical potential on one location of the scalp, all three dimensions - frequency, time and amplitude (expressed as ERD or ERS) - are needed. To study the topographical distribution of brain electrical potentials, multichannel EEG data must be analyzed.

3. Frequency band and ERD

One characteristic feature of the ERD is that the desynchronization may be limited to individual alpha components and not have any effect on other alpha components in the EEG. The existence of rhythms with similar frequencies where one is reactive and the other is not, is documented in Fig. 2. The data are from a voluntary movement experiment, where the EEG was recorded 4 sec before and 2 sec after movement onset. The 1-sec power spectra display two peaks, one of which was unchanged during movement; the other was diminished. This can be interpreted to mean that at least two rhythms are generated within the central region (bipolar EEG recording from right central region), whereby only one is reactive to finger movements. From the ERD time course (Fig. 2, lower panel), it can be seen that the desynchronization in the 12-14 Hz band started about 2 sec before movement onset and recurred thereafter. The 13.5 Hz

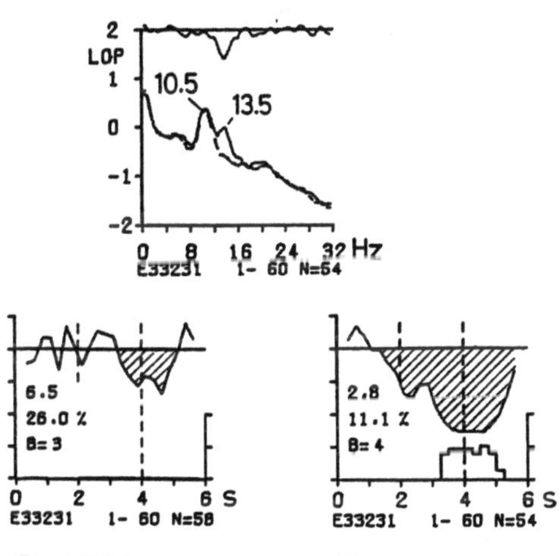

Fig. 2: Example of two logarithmic 1-sec spectra, one calculated before (full line) and one during (dotted line) hand movements. The ERD time course for 2 frequency bands are plotted below. The numbers indicate the peak frequency of the two peaks.

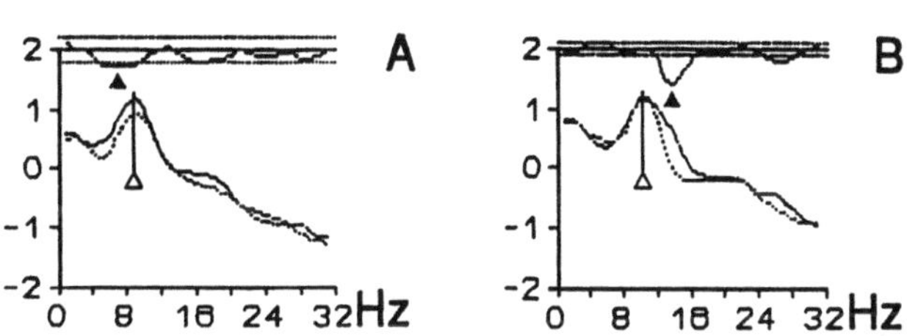

Fig. 3: Examples of 1-sec spectra from a reading experiment displaying desynchronization of lower (A) and upper (B) alpha components. Note the differences between maximal reactivity and peak frequency. Data from two subjects.

rhythm in this example can be considered to be an intrinsic sensorimotor rhythm, or mu rhythm, which is desynchronized whenever sensorimotor areas become activated during the course of planning a movement [16]. The 10.5 Hz rhythm is unaffected by movement and thus not related to sensorimotor activation.

Very often, the spectral alpha peak is broad during rest, then becomes smaller and demonstrates a small peak frequency shift during activation. This can be interpreted to mean that either upper or lower alpha components generated in underlying cortical tissue are affected by cortical activation. When lower frequency components are desynchronized, the frequency increases (Fig. 3A), and when upper frequency components are affected, the frequency decreases (Fig. 3B). However, an event-related alpha frequency shift found in scalp recordings does not necessarily mean that the frequency of alpha components is increased or decreased.

4. Cortical activation and ERD

Topographical ERD studies in the past few years have given strong evidence that there is a close relationship between topographical localization of the ERD on the scalp and excitation or activation of underlying cortical structures [11,17,18]. Typical examples are given in Fig. 4.

During voluntary finger movement, the ERD is localized to the electrodes C_3 and C_4 (Fig. 4A). This bilateral pattern during one-sided hand movement is not surprising, because blood flow measurements reveal a similar bilateral pattern [19]. The movement of one hand apparently involves not only the contralateral hemisphere but also engages the ipsilateral side.

Reading words results in an occipital localized ERD when upper alpha components are analyzed (Fig. 4B). The ERD is less localized when lower components are considered [20]. During planning of speech, wide areas of the left hemisphere, including both speech centers and the motor cortex; display ERD (Fig. 4C). In addition, the mid-precentral area can display a localized ERD, indicating perhaps the activation of the SMA. Feedback during a timing experiment results in a right parietal localized ERD (Fig. 4D), indicating that the right parietal cortex is needed when performing a time estimation task.

These results obtained in different subjects and group studies give strong evidence that the participation or involvement of a cortical area at a special moment in time in a special task can affect the generation of its intrinsic rhythmic activity and can be visualized on the scalp by a localized ERD. The correlation between anatomical substrate and the ERD pattern supports the hypothesis of area-specific intrinsic or resting rhythms as formulated by Kuhlman [16].

It is of interest that a type of preactivation or priming can also be visualized with ERD mapping. In a visual judgement task with a warning stimulus, the imperative stimulus was

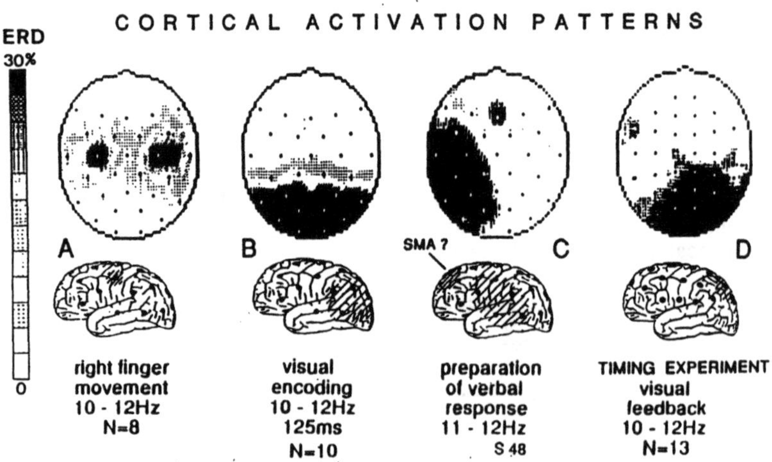

Fig. 4: ERD maps calculated during different tasks, such as voluntary finger movements (A), reading of words (B), prior to speech (C), and during visual feedback in a timing experiment (D). Each map represents a time interval of 125 msec. Maps A, B and D are grand average maps over several subjects, map C represents data from a single subject. Black coloring marks the areas with the largest alpha power decrease or ERD; analyzed frequency bands are indicated.

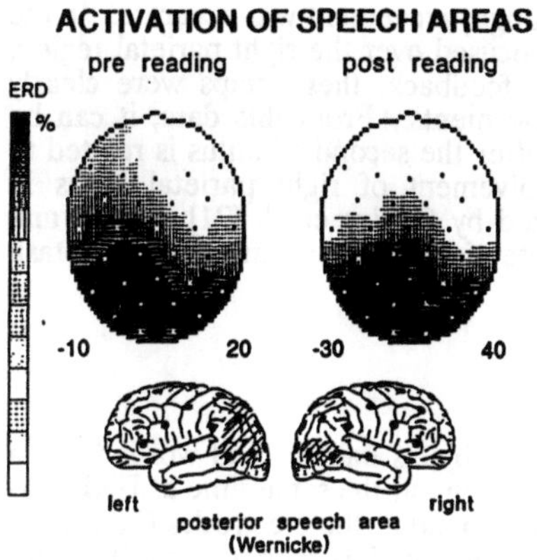

Fig. 5: Grand average ERD maps (10 subjects) calculated in intervals of 375 msec preceding the visual stimulus and during preparation for the verbal response (500 msec post-stimulus). Black coloring marks the areas with maximal alpha power decrease or ERD. Each map represents a time interval of 125 msec. Note the different scales for the maps: the range of the scale is indicated next to the map.

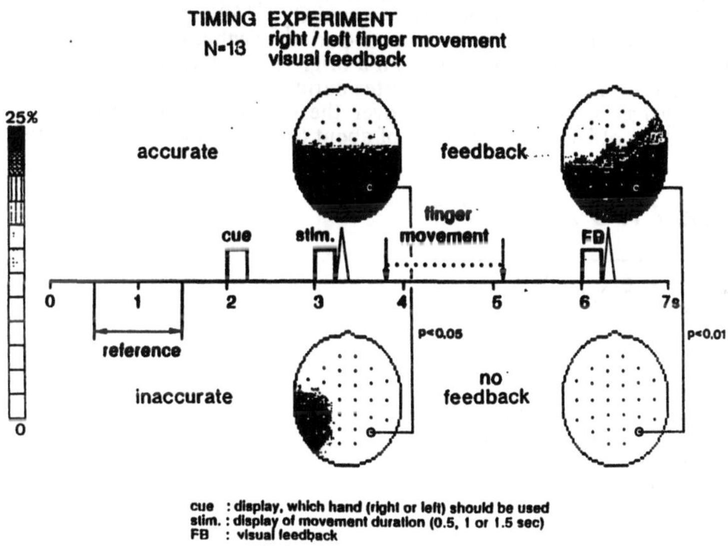

Fig. 6: Grand average ERD maps (N=13) from a movement timing experiment with a visual feedback. ERD maps calculated after the stimulus which indicated the desired time of movement (0.5 or 1.3 sec), and during visual feedback which revealed the actual duration. The ERD map after the stimulus corresponds to "accurate" and "non-accurate" movement and show a significant ($p < 0.05$, Wilcoxon text) difference in the right parietal region. Both the maps calculated as a control measure - without visual feedback - and the maps calculated with visual feedback revealed a significant ($p < 0.01$) difference in the right parietal region. Note the significant ERD difference over the right parietal region before movement and the involvement of the right parietal side during visual feedback.

followed by a verbal response [18]. The map calculated before the visual stimulus displayed the same topographical ERD pattern as the map calculated prior to the verbal response (Fig. 5).

Another example is from a timing experiment (Fig. 6). The procedure was as follows: a cue stimulus indicated the side of finger movement (left or right), and a stimulus one second later indicated the desired duration of finger movement. The movement response was followed by visual feedback indicating whether or not the duration of finger movement was accurate. In the case of accurate movements, similar maps, with an ERD focused over the right parietal region, were found after the second stimulus and during visual feedback; these maps were clearly different from the pattern found during non-accurate movements. From this data, it can be speculated that a preactivation of the right parietal area after the second stimulus is related to the performance of an accurate movement. The involvement of right parietal areas in performing accurate movements was recently demonstrated by Gevins et al. [21]. Brunia and Damen [22] reported right-sided preponderance of slow potential shifts in a time estimation task with the largest amplitude increase over the parietal areas.

5. Event-related synchronization

Analyzing the data from three different studies where the subjects had to read words from a computer-controlled display, we made an interesting observation: at the same time as ERD was maximal over the occipital lobe, a strictly localized ERS was found at the electrodes C_3 and C_4 overlying the sensorimotor cortex [20]. The results from two studies with slightly different experimental paradigms are displayed in Fig. 7. Their time courses indicate that the extrema of central ERS and occipital ERD behave coherently. This can be interpreted to mean that excited cortical structures in the visual cortex are found in parallel with inhibited cortical structures in the sensorimotor cortex during a purely visual task without any motor response.

The maps calculated during visual encoding with a bilateral ERS pattern focused to electrodes C_3 and C_4 look like mirror images of the ERD maps found during voluntary finger movement with a central localized ERD on the electrodes C_3 and C_4 [17,20].

In experiments with monkeys, it was shown that a somato-parietal rhythm (corresponding to the mu rhythm in humans) can be elicited after turning on a bright light [23]. This is probably analogous to the ERS in humans over the sensorimotor cortex during visual encoding.

Synchronized alpha activity has also been reported over the occipital region after photic stimulation [12, 13, 24]. This type of ERS had a latency of about 650 msec and was found after

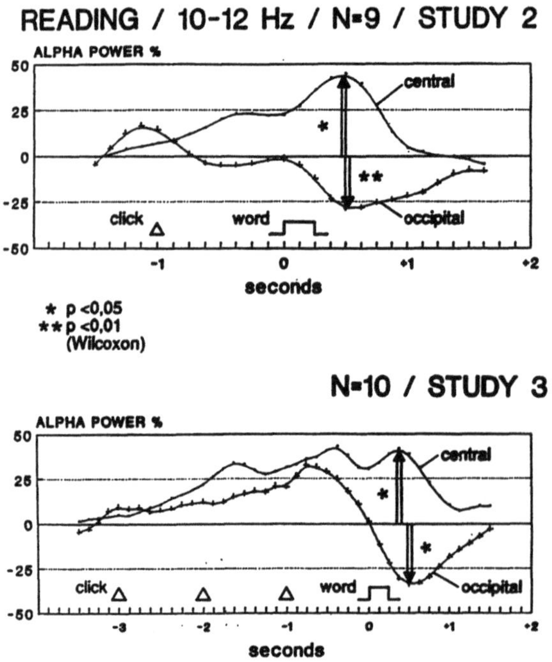

Fig. 7: Grand average ERD time courses calculated for electrodes C_4 and TPO_2 (between O_2 and T_6) in a reading task *without* any motor response. Note the contrary behavior of central and occipital rhythms during visual encoding. Data from two different experiments are displayed.

desynchronization. This means that the occipital alpha rhythm first exhibited an amplitude attenuation followed by amplitude enhancement, where the magnitude of attenuation and enhancement depended on the amplitude of the alpha rhythm at the time of stimulation.

Another type of ERS was reported in comatose patients [25]. Here, it is of interest that the ERS in the comatose state changed to ERD after clinical recovery when the level of consciousness was reached. This means that the same physical stimulus can elicit ERS or ERD, depending on the state of consciousness.

6. Conclusions

The ERD is a phasic phenomenon found in the alpha band and is different for lower and upper alpha components [26]. In general, upper alpha band ERD is more localized and short-lasting than lower alpha band ERD, and is therefore more task-specific. There is evidence from a series of experiments that ERD reflects a type of increased cortical excitability or activation. ERD is also found prior to stimulation or voluntary movement. This means that ERD is not exclusively a correlate of sensory processing or execution of a motor program, but can also indicate preactivation or priming of such cortical areas needed in a special task. ERD can be used as an indication of localized brain activation in cognitive tasks, but caution must be taken, because factors like frequency band, expectancy conditions, stimulus type, and localization of recording electrodes can influence the ERD [18]. It can be speculated that when a cortical area is not involved in processing or does not expect to process input, as e.g. the motor cortex in a purely visual task, or the visual cortex during voluntary movement, this cortical area is inhibited or in an 'idling' state, and as a result of this, displays synchronized activity. The intrinsic rhythm of the sensorimotor cortex is the mu rhythm; this mu rhythm is blocked or desynchronized over both hemispheres during one-sided movement [27] and maximally synchronized during visual encoding. It is of interest to note that the states of ERD and ERS have about the same duration and onset. This means that ERS reflects an active process in one area and is linked to the ERD observed in another cortical area. Whether areas displaying ERS reflect a decrease in metabolism and areas with ERD reflect an increase of metabolism can only be speculated and not proven, because measurement of glucose and oxygen metabolism with positron emission tomography (PET) have a poor time resolution and therefore cannot be used to study cortical activation patterns within fractions of seconds.

Acknowledgements: The author is indebted to Mag. Clara Kirschner for preparing the manuscript, to W. Mohl for his assistance with data acquisition and processing, and to Dipl.-Ing. M. Peltoranta for supplying the data for the non-stationary power spectrum in Fig. 1.

References

1. W.J. Freeman, B. Baird: Behav. Neurosci., **101**, 393 (1987)
2. J.J. Bouyer, M.F. Montaron, A. Rougeul-Buser, P. Buser: In *Rhythmic EEG Activities and Cortical Functioning*, ed. by G. Pfurtscheller, P. Buser, F.H. Lopes da Silva, H. Petsche, Developments in Neuroscience, Vol. 10 (Elsevier, Amsterdam 1980), p. 63
3. R. Galambos: In *The Central Nervous System and Behavior*, ed. by H. Brazier (Josiah Macy Jr. Foundation, New York 1958) p. 375
4. C.M. Gray, W. Singer: Proc. Natl. Acad. Sci. USA **86**, 1698 (1989)
5. D. Giannitrapani: Electroenceph. Clin. Neurophysiol. **27**, 480 (1969)
6. J. DeFrance, D.E. Sheer: In *The EEG of Mental Activities*, ed. by D. Giannitrapani, Murri (Karger, Basel 1988) p. 153
7. J.P. Banquet: Electroenceph. Clin. Neurophysiol. **35**, 143 (1973)
8. D.E. Sheer: In *Behavior and Brain Electrical Activity*, ed. by N. Burch, H.L. Altshuler (Plenum, New York 1970) p. 325
9. T. Mulholland: In *Attention in Neurophysiology*, ed. by C.R. Evans, T.B. Mulholland (Butterworth, London 1969) p. 100
10. G. Pfurtscheller, A. Aranibar: Electroenceph. Clin. Neurophysiol. **42**, 817 (1977)
11. G. Pfurtscheller, W. Klimesch: In *Topographic Brain Mapping of EEG and Evoked Potentials*, ed. by K. Maurer (Springer, Berlin 1989) p. 303
12. T. Nogawa, K. Katayama, Y. Tabata, T. Ohshio, T. Kawahara: Electroenceph. Clin. Neurophysiol. **40**, 78 (1976)

13. J. Tatsuno, J. Mori, H. Ashida, E. Maru: In *Rhythmic EEG Activities and Cortical Functioning*, ed. by G. Pfurtscheller, P. Buser, F.H. Lopes da Silva, H. Petsche, Developments in Neuroscience, Vol. 10 (Elsevier, Amsterdam 1980) p. 249
14. H.J. Reitboeck, M. Pabst, R. Eckhorn: In *Computer Simulation in Brain Science*, ed. by M.J. Cotterill (Cambridge U. Press, Cambridge 1988) p.479
15. N. Kawabata: Kybernetik **12**, 40 (1972)
16. W.N. Kuhlman: In *Rhythmic EEG Activities and Cortical Functioning*, ed. by G. Pfurtscheller, P. Buser, F.H. Lopes da Silva, H. Petsche, Developments in Neuroscience, Vol. 10 (Elsevier, Amsterdam 1980), p. 105
17. G. Pfurtscheller, A. Berghold: Electroenceph. Clin. Neurophysiol. **72**, 250 (1989)
18. W. Klimesch, G. Pfurtscheller, W. Mohl, H. Schimke: Int. J. Psychophysiol. **8**, 297 (1990)
19. P.E. Roland, L. Widen: In *Functional Brain Imaging*, ed. by G. Pfurtscheller and F.H. Lopes da Silva (Huber, Toronto 1988) p. 213.
20. G. Pfurtscheller, W. Klimesch: In *Induced Rhythms of the Brain*, ed. by E. Basar, T. Bullock (Boston, Birkhäuser 1991, in press)
21. A.S. Gevins, B.A. Cutillo, S.L. Bressler, N.H. Morgan, R.M. White, J.Illes, D.S. Greer: Electroenceph. Clin. Neurophysiol. **74**, 147 (1989)
22. C.H.M. Brunia, E.J.P. Damen: Electroenceph. Clin. Neurophysiol. **69**, 234 (1988)
23. A. Rouguel, J.J. Bouyer, L. Dedet, O. Debray: Electroenceph. Clin. Neurophysiol. **46**, 310 (1979)
24. L.K. Morrell: Electroenceph. Clin. Neurophysiol. **21**, 552 (1966)
25. G. Pfurtscheller, G. Schwarz, B. Pfurtscheller, W. List: Electroenceph. Clin. Neurophysiol. **56**, 114 (1983)
26. G. Pfurtscheller, W. Klimesch: Brain Topogr. (1990, in press)
27. G.E. Chatrian: In *Handbook of Electroencephalography and Clinical Neurophysiology*, ed. by A. Remond, Vol. 6, (Elsevier, Amsterdam 1976) p. 46.

Part VII

Rhythms in Perception

Psychological Modification and Synergetic Modelling of Perceptual Oscillations

P. Kruse, M. Stadler, and D. Strüber

Department of Psychology, University of Bremen,
W-2800 Bremen 33, Fed. Rep. of Germany

Abstract. The rapidity, ease, and temporal stability of actual genetical order formation in perception suggests a general theoretical concept in which the perceptual process is viewed as a more or less direct informational input. The aspect of autonomous order formation in perception and the question for inner-systemic tendencies which govern the emergence of relation, form, and meaning are lost from view. Following the historical line of Gestalt psychology some old phenomena are revisited which gain special importance in investigating and modelling perceptual processes on the background of self-organization theory. On the basis of spontaneous perceptual oscillations and reorganizations some theoretical and empirical aspects of investigating autonomous order formation in perception are discussed and illustrated. The idea is forwarded that these perceptual phenomena and possibly perception in general can be described and understood best by describing and understanding the underlying dynamics of the cognitive system. Taking the example of perceptual multistability the approach of synergetics is favoured as a promising tool for further investigation.

1. Introduction

> "Now, one of the main tasks which psychology has to solve consists in the discovery of those functional relations which are responsible for the occurrence and the characteristics of our experiences. We want to know not merely *what* happens in mental life but also *how* and *why* it happens. And from this point of view any number of discoveries are possible in psychology. Moreover in this respect the work of psychology seems to be at least as hard as is that of any other science. Curiously enough, one of the principal reasons for this is precisely the condition which seems at first to constitute a facilitating factor, namely, the extreme familiarity of psychological experiences as such. The reason is this: hidden facts of dependence will seldom be revealed by accident. They will be discovered when people ask questions about observed facts. And people do not tend to ask any questions about facts with which they are thoroughly acquainted; they ask questions about unusual events, about things which differ strictly from familiar facts. Since we are, at least as adults, entirely familiar with most psychological data as such, there will be few occasions on which these experiences strike us as strange, and thus evoke a questioning attitude. This, I believe, is one of the major difficulties with which psychology has been confronted up to the present time" [1].

This introductory statement of the Gestalt psychologist Wolfgang Köhler is to the point especially for the psychological investigation of perception. As a rule perceptual processes seem to be self-evident and the question why things look the way they do is too easily

Eds.: H. Haken and H.P. Koepchen

answered in a more or less trivial way. Things look as they do because they are as they are. If one understands perception as information processing one has to look for properly tuned filtermechanisms and processing algorithms [2] and if one understands perception as a direct input of relevant cues for behavioral coordination one has to look for critical variables in the optical flow [3]. In these views the possibility of autonomous order formation in the cognitive system becomes lost from sight. It was one major contribution of Gestalt psychology to demonstrate that inner-systemic tendencies generally play an important role and that understanding cognitive processes is necessarily dependent on understanding the involved intrinsic dynamics of the cognitive system itself [4]. In perception our experience is predominated by the rapidity, ease, and temporal stability of actual genetical order formation. The efficiency of the process often prevents the questioning attitude needed to further investigate the *how* and *why* of emerging order.

One way out is to look for unusual events, for things which differ strictly from familiar facts, as Köhler formulated in the statement cited at the beginning. In perception this means e.g. to look for conditions under which (i) a random or confused stimulus leads to an orderly or meaningful percept, (ii) a constant stimulus leads to a varying percept, (iii) a steadily varying stimulus leads to sudden organization or reorganization in perception [5]. In the perspective of understanding perception as a more or less direct process of informational input, phenomena of such a kind clearly have to be evaluated as unusual or even as exceptional slips of the system. In the following we recall some old and well-known perceptual phenomena in which constant stimuli lead to varying percepts and steadily varying stimulus aspects lead to sudden perceptual reorganization. The central importance of these seemingly exceptional phenomena for understanding perceptual processes will be underlined. Possibilities and problems of investigating and modelling the phenomena on the background of self-organization concepts will be dicussed and illustrated by empirical results.

2. Spontaneous oscillations in perception

The existence of spontaneous oscillations has been demonstrated for various levels of perceptual processing and has been investigated since the early days of perceptual research. One basic phenomenon is the so-called Troxler effect [6]. A spot, slightly brighter or darker than the surrounding field, is presented extra-foveally. After some time the spot fades and reappears, producing the impression of periodic fluctuation of the stimulus. Urbantschitsch [7] demonstrated the rhythmical fading of minimal auditory sensations and opened up the phenomenon to systematic investigation. There was a long controversy about the origin of the spontaneous changes in minimal sensory experience [8]. The most commonly accepted theory was to explain the phenomenon as **fluctuation of attention**. To view the rhythmical changes as generated in more central parts of the nervous system is additionally supported by a finding of Stadler and Erke [9]. They were able to show comparable rhythmical changes in the perceived size of objects.

Another phenomenon to be mentioned in this context is so-called **binocular rivalry**. When two eyes view discrepant monocular stimuli, stable single vision gives way to alternating periods of monocular dominance. When e.g. a white area is viewed with a red filter in front of one eye and a green filter in front of the other, the perceived colour of the area changes continuously as a consequence of the changing ocular dominance. Rivalry was first described more than two centuries ago [10] and was introduced to perceptual research by

Wheatstone at the beginning of the last century [11]. The spontaneous alternation is modelled in connection with the process of binocular fusion. Therefore the phenomenon also has to be regarded as centrally located [12].

The most well-known and most fascinating phenomenon of spontaneous oscillations in perception is the phenomenon of **multistability**. In a variety of physically stable stimulus situations perception changes continuously and without external triggering between two or more distinct perceptual alternatives. Such ambiguous stimulus situations can be constructed for different senses and for different levels of complexity. In visual perception multistability can be demonstrated from the basic object constitution up to the ambiguity of meaning. The phenomenon was first recognized by Porterfield [13] and the beginning of more systematic research coincides with the general starting point of modern scientific psychology which dates back to the work of Wilhelm Wundt [14]. Unless Wundt preferred peripheral explanations, the phenomenon is clearly of central origin. The most accepted theories describe the spontaneous changes between different perceptual alternatives as a result of underlying processes of saturation or adaptation [15]. Because of the quantitative similarity of the measured periodicity, sometimes a common underlying process has been suggested for the rhythmical fading of minimal sensory experiences, the fluctuation of perceived size of objects, and perceptual multistability (see, e.g., [9]).

All phenomena mentioned share a history of about a century or more of scientific recognition. Indeed, multistability is probably the most frequently investigated phenomenon in perceptual psychology. But in spite of the long and intense scientific preoccupation a satisfying theoretical integration is still lacking for all phenomena. One possible explanation for this remarkable discrepancy can be seen in the general acceptance of the information processing model of perception. As stated at the beginning, understanding perception as a process of informational input necessarily leads to a position in which any form of autonomous order formation in the perceptual system has to be evaluated as a spectacular or marginal exception. But instead of this the phenomena can be understood as a general hint at the basic autonomy of order formation in perception. There is a lot of evidence e.g. for suggesting that in principle any stimulus situation has to be regarded as ambiguous and that therefore any perception is multistable [16]. On the basis of the known empirical results concerning the divergent phenomena of spontaneous oscillation in perception, there is no convincing evidence for a common underlying mechanism. But the oscillations can be seen as an indicator of the presence of neuronal self-organization on different levels and in different parts of the perceptual system. In such a view the phenomena gain the importance of paradigmatical tools to investigate the dynamics underlying the process of perceptual order formation. Levelt e.g. formulated for the methodological importance of binocular rivalry "...that the eyes when presented with ... non-fusing patterns relinquish their secret laws of interaction" [17]. Basically the same idea was mentioned for perceptual multistability by Köhler fifty years ago: "Stationary visual percepts, a tree, a stone, or a book are as a rule extremely reticent as to the nature of the neural events which underlie their existence. We may hope to learn more about brain correlates if we turn to instances in which percept processes seem to be in a more active state. This is the case with `reversible figures´" [18]. Consequently the phenomena of spontaneous oscillation in perception are of special interest for a self-organization theory of cognition.

3. Synergetic modelling of perceptual oscillations

Self-organizing networks have already been designed to simulate binocular rivalry [12] and perceptual multistability [19, 20]. But up to now these connectionist or neural network models meet only a few aspects of the complex dynamics of the phenomena. The approach which comes closest to the different properties of perceptual multistability has been developed recently on the basis of the mathematical theory of synergetic systems as introduced by Hermann Haken [21]. Using coupled differential equations Ditzinger and Haken [22, 23] were able to simulate the temporal characteristics of perceptual multistability with excellent agreement between the mathematical data and the results of existing psycho-physical experiments. This agreement includes the phenomenon of hysteresis, the perceived changes of ambiguous patterns with two or more perceptual alternatives, and the distribution function of reversion times. For a detailed understanding and theoretical integration of the phenomenon of perceptual multistability a close cooperation between synergetic modelling and psychological investigation seems to be very promising.

The fruitfulness of a comparable cooperation has already been demonstrated by Haken, Kelso, and co-workers [24-26] investigating the emergence and change of patterns of coordination in biological movement. For the nonequilibrium phase transition between two patterns of movement coordination the existence of critical fluctuations and slowing down of relaxation time was predicted and empirically confirmed. A detailed stability analysis was performed by using the switching time between the two patterns [27] or the summation of different potentials [26] as a way to estimate the intrinsic dynamics of the process. This approach was sometimes called "phenomenological synergetics".

As mentioned before, first steps have already been taken towards a comparable analysis of the intrinsic dynamics underlying the process of perceptual multistability. But the characteristics of the phenomenon cause a number of problems for a detailed stability analysis in the sense of the approach of phenomenological synergetics. One problem is caused by the heterogeneity of perceptual multistability. There is a great variety of different ambiguous patterns on different levels of complexity. As shown by Gräser [28], the temporal characteristics of the reversion process vary considerably between the patterns. To avoid difficulties, the cooperation between synergetic modelling and empirical investigation always has to be restricted to one and the same category of perceptual multistability. Another more central problem is elicited by the fact that in perception the emergence and change of stable order is far too quick to measure the presumed nonequilibrium phase transitions themselves. The stationary solution to which the system relaxes predominates in conscious experience. In the case of multistability the system seems to pass from stability to stability without any intermediate instability. The direct demonstration of critical fluctuations or critical slowing down of relaxation time, typical for phase transitions, and the quantification of switching time as a measure of the underlying dynamics seems to be almost impossible. Concerning perceptual multistability only hysteresis can be demonstrated and measured without fundamental difficulties [29].

4. Modelling and modification of perceptual multistability

Reviewing the existing psychological results concerning perceptual multistability, ambiguous patterns which are constructed on the basis of the phenomenon of apparent

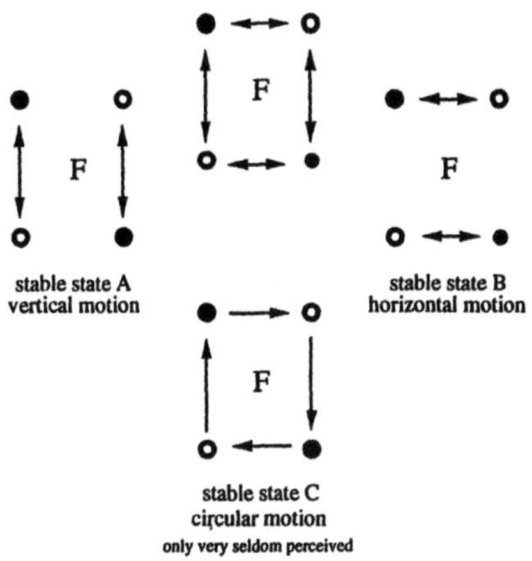

depending on the frequency of flashing lights sometimes no motion can be seen (succession or simultaneity)

Fig. 1. Stroboscopic alternative motion (SAM, F : fixation point).

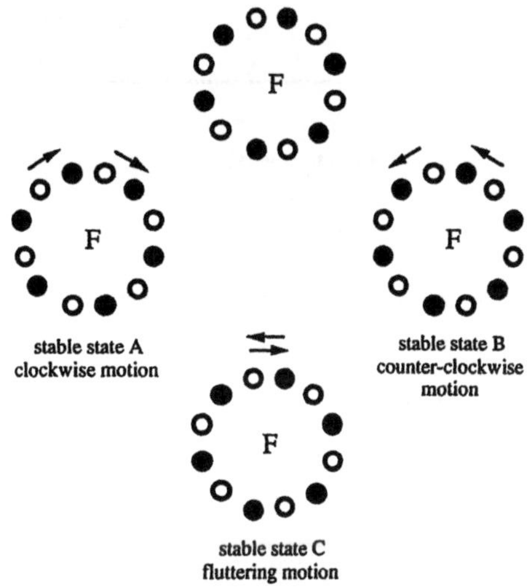

depending on the frequency of flashing lights sometimes no motion can be seen (succession or simultaneity)

Fig. 2. Circular alternative motion (CAM, F : fixation point).

motion (AM) seem to be most adequate as an empirical basis for the synergetic modelling. Ambiguous AM displays are extremely reliable in their perceptual characteristics and easily modifiable in various ways [15]. In the following, two different ambiguous AM displays are used. In the so-called stroboscopic alternative motion (SAM) four point-lights are positioned in the form of an upright rectangle. When one pair of point-lights corresponding to diagonal corners is presented in alternation with the other pair and when the speed of alternation is properly tuned, three different perceptual alternatives or paths of AM can be seen (see figure 1). The probability of the different perceptual alternatives depends on the ratio between the horizontal and vertical distances. The point of equal probability between the horizontal and the vertical motion (stable states B and A) usually is aproximately around the ratio of the golden section. Varying the ratio of horizontal and vertical distances easily allows a demonstration of the effect of hysteresis or objective set [30, 31]. The second AM display is quite similar. A larger number of point-lights are positioned on a circle. The even number of point-lights is divided into two groups by arranging direct neighbours into different groups. Again one group of point-lights is presented in alternation with the other. When the frequency of alternation is in a range which allows the perception of AM three different stable states can be perceived (see figure 2). The clockwise motion is usually predominant (stable state A). But unlike the conditions in the stroboscopic alternative motion, where the third alternative (circular motion) has only a low probability, all three alternatives are experienced by most of the perceivers without previous instruction. This second ambiguous AM display we will call circular alternative motion (CAM).

Viewing SAM and CAM a continuous change between the different perceptual alternatives is perceived. In figure 3 the oscillation between different stable states is schematically described for a bistable display and some parameters to measure the reversion process are presented. Usually the rate of apparent change (RAC), the speed with which an

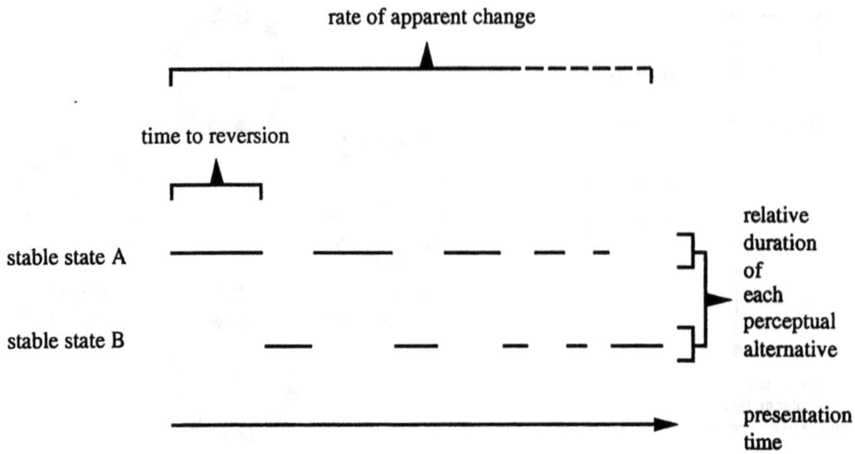

Fig. 3. Possible parameters for measuring the reversion process in multistable perception. The rate of apparent change (RAC) counts the frequency of perceived changes during a limited time interval. The relative duration counts the time each alternative is seen during a limited time interval.

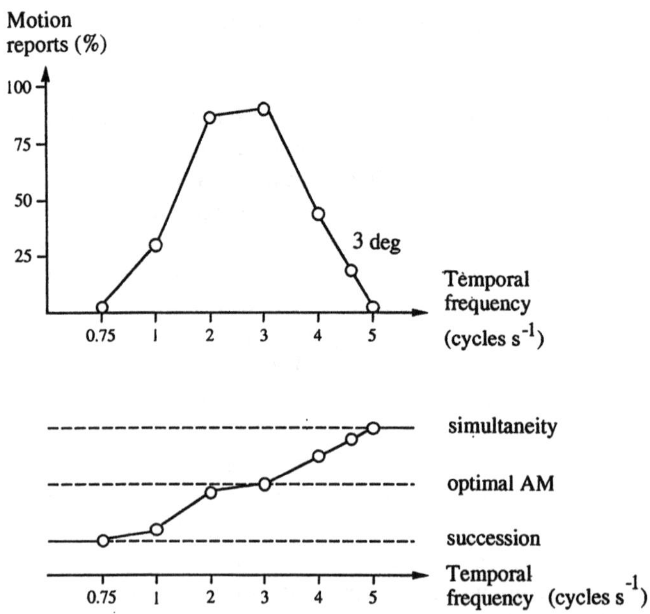

Fig. 4. AM judgements as a function of temporal frequency using two alternating point-lights (spatial separation = 3 deg of visual angle, data from [32]).

individual perceptual system shifts between the possible alternatives, varies from person to person. Besides the phase transition between different stable states a more basic discontinuity or spontaneous order formation exists concerning AM which is closely connected to the described multistability. If one uses only two point-lights spatially separated and alternating, appearence and disappearence of AM occurs (see figure 4). Starting with a very low frequency

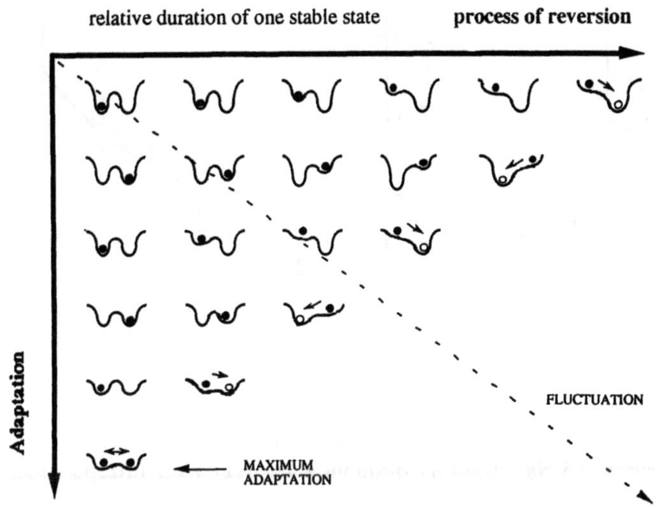

Fig. 5. Development of potential landscape during the perception of a bistable pattern. Successive shortening of the relative duration of one perceptual alternative up to the point of maximum adaptation.

of alternation the two point-lights are perceived as totally independent (succession). When the frequency is increased the clear impression of a moving spot appears (optimal AM). When the frequency is increased further the repetitive AM disappears again and the two point-lights are seen simulaneously (simultaneity). It will be an interesting question for further investigation to evaluate whether this discontinuity can be understood as a nonequilibrium phase transition too. But even in the area of optimal AM a trained observer can experience a fluctuation in the quality of the perceived motion. This fluctuation hints at an underlying process of adaptation which can be seen as responsible for the perceptual oscillation viewing SAM or CAM.

When the perceptual system has relaxed to one stable state a process of adaptation starts which changes the potential landscape of the underlying dynamics. In a bistable situation e.g. (see figure 5) the valley which describes stable state A is continuously flattening when the particle which describes the actual perception stays in this alternative. Depending on innersystemic fluctuations the particle changes to stable state B after a given time. When the adaptation process is faster than the recovery, the time to reversion becomes shorter and shorter until a point is reached where the change between the stable states is directly governed by innersystemic fluctuations. This model leads to concrete empirical expectations. During prolonged observation of an ambiguous pattern the time to reversion first is continuously shortened for all perceptual alternatives and therefore RAC increases to a point of maximum adaptation. This linear increase has been empirically confirmed in our own experiments using the SAM (see figure 6). Additionally we found that there is a strong positive correlation between stimulus frequency and RAC (see figure 7) which indicates that the adaptative effect is the stronger the higher the frequency of the alternating pairs of point-lights of the SAM. After the point of maximum adaptation the perceptual change should directly reflect the innersystemic fluctuations. Therefore it is expected that the distribution function of the

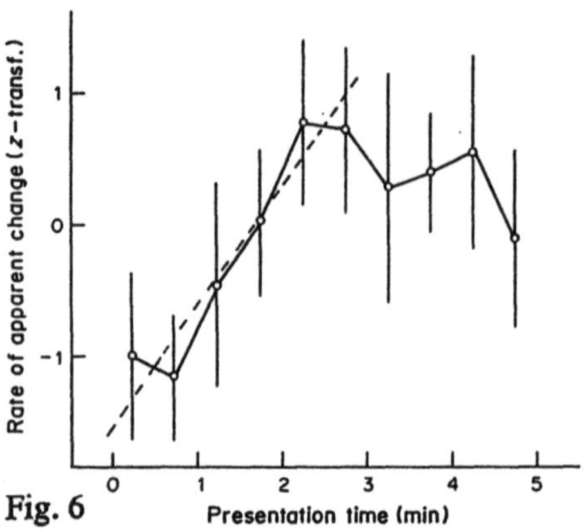

Fig. 6

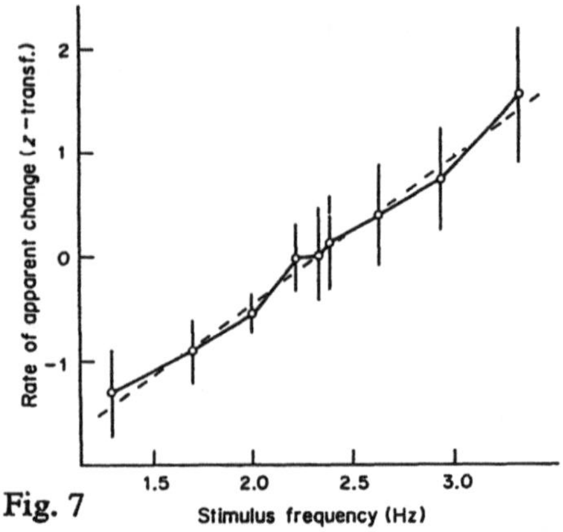

Fig. 7

Fig. 6. The increase of rate of apparent change with presentation time of SAM. Polygon of the mean values (30-sec intervals, 13 Ss, stimulus frequency 2.3 Hz) with interpersonal standard deviations. (From [33]).

Fig. 7. The increase of rate of apparent change over SAM frequency. Mean values (13 Ss) with interpersonal standard deviations (from Kruse et al. [33]).

Fig. 8

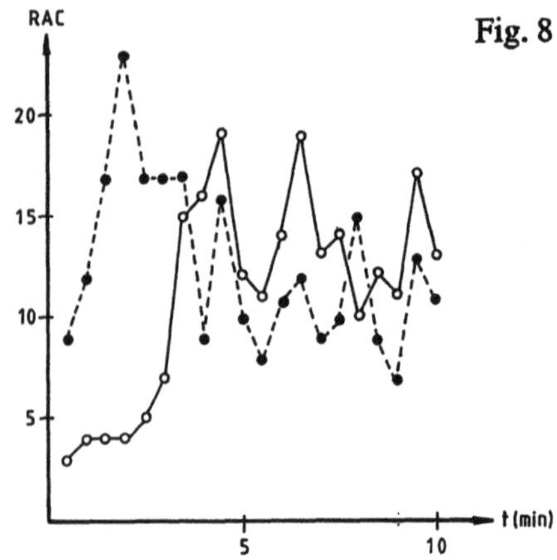

Fig. 9

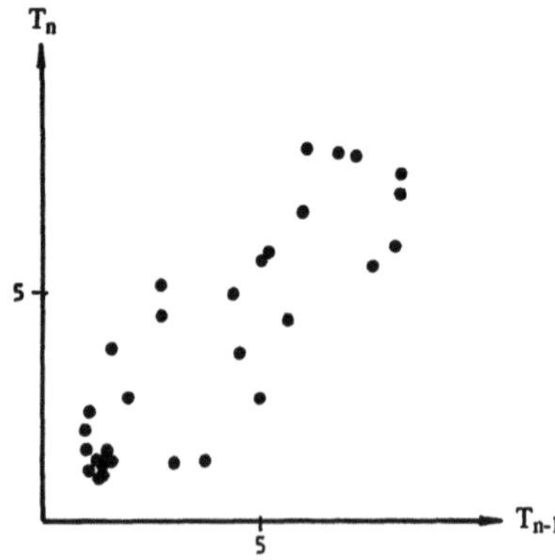

Fig. 8. Development of rate of apparent change during 10-minute presentation of SAM (filled circles and dashed lines : S_1, open circles and solid lines : S_2, 30-sec intervals).

Fig. 9. Return map of horizontal times to reversion (sec) up to the point of maximum adaptation during perception of SAM (first 5 minutes data of S_2).

different times to reversion or switching times after the point of maximum adaptation reflects a more or less stochastic process [34].

To analyse this question for each individual measurement the initial adaptation process has to be separated. In figure 8 we see the reversion process (RAC) of two subjects (Ss) viewing the SAM for 10 minutes. S_1 reaches the point of maximum adaptation after only 2 minutes whereas S_2 reaches the point only after 5 minutes. To demonstrate the characteristics of the processes one can use the presentation form of return maps [35]. In return maps a

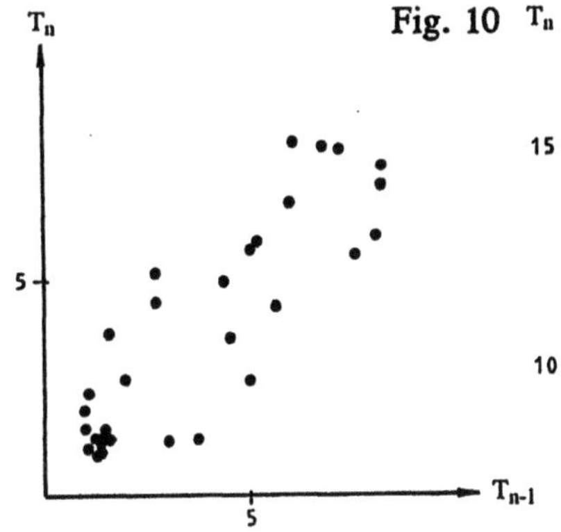

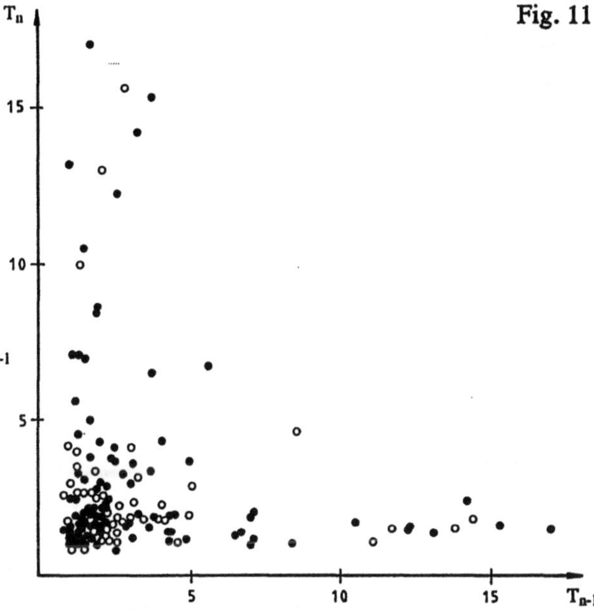

Fig. 10. Return map of horizontal times to reversion (sec) after the point of maximum adaptation during perception of SAM (last 5 minutes data of S_2).

Fig. 11. Return map of times to reversion (sec) separately plotted for the horizontal motion alternative (filled circles) and the vertical alternative (open circles) during 10-minute perception of SAM (data of S_1).

value T_n of a time series is always plotted against the previous one T_{n-1}. Time series one can use in the context of ambiguous patterns, for example, consist of the different times to reversion for one perceptual alternative. In the presentation form of the return map a linear decrease of these times - as expected during the adaptation phase - leads to a more or less distributed cloud of points with a clear diagonal orientation. A stochastic process - as expected after the point of maximum adaptation - ends up with no functional relationship of the points. Figure 9 shows the return map of the times to reversion of the long adaptation phase of S_2 only for the horizontal motion (stable state B). The data are in good agreement with the expectation. Figure 10 shows the same data after the maximum point of adaptation. In contradiction to the expectation these data clearly have a functional relationship. It is indicated that after the point of maximum adaptation there is no stochastic process. Looking at the time series itself shows that during prolonged observation and after the point of maximum adaptation some short times to reversion are always followed by one single long time. Producing the return map for the whole time series of S_1 separately for the horizontal (stable state B) and vertical alternative (stable state A) confirms the results (figure 11). Because of the short adaptation phase of S_1 separation of the initial adaptation was not necessary. Compared to a large number of time series from the same and other Ss the results proved to be typical. Obviously the characteristics of the underlying dynamics of the spontaneous oscillations of perceptual multistability need further investigation.

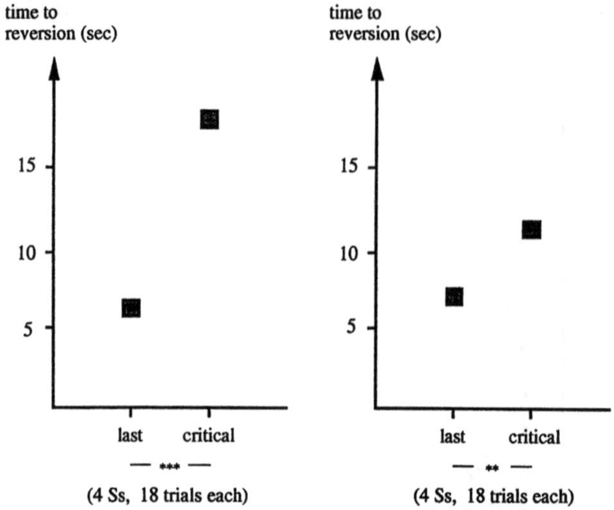

Fig. 12. Effect of contextual presentation of unambiguous AM on perception of SAM (vertical bias left, horizontal bias right). Comparison between the last time to reversion before and the first time to reversion after adding the bias.

One idea for analyzing the potential landscape of the underlying dynamics more closely is to actively change or bias the potential landscape by adding a modifying influence. In a coarse categorization the possible ways of modifying the dynamics of the reversion process in multistable perception can be subsumed under the well-known factors of perceptual organization introduced and investigated by Gestalt theory: e.g. vicinity, similarity, closure, symmetry, continuity, common motion, meaning, and context. As already described the factor of vicinity plays a major role in demonstrating hysteresis on the basis of the SAM. The determining influence of context on the perception of one SAM has been shown by Ramachandran and Anstis [36]. When one SAM is surrounded by a matrix of other SAMs, the perceived motion alternative is always the same for all ambiguous displays. When one SAM is accompanied by clearly horizontal or clearly vertical AMs, the perceived direction of the SAM is strongly biased in the direction of the unambiguous motion. The strength of the biases also depends on the distances between the ambiguous and unambiguous displays. Even when the SAM has already been presented for a long time and the possible motion alternatives are adapted, the biases are still able to force the ambiguous display in one direction. Figure 12 shows the results of an experiment where at the beginning a SAM is presented and after some time clearly horizontal or clearly vertical motions are added. The biases always cause a significant increase of the time to reversion after the intervention. Using the factors of context and common motion the potential landscape is changed in favour of one perceptual alternative. Another example demonstrates the influence of the factor of similarity. In the ambiguous display of the CAM a bias is introduced by adding a frame around one of the alternating point-lights. After every cycle of alternation the frame is moved one point-light in the counter-clockwise direction. As a result a frame is perceived which clearly moves counter-clockwise. In principle the CAM is unaffected by this. The point-lights of the CAM still can elicit the described three motion alternatives. But to see the clockwise or the fluttering motion, the perceptual system has to accept a change of the shape

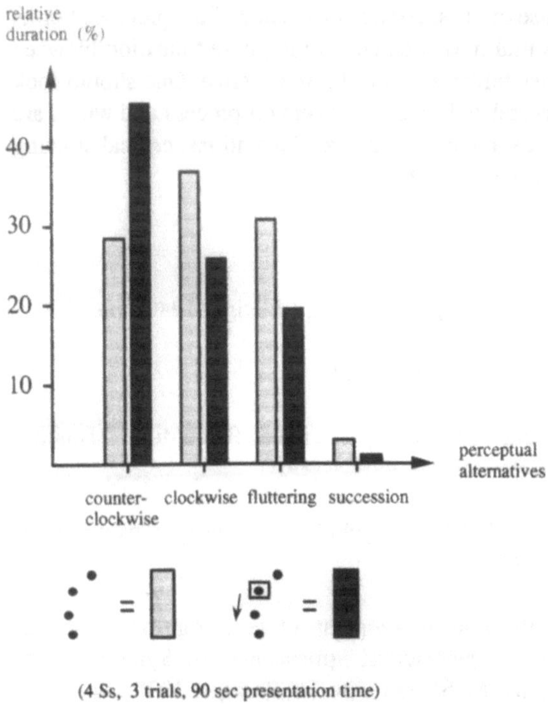

Fig. 13. Effect of bias of similarity or figural identity on perception of CAM (unbiased relative duration : left bars, biased relative duration : right bars).

of one part of the pattern. In figure 13 the effect of this bias of similarity is shown. Even introducing a bias by adding meaning in the form of little arrows instead of the point-lights is able to significantly change the probability of the perceptual alternatives of the CAM. When arrows are used which point to the counter-clockwise direction the relative duration of this motion alternative is enhanced compared to the unbiased CAM. The same can easily be done with the SAM. A fascinating manifoldness of possibilities of gradually biasing the ambiguous AM displays can be created.

5. Conclusions

Spontaneous oscillations in perception underline the aspect of autonomous order formation in cognition. Phenomena like the Troxler effect, binocular rivalry, or perceptual multistability gain the importance of paradigmatical tools for investigating the intrinsic dynamics of perceptual processes on various levels of complexity. A direct cooperation between synergetic modelling and psychological investigation promises important insights in the process of cognitive order formation. Concerning perception the use of ambiguous apparent motion displays seems to be especially suitable. For synergetic modelling of multistable perception a more detailed psycho-physical investigation of the characteristics of the reversion process is needed. To perform a detailed stability analysis of the intrinsic dynamics of the reversion process, the possibilities of biasing the different perceptual alternatives of the ambiguous patterns may be a fruitful approach. Because of the speed and efficiency of the process of order formation in perception, a direct behavioural measurement of the phase transition between different perceptual alternatives in multistability seems to be impossible. One should look for physiological parameters which are correlated with the reversion process and which are fast enough to allow, for example, the measurement of critical fluctuations, critical slowing down of relaxation time, or the measurement of switching time.

Reference notes

[1] W. Köhler: Dynamics in Psychology, (New York, Liveright 1940), pp.5f

[2] D. Marr: Vision, (San Francisco, Freeman 1982)

[3] J.J. Gibson: The Ecological Approach to Visual Perception, (Boston, Houghton Mifflin 1979)

[4] See e.g. W. Köhler: The Task of Gestalt Psychology, (Princeton, New Jersey, Princeton University Press 1969)

[5] M. Stadler, P. Kruse: The self-organization perspective in cognition research: historical remarks and new experimental approaches, in Synergetics of Cognition, ed. by H. Haken and M. Stadler, (Berlin, Springer 1990), pp. 32-52

[6] D. Troxler: Ophthal. Bibliothek. 2, 51-53 (1804)

[7] V. Urbantschitsch: Centralbl. med. Wiss. 13, 625-628 (1875)

[8] J.P. Guilford: Amer. J. Psychol. 38, 534-538 (1927)

[9] M. Stadler, H. Erke: Vision Res. 8, 1081-1092 (1968)

[10] E.-F. Dutour: Math. Phys. Pres. dDvers Sav. 3, 514-530 (1760)

[11] C. Wheatstone: Phil. Trans. Royal Soc. London 128, 371-394 (1838)

[12] R. Blake: Psychol. Rev. 96, 143-167 (1989)

[13] W. Porterfield: A Treatize on the Eye, the Manner and Phenomena of Vision, (Edinburgh, Hamilton and Balfour 1795)

[14] W. Wundt: Philosoph. Stud. 14, 1-118 (1898)

[15] P. Kruse: Delfin 11, 35-57 (1988)

[16] P. Kruse, M. Stadler: Stability and instability in cognitive systems: multistability, suggestion and psychosomatic interaction, in Synergetics of Cognition, ed. by H. Haken, M. Stadler (Berlin, Springer 1990)

[17] W.J.M. Levelt: On Binocular Rivalry, (The Hague, Mounton 1968), p.2

[18] See [1], pp. 67f

[19] A.H. Kawamoto, J.A. Anderson: Acta Psychol. 59, 35-65 (1985)

[20] P.K.Kienker, T.J. Sejnowski, G.E. Hinton, L.E. Schumacher. Perception 15, 197-216 (1986)

[21] H. Haken: Synergetics. An Introduction, (Berlin, Springer 1977)

[22] T. Ditzinger, H. Haken: Biol. Cybern. 61, 279-287 (1989)

[23] T. Ditzinger, H. Haken: Biol. Cybern. (in press)

[24] H. Haken, J.A.S. Kels, H. Bunz: Biol. Cybern. 51, 347-356 (1985)

[25] J.A.S. Kelso, G. Schöner: Springer Proc. Phys. 19, 224 (1987)

[26] J.A.S.Kelso: Phase transitions: foundations of behavior, in Synergetics of Cognition, ed. by H. Haken, M. Stadler (Berlin, Springer 1990)

[27] See e.g. G. Schöner, H. Haken, J.A.S. Kelso: Biol. Cybern. 53, 247-257 (1986)

[28] H. Gräser: Spontane Reversionsprozesse in der Figuralwahrnehmung, (Doct. Thesis, Trier 1977)

[29] T. Poston, I. Stewart: Behav. Sci. 23, 318-334 (1978)

[30] W. Metzger: Psychologie, (Dresden, Steinkopff 1941)

[31] H.S. Hock, J.A.S. Kelso, G. Schöner: Bistability, hysteresis, and loss of temporal stability in the perceptual organization of apparent motion (in prep.)

[32] D. Finlay, T. Caelli: Perception 8, 595-602 (1979)

[33] P. Kruse, M. Stadler, T. Wehner: Vision Res. 26, 327-335 (1986)

[34] A. Borsellino, F. Carlini, M. Riani, M. T. Tuccio, A. DeMarco, P. Penengo, A. Trabucco: Perception 11, 263-273 (1982)

[35] G. Schöner: Dynamic pattern theory of perceptual organization: An example from apparent movement (in prep.)

[36] V.S. Ramachandran, S.M. Anstis. Perception 14, 135-143 (1985)

Part VIII

Aspects of Systems Theory

Spatio-Temporal EEG Patterns

R. Friedrich, A. Fuchs, and H. Haken

Institut für Theoretische Physik und Synergetik,
Universität Stuttgart, Pfaffenwaldring 57/IV,
W-7000 Stuttgart 80, Fed. Rep. of Germany

We present recent results obtained from a synergetic analysis of spatio-temporal EEG patterns.

1 Introduction

Systems far from equilibrium exhibit the spontaneous emergence of spatial, temporal, and spatio-temporal patterns by the mechanism of self-organization [1], [2], [3]. Recently, considerable interest has been addressed to the study of systems which form coherent spatial structures undergoing a complex and chaotic temporal evolution. Examples may be found e.g. in the fields of hydrodynamic instabilities or chemical reactions. It has been recognized that similar phenomena can also be observed in biological systems. The present paper is devoted to the examination of spatio-temporal patterns connected with the electrical activity of the brain measured in the form of electroencephalograms (EEG).

In the case of brain waves, research interest has predominantly been devoted to the properties of the time signal obtained by measuring the potential difference between two locations on the scalp. As a basic problem the question has been raised whether the underlying dynamics is related with a low dimensional chaotic attractor. This question has been investigated on the basis of measuring the metric properties of strange attractors like correlation dimensions or Lyapunov exponents [7], [9], [10]. Apparently, the fact that low dimensional temporal behaviour in a complex system like the brain usually exhibits coherent spatial patterns has not yet been sufficiently recognized in the literature dealing with dynamical properties of the EEG. In the neurophysiological literature, however, it is well established that under certain circumstances the EEG patterns appear to be spatially well-ordered [11].

The observation that the spatio-temporal behaviour of the EEG exhibits coherent spatial patterns which undergo a complex temporal evolution has led us to perform a 'macroscopic treatment' of these patterns in close analogy to pattern formation in nonequilibrium systems [4], [5], [6]. We have analysed several data sets provided by D. Lehmann. So far we have focused our interest on the investigation of α-waves, which are detectable on the scalp of a person at rest keeping his eyes closed, as well as brain waves occurring during an epileptic seizure. We remind the reader that the α-rhythm exhibits more or less pronounced regular oscillations with

Springer Series in Synergetics, Vol. 55 **Rhythms in Physiological Systems**
Eds.: H. Haken and H.P. Koepchen

a frequency of about 10 Hz, whose intensity is irregularly modulated. The oscillations are associated with wavelike spatial behaviour (see the discussion below). If the person carries out some activity, such as opening his eyes, the α-waves are blocked and a disorganization of the wave patterns takes place. The spatio-temporal patterns arising during an epileptic seizure are quite different although they share the property of being spatially coherent.

The present article summarizes recent results in the synergetic analysis of spatio-temporal EEG patterns. In the next section we shall outline our approach in some mathematical detail. Then we shall derive a model for the case of epileptic EEGs which accounts for the dynamical and spatial properties. The last section is devoted to α-waves. We discuss several models and introduce quantities which allow a classification of the spatial characteristics of the waves and yield considerable insights into their dynamical properties.

2 Analysing spatio-temporal EEG patterns

The basic ideas underlying the present analysis of brain waves are taken from synergetics [1], [2], [3]. Let us mention the following main points:

a) The EEG signals are generated by the cooperative behaviour of neural activity. Thus it is necessary to analyse the spatio-temporal behaviour which is obtained by measuring the signals at a mesh of grid points uniformly distributed over the scalp.

b) It is well-known from the examination of synergetic systems that only a few degrees of freedom may govern the dynamical behaviour of a complex system. These degrees of freedom may be intimately related to spatial patterns of the system.

An analysis of spatio-temporal patterns therefore aims at detecting the relevant degrees of freedom and the corresponding spatial patterns. Let us explain our approach in some mathematical detail. The starting point is the assumption that the dynamics of the system, which is described by a state vector $\mathbf{q}(\mathbf{r}, t)$, is governed by a nonlinear evolution equation of the type

$$\dot{\mathbf{q}}(\mathbf{r}, t) = \mathbf{N}[\mathbf{q}(\mathbf{r}, t), \nabla, \sigma] + fluctuations. \tag{1}$$

It has been well established on the basis of the synergetic theory of self-organization [1], [2], [3] that the state vector $\mathbf{q}(\mathbf{r}, t)$ of a nonequilibrium system close to a critical point has the form

$$\mathbf{q}(\mathbf{r}, t) = \sum_i u_i(t)\mathbf{v}_i^u(\mathbf{r}) + \sum_i s_i[u_j(t)]\mathbf{v}_i^s(\mathbf{r}). \tag{2}$$

The main point of the representation of the state vector $\mathbf{q}$ is the fact that the temporal evolution is solely determined by the *order parameters* $u_i(t)$:

$$s_i(t) = s_i[u_j(t)]. \tag{3}$$

Usually, only a few order parameters $u_i(t)$ are highly excited. Their dynamics is determined by a (low dimensional) dynamical system, the *order parameter equations*:

$$\dot{u}_i(t) = p_i[u_j(t)] + \mathit{fluctuations}. \tag{4}$$

Equations (2)-(4) represent the mathematical formulation of the *slaving principle* of synergetics. This principle lies at the heart of our understanding of the formation of spatio-temporal patterns in complex systems.

The close analogy of brain waves with spatio-temporal patterns in synergetic systems suggests an analysis of EEG patterns based on the synergetic theory of self-organization. An analysis of spatio-temporal patterns has to determine the spatial modes $\mathbf{v}_i^u(\mathbf{r})$ and the dynamics underlying the temporal evolution of their amplitudes. A systematic approach for the case of nonequilibium phase transitions has been devised in [3]. In the present situation the dynamics seems to contain chaotic elements. In that case a systematic analysis is not attainable at present. Nevertheless we shall pursue a 'pragmatic' approach which is based on the decomposition of the patterns using the Karhunen-Loève method. Let us briefly discuss this method (see e.g. [3]). The potentials $V_j(t)$ measured at grid points $\mathbf{x}_j$ are lumped together to a vector $\mathbf{V}(t)$

$$\{V_j(t)\} = \mathbf{V}(t). \tag{5}$$

In the next step the correlation matrix M_{jk} is determined (T denotes the length of the time signal):

$$M_{jk} = \frac{1}{T}\int_0^T V_j(\tau)V_k(\tau)d\tau. \tag{6}$$

The vector $\mathbf{V}(t)$ is then represented as a superposition

$$\mathbf{V}(t) = \sum_\alpha \eta_\alpha(t)\, \mathbf{v}_\alpha, \tag{7}$$

where the vectors $\mathbf{v}_\alpha$ are the normalized eigenvectors of the correlation matrix M:

$$M\mathbf{v}_\alpha = \lambda_\alpha \mathbf{v}_\alpha. \tag{8}$$

The amplitudes $\eta_\alpha(t)$ of the expansion are determined by the scalar product

$$\eta_\alpha(t) = \mathbf{v}_\alpha(t) \cdot \mathbf{V}(t). \tag{9}$$

The eigenvalues λ_α turn out to be given by

$$\lambda_\alpha = \frac{1}{T}\int_0^T \eta_\alpha(\tau)^2 d\tau, \tag{10}$$

i.e. they are a measure of the mean intensity of the contribution of the mode $\mathbf{v}_\alpha$ to the total pattern. Thus it is tempting to neglect modes with small values of the eigenvalues λ_α especially if there is a wide gap in the spectrum of the eigenvalues separating the state space into a highly excited part and a remainder. This separation is similar to the representation of the state vector if the modes contained in the remainder are enslaved by the highly excited modes. In the case where the highly excited part is spanned by only a few modes a low dimensional reconstruction of the dynamics becomes attainable.

The second step of the analysis has to deal with the dynamics, i.e. to establish the order parameter equations on the basis of the low dimensional representation of the dynamics. It is assumed that the vector field $\mathbf{p}[\mathbf{u}]$ can be approximated by low order polynomials in $u_i(t)$:

$$p_i[\mathbf{u}] = \sum_j \Lambda_{i;j} u_j + \sum_{j,k} \alpha_{i;j,k} u_j u_k + \sum_{j,k,l} \alpha_{i;j,k,l} u_j u_k u_l + ... \tag{11}$$

This is possible close to instability. A reconstruction of the dynamics then has to determine the coefficients $\Lambda_{ij}, \alpha_{i;j,k}, ...$ from the experimentally derived time series of the amplitudes $u_i(t)$. This is a complicated task due to the fact that the number of unknown coefficients turns out to be large even for only a few order parameters $u_i(t)$. However, two tools facilitate the modeling of the dynamics. First, symmetry considerations restrict the class of vector fields which are expected to describe the dynamics. If the evolution equation is invariant with respect to a group of symmetry operations g the order parameter equations have to be equivariant with respect to representations $D(g)$ of the symmetry group restricted to the subspace spanned by the unstable modes $v_\alpha(\mathbf{r})$:

$$\mathbf{p}[D(g)\mathbf{u}(t)] = D(g)\mathbf{p}[\mathbf{u}(t)]. \tag{12}$$

The main simplification introduced by the symmetry concept is due to the fact that the equivariance condition reduces the number of the above unknown coefficients considerably. This rather general result is based on the Hilbert-Weyl theorem, according to which the polynomial vector field $\mathbf{p}[\mathbf{u}]$ is a superposition of a few generating polynomials $\mathbf{p}_\mu[\mathbf{u}]$:

$$\mathbf{p}[\mathbf{u}] = \sum_\mu a_\mu[I_\nu]\mathbf{p}^\mu[\mathbf{u}]. \tag{13}$$

Here the quantities a_μ are functions of the set of generating polynomial invariants I_ν. A truncation of the polynomials at low orders yields a vector field $\mathbf{p}[\mathbf{u}]$ which depends only on a few coefficients. We further note that the set of generating invariants I_ν obey a closed set of evolution equations, which can be derived from the order parameter equations:

$$\frac{d}{dt} I_\nu(t) = H_\nu[I_1, I_2, ..., I_{\nu_0}] \qquad \nu = 1, ..., \nu_0. \tag{14}$$

For an extended discussion of the technical details we refer the reader to [12].

The second tool that facilitates the modeling of the dynamics is the notion of *normal forms* (see [13], [14]). Normal form theory has successfully been used to classify a great variety of instabilities with higher dimensional unstable subspaces. One may expect to identify experimentally observed phase portraits with the phase portraits of a suitable normal form.

3 Analysis of an EEG of an epileptic seizure

Figure 1 shows an example of the temporal evolution of the EEG potential derived during an epileptic seizure. The measurements taken at 25 different points on the scalp correspond to the individual boxes of fig.1. The potentials were measured with respect to one reference electrode. It is remarkable that the time signal seems to be rather coherent and nearly time periodic. Figure 2 exhibits a three dimensional reconstruction of the attractor from the time signal $H(t)$ derived between two points. The three dimensional space is spanned by $H(t), \dot{H}(t), \ddot{H}(t)$. The reconstruction yields a characteristic geometry of the attractor and indicates that the dynamics is in fact low dimensional. An estimate of the correlation dimension [8] yields a value between two and three. At this point we would like to emphasize that a more accurate determination of the dimension is practically impossible due to the apparent nonstationarity of the time signal. We mention that similar results have already been obtained using data derived from a different source by Babloyantz and Destexhe [7], who give a correlation dimension of 2.05 ± 0.09.

Figure 3 exhibits the temporal evolution of the spatial patterns during one 'cycle'. The patterns consist of two regions with different polarity. In the first time segment the pattern exhibits small oscillations around a mean orientation (patterns 1-10), which have their counterpart in the small oscillations ('spikes') of the time signals. Then the orientation of the pattern changes by about 180^0, subsequently performing small oscillations. A second reversal drives the system back to a pattern

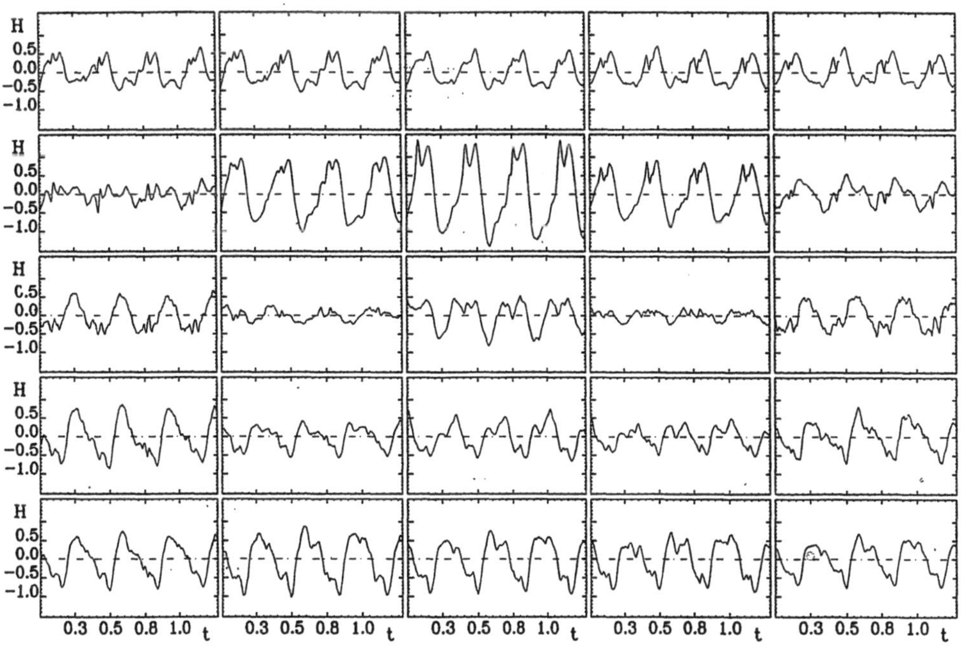

Fig. 1: Time signals taken at 25 positions (epileptic seizure).

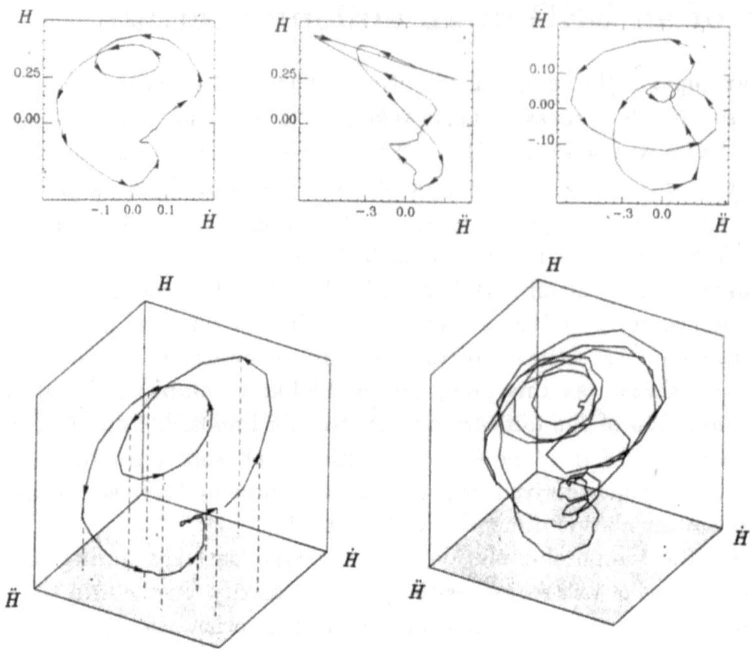

Fig. 2: Reconstruction of the attractor from the time signal between two different points on the skull (epileptic seizure).

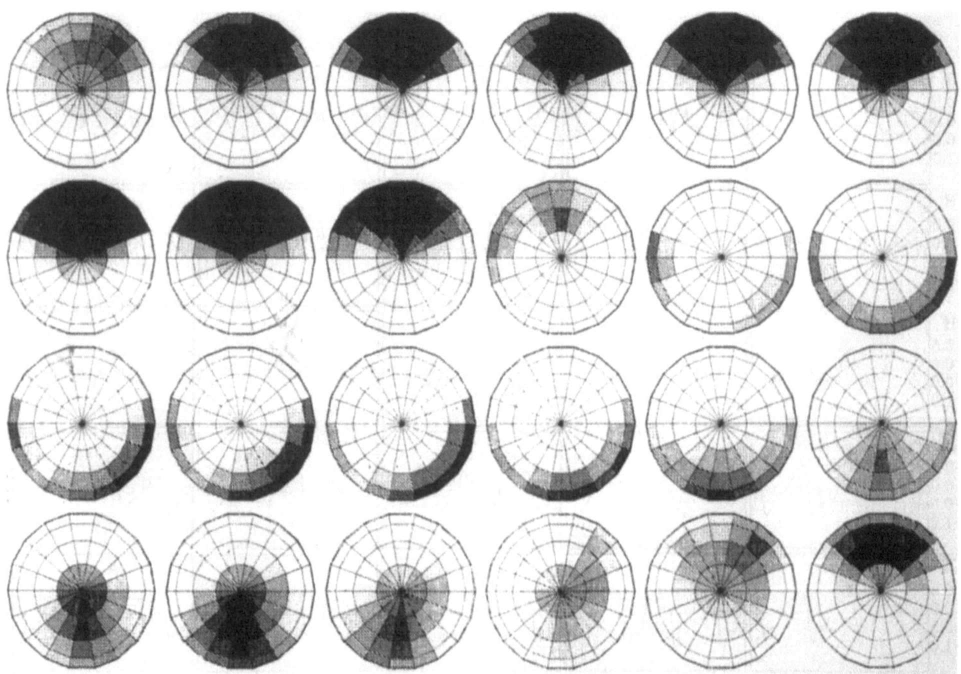

Fig. 3: Spatio-temporal patterns for epileptic seizure. The sequence corresponds to one 'cycle'.

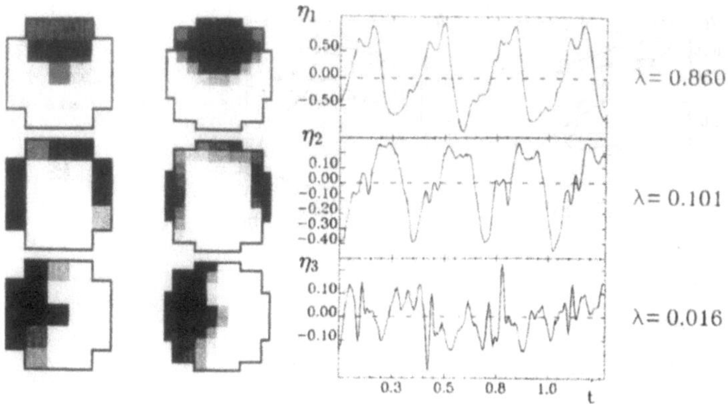

Fig. 4: Spatial modes $\mathbf{v}_i$ obtained from a Karhunen-Loève expansion (epileptic seizure).

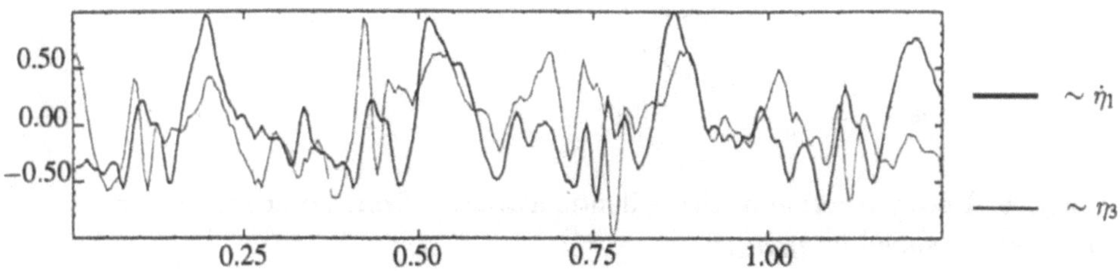

Fig. 5: Time signals of $\dot{\eta}_1$ (bold) and η_3 (light).

which is roughly similar to the initial one. The reversal of the polarity corresponds to the 'relaxation wave' (large oscillation) in the time signal $H(t)$. This significant spatial behaviour indicates the coherence of neuronal activity and shows that the single neuronal subunits are enslaved by the collective behaviour.

In order to analyse the behaviour we have performed a Karhunen-Loève decomposition of the spatial patterns. The results are exhibited in fig. 4. As it turns out, the first three modes cover about 97 per cent of the total potential $V(t)$. The contributions of higher modes are not only negligible in the time mean, as follows from the Karhunen-Loève method, but also at each time instant, as a reconstruction of the patterns from the first three modes shows. From the point of symmetry, the essential structure is provided by the modes connected with the amplitudes η_1 and η_3 (compare fig. 4), whereas the mode connected with η_2 is of a different kind of symmetry. (In terms of Bessel functions, the modes 1 and 3 correspond to m = 1, whereas the mode 2 corresponds, roughly, to m = 2.) A closer inspection of the experimental data reveals that η_3 is proportional to the time derivative $\dot{\eta}_1$. This important result is demonstrated in fig. 5.

It is now evident that these experimental observations suggest that one should reconstruct the attractor by means of the amplitude η_1 and its time derivatives. Since we may embed the signal H(t) in a three dimensional space (see fig. 2) and since the estimation of the correlation dimension yields a value smaller than three we may reconstruct the attractor in the space spanned by $\eta_1, \dot{\eta}_1, \ddot{\eta}_1$. This reconstruction

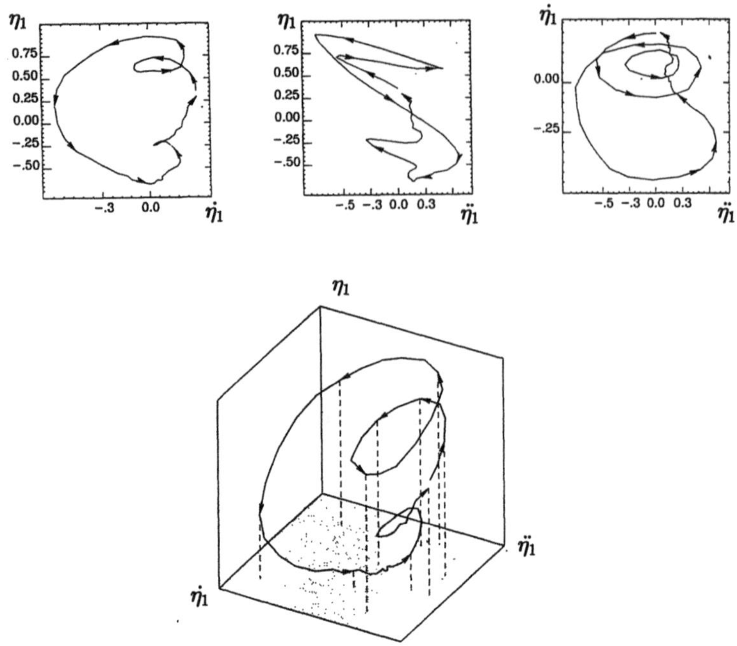

Fig. 6: Reconstruction of the epileptic attractor from experimental time series in the space spanned by $\eta_1, \dot\eta_1, \ddot\eta_1$.

is exhibited in fig. 6. It appears to be similar to the one obtained from a single time signal H(t). However, the present reconstruction yields a different interpretation: Since the amplitude η_1 is the amplitude of the most dominant Karhunen-Loève mode $\mathbf{v}_1$ and from η_1 we obtain by time differentiation the amplitude of the mode $\mathbf{v}_3$, the representation contains the information of the temporal and the spatial properties of the EEG waves during an epileptic seizure.

Let us now examine the temporal dynamics of the amplitude η_1. The geometry of the phase space suggests that the dynamics is generated by an attractor of the Shilnikov-type [14]. This attractor is organized around a fixed point with a stable one dimensional manifold and a transverse two dimensional unstable manifold on which the trajectories are spiraling away eventually approaching the fixed point again along the stable manifold. We remind the reader that a class of dynamical systems exhibiting Shilnikov type attractors has been discussed by Arnéodo, Coullet, and Spiegel [15]. This class of dynamical systems arises as a normal form of a codimension three instability whose linear behaviour is governed by

$$\dot A_1 = A_2, \tag{15}$$

$$\dot A_2 = A_3, \tag{16}$$

$$\dot A_3 = -\mu_0 A_1 - \mu_1 A_2 - \mu_3 A_3, \tag{17}$$

where A_i denotes the amplitudes of the three modes. The normal form, which takes into account the nonlinear interaction among the modes, reads

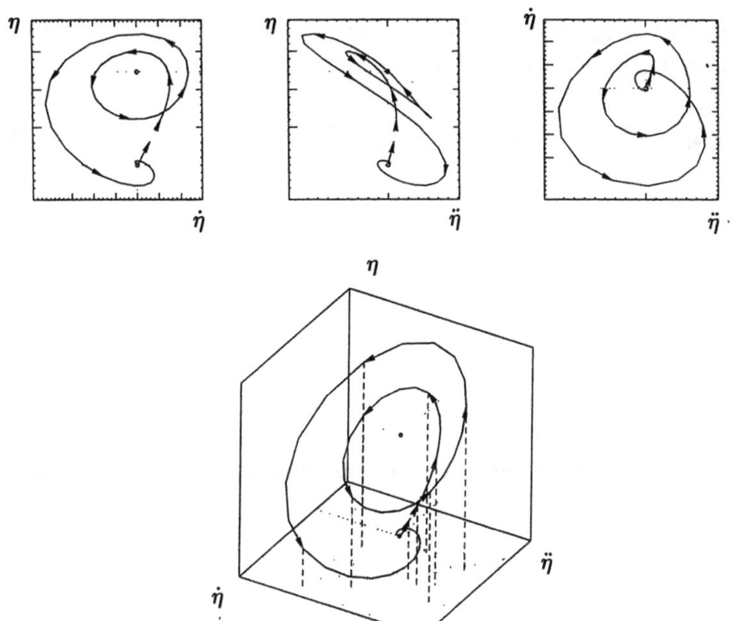

Fig. 7: Reconstruction of the attractor for epileptic EEG as obtained from the model equation (19).

$$\begin{aligned}\frac{d^3}{dt^3}\eta(t) &+ \mu_2\frac{d^2}{dt^2}\eta(t) + \mu_1\frac{d}{dt}\eta(t) + \mu_0\eta(t) \\ &= k_1\eta^3 + k_2\eta^2\dot\eta + k_3\eta^2\ddot\eta + k_4\eta\dot\eta^2 + k_5\dot\eta^3 + k_6\eta\ddot\eta^2, \qquad (18)\end{aligned}$$

where μ_0, μ_1, μ_2, and $k_1, ..., k_6$ are constant coefficients, which are determined by the specific properties of the system. The normal form holds for systems which possess the additional symmetry $\eta \to -\eta$. It is well-known that this normal form generates attractors of Shilnikov type.

We use a similar differential equation to model the dynamics of the amplitude $\eta_1(t)$. We mention that it is possible to optimize the values of the coefficients in this normal form in such a way that the distance between the experimentally obtained trajectories and the solutions of the differential equation is minimized. Figure 7 shows that a rough approximation based on the differential equation

$$\frac{d^3}{dt^3}\eta(t) + \mu_2\frac{d^2}{dt^2}\eta(t) + \mu_1\frac{d}{dt}\eta(t) + \mu_0\eta(t) = c - b\eta(t)^2 - a\eta(t)^3 \qquad (19)$$

already yields an accurate representation of the attractor. We may expect that a still better agreement can be found when other terms are included.

Let us explain why this class of dynamical systems applies to the case of epileptic brain waves. First, we have experimentally obtained the relation

$$\frac{d}{dt}\eta_1(t) = \eta_3(t). \qquad (20)$$

This serves as a first evolution equation. A further evolution equation is assumed to hold for the amplitude $\eta_3(t)$, which we may write in the form

$$\frac{d^2}{dt^2}\eta_3(t) = -\mu_2\frac{d}{dt}\eta_3(t) - \mu_1\eta_3(t) - \mu_0\eta_1(t) + f[\eta_1(t), \eta_3(t), \dot{\eta}_3(t)]. \qquad (21)$$

Here $f[\eta_1(t), \eta_3(t), \dot{\eta}_3(t)]$ is a nonlinear function of the amplitudes $\eta_1(t), \eta_3(t), \dot{\eta}_3(t)$. Using a second order differential equation allows an oscillatory behaviour of the amplitude $\eta_3(t)$ in the case where the coupling to the amplitude $\eta_1(t)$ is neglected. That means that the mode $\mathbf{v_3}$ can perform a standing wave motion. Eliminating the amplitude $\eta_3(t)$ and approximating the nonlinear function $f[\eta_1, \dot{\eta}_1, \ddot{\eta}_1]$ in a suitable way we end up with the evolution equation (19) for the amplitude $\eta_1(t)$ which, as we have seen, yields a reasonable approximation of the dynamics. For the sake of completeness we mention that the function $f[\eta_1(t), \dot{\eta}_1(t), \ddot{\eta}_1(t)]$ may depend on higher derivatives of $\eta_1(t)$, provided the values of $\mu_i, i = 0, ..., 2$ are small. That means that the system is close to a codimension three instability and is essentially described by the above normal form.

Let us finally indicate that the main characteristics of the spatio-temporal behaviour of epileptic brain waves is now reproduced in terms of the modes with the amplitudes $\eta_1(t)$ and $\eta_3(t)$ together with the dynamical system governing their time evolution. The azimuthal spatial structures of the two modes roughly have the form

$$\begin{aligned} v_1(r,\phi) &\sim \sin(\phi+\alpha), \\ v_3(r,\phi) &\sim \cos(\phi+\alpha). \end{aligned} \qquad (22)$$

The pattern is a superposition of these two modes

$$V(\phi,t) \sim \eta_1(t)\sin(\phi+\alpha) + \eta_3(t)\cos(\phi+\alpha). \qquad (23)$$

(We map the patterns onto a circular disk and introduce the cylindrical coordinates r and ϕ.) The first, dominant contribution mainly describes the observed reversals of the pattern ('relaxation wave') whereas the second contribution accounts for the rotational motion of the pattern ('spikes'). The dynamics accurately describes the temporal evolution. A successful reconstruction of the epileptic EEG wave on the macroscopic level of description has been achieved. Let us finally mention that we have analysed a second data set. For that case similar results have been obtained, indicating that the main characteristics of the epileptic waves we have described above are not singular.

4 Analysis of α-waves

We have analysed α-EEG time series along similar lines. Partial results have already been published in [5]. In comparison to the dynamics of the epileptic EEG patterns the temporal evolution is much more involved. In fact, the estimations of the fractal dimension performed by Babloyantz et al. [7], as well as our own, yield a dimension of larger than six, at least. We want to point out that one should take into account the possibility that the dynamics is not low dimensional, since the number of data points, which have to be used to estimate such relatively high fractal dimensions (see [16]), is not sufficient. Additionally, one should bear in mind that the dynamical behaviour may be strongly influenced by fluctuations, i.e. dynamical processes varying on a faster time scale than the time scale of the α-rhythm. Furthermore, the dynamics cannot be expected to be stationary over a period sufficiently long so that the mathematical construct 'attractor' may not be an adequate tool for describing the dynamics. However, certainly time segments exist during which deterministic elements show up in the spatio-temporal dynamics. As ordered structures they are generated by self-organization and are therefore expected to obey low dimensional dynamics.

Let us first phenomenologically describe the spatio-temporal behaviour of α-waves, which is found by an inspection of the evolution of the waves. The most prominent features of the α-wave patterns are shown in fig. 8. The patterns consist, roughly speaking, of two regions of opposite polarity. It is possible to find temporal sequences were the dynamics undergoes a rotating wave motion (either clockwise or counterclockwise), or a standing wave motion. The complicated temporal evolution is manifested in an aperiodic appearance and an aperiodically varying length of the three sequences. Furthermore, there are rarely occurring sequences during which the motion cannot be attributed to one of these three types of motion. The phenomenological features can be quantitatively examined by means of an analysis based on a Karhunen-Loève expansion. The obtained modes and the time series of their amplitudes are exhibited in fig. 9. The spatial structure of the modes is striking, especially with respect to their variation in azimuthal direction, which is apparently given by

$$\begin{aligned} v_1 &\sim v_1(r)\cos(\phi+\beta), \\ v_2 &\sim v_2(r)\sin(\phi+\beta), \\ v_3 &\sim v_3(r), \\ v_4 &\sim v_4(r)\cos(2\phi+\gamma), \\ v_5 &\sim v_5(r)\sin(2\phi+\gamma). \end{aligned} \tag{24}$$

(We map the patterns onto a circular disk and introduce the cylindrical coordinates r, ϕ.) It is convenient to introduce the following complex amplitudes:

$$\xi_0(t) = \eta_3(t) \qquad ,\xi_1(t) = \eta_1(t) + i\eta_2(t) \qquad ,\xi_2(t) = \eta_4(t) + i\eta_5(t). \tag{25}$$

These amplitudes belong to complex modes with an azimuthal dependence characterized by $e^{im\phi}$. Furthermore, we may decompose $\xi_1(t)$ and $\xi_2(t)$ into moduli and phases $\psi_1(t)$ and $\psi_2(t)$. Then the azimuthal variation of the pattern is given by

$$\begin{aligned} V(r,\phi,t) &= \xi_0(t)v_0(r) + v_1(r)|\xi_1(t)|\sin(\phi+\psi_1(t)) \\ &+ v_2(r)|\xi_2(t)|\sin(2\phi+\psi_2(t)). \end{aligned} \tag{26}$$

(We have assumed that the radial dependence of $v_1(r)$ and $v_2(r)$ as well as $v_4(r)$ and $v_5(r)$ is similar). The result is depicted in fig. 10. Especially the behaviour of the

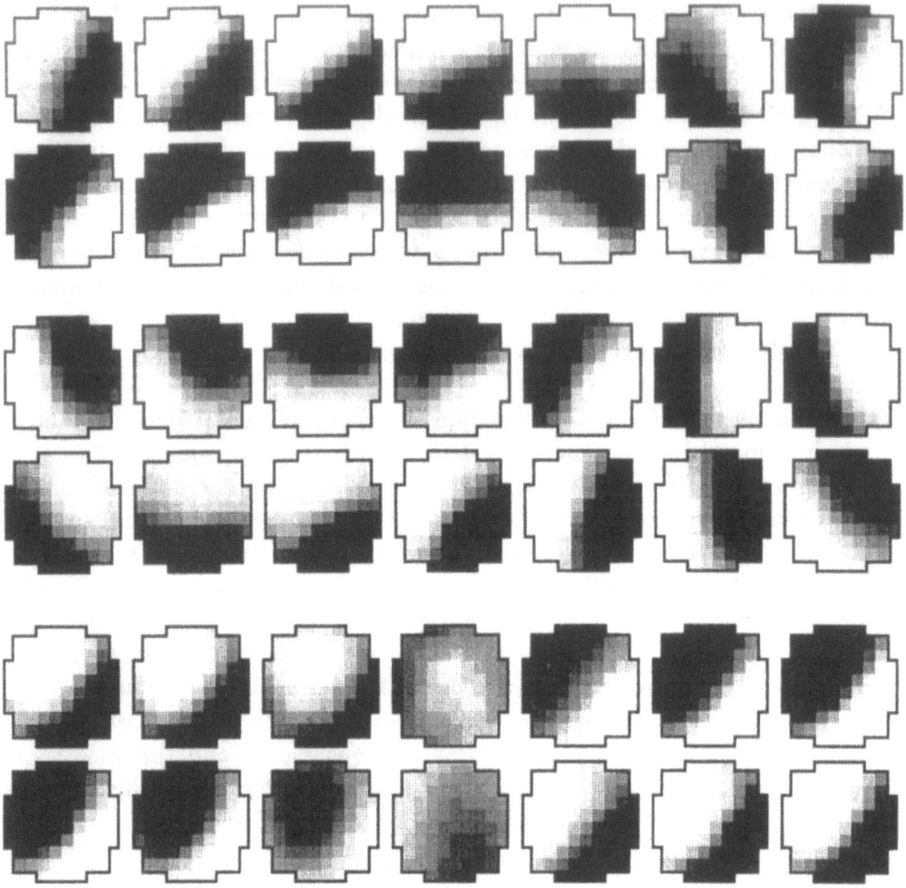

Fig. 8: Typical sequences occurring in the temporal evolution of α-waves:
(a) clockwise rotating wave motion;
(b) counterclockwise rotating wave motion;
(c) standing wave motion.

phases is interesting: the phases exhibit a kind of diffusion process. However, it has not yet been possible to decide whether this process is generated by noise (i.e. degrees of freedom varying on a much faster time scale), like Brownian motion, whether it is a result of chaotic dynamics, or whether it is due to a change of the state of the 'system'. In the following we shall report on efforts to model the spatio-temporal patterns of the α-waves. This modelling aims at an identification of elements of deterministic dynamics in the α-waves. Although the accuracy of the models are less evident as compared to the case of epileptic waves they nevertheless contribute significantly to our understanding of the spatio-temporal behaviour of α-waves.

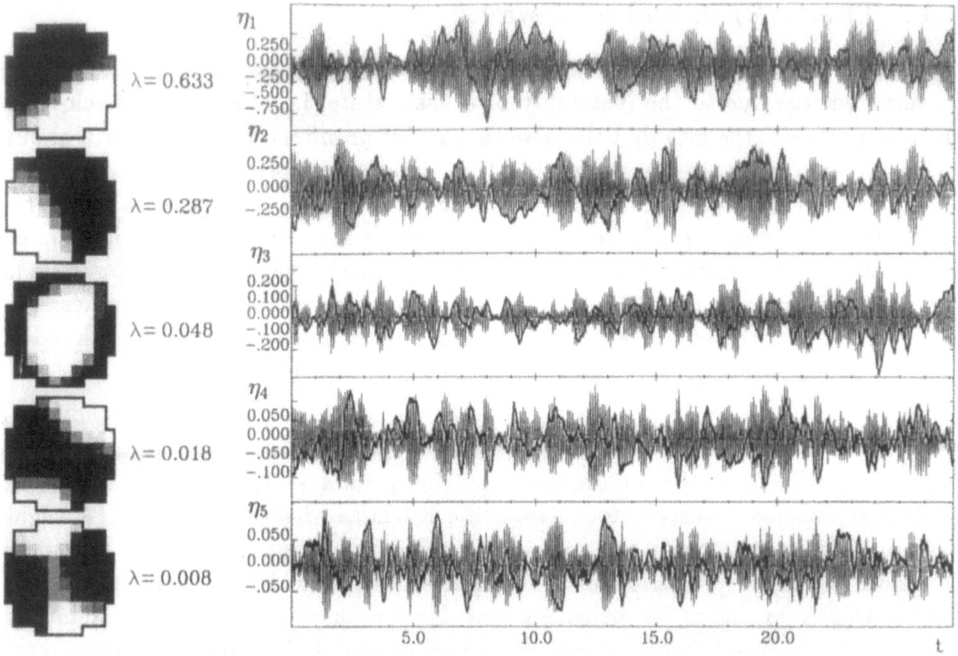

Fig. 9: Results of Karhunen-Loève expansion for the case of α-waves.

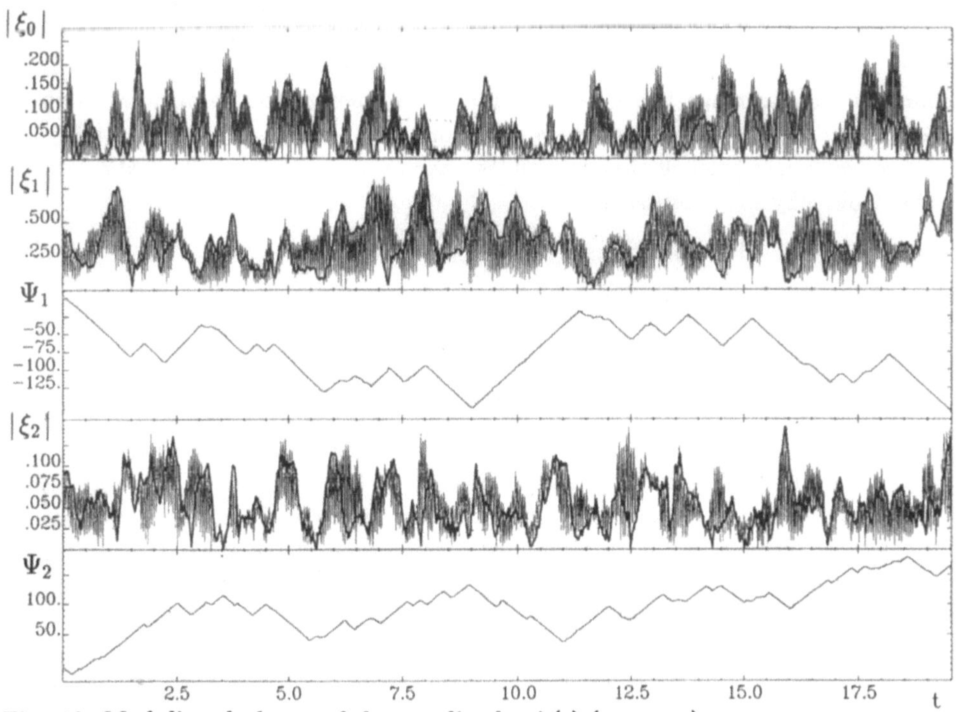

Fig. 10: Moduli and phases of the amplitudes $\xi_i(t)$ (α-waves).

4.1 The O(2)-model of α-waves

In order to model the α-waves along the lines of synergetics we assume that the patterns emerge due to the instability of a basic state, i.e. the system is close to criticality. Due to the fact that the α-waves exhibit pronounced oscillations of about 10 Hz one is further led to assume that this instability is oscillatory. Assumptions on symmetries allow one to make more detailed statements. The patterns are represented on a circular disk. Therefore, it is natural to assume that the basic state as well as the whole system exhibits the symmetry of the circular disk; i.e. the system is invariant with respect to rotations in the azimuthal direction as well as an inversion $\phi \rightarrow -\phi$ (O(2)-symmetry). From this assumption it immediately follows that the normal modes of the system exhibit the following spatial form:

$$v(r,\phi) = v_{jm}(r)\exp(im\phi), \qquad m = 0, \pm 1, \pm 2, ... \tag{27}$$

It is remarkable that the assumption of O(2)-symmetry already yields the same form of the normal modes as found from the Karhunen-Loève expansion. (A slight asymmetry is manifested in the values of the eigenvalues, which differ for two pairs of modes $\mathbf{v}_1$ and $\mathbf{v}_2$, $\mathbf{v}_4$ and $\mathbf{v}_5$. The eigenvalues should be equal in the case of perfect O(2)-symmetry.) We point out that this fact justifies our assumption on the underlying symmetry, since a strong distortion of the SO(2)-symmetry would show up in a different azimuthal spatial dependence. The assumption of an oscillatory instability in the presence of O(2)-symmetry has to be completed by a specification of the number of unstable modes, i.e. the number of order parameters.

To start with the simplest case let us first assume that the instability is induced by the modes with $m = \pm 1$, i.e. we take into account only the two most dominant modes of the Karhunen-Loève expansion. Let us denote the amplitudes of these modes by ξ_1, ξ_{-1}. (We remind the reader that $\xi_1(t)$ and $\xi_{-1}(t)$ are essentially the positive frequency parts of the amplitudes defined in eq. (25) .) Thus, the pattern is determined by the expression

$$V(\mathbf{r},t) = \xi_1(t)v_1(r)e^{i\phi} + \xi_{-1}(t)v_{-1}(r)e^{-i\phi} + c.c. \tag{28}$$

The order parameter equation takes the form (up to cubic terms):

$$\begin{aligned} \dot{\xi}_1(t) &= \xi_1(t)[\lambda_1 - A(|\xi_1(t)|^2 + |\xi_{-1}(t)|^2) - B|\xi_{-1}(t)|^2], \\ \dot{\xi}_{-1}(t) &= \xi_{-1}(t)[\lambda_1 - A(|\xi_1(t)|^2 + |\xi_{-1}(t)|^2) - B|\xi_1(t)|^2]. \end{aligned} \tag{29}$$

The structure of the order parameter equations is solely determined by the assumption of an oscillatory instability in the presence of O(2)-symmetry. Specific properties of the system are contained in the complex values A, B of the coefficients of the nonlinear terms as well as the complex coefficient λ_1 of the linear terms. Due to symmetry the order parameter equations have to be equivariant with respect to the transformations (invariance with respect to the transformations $\phi \rightarrow \phi + \beta$)

$$\xi_m \rightarrow \xi_m e^{im\beta}, \qquad \xi_{-m} \rightarrow \xi_{-m} e^{-im\beta} \tag{30}$$

and (invariance with respect to the transformation $\phi \rightarrow -\phi$)

$$\xi_m(t) \to \xi_{-m}(t), \qquad \xi_{-m}(t) \to \xi_m(t). \tag{31}$$

A third type of equivariance, namely

$$\xi_m(t) \to \xi_m(t)e^{i\omega t}, \qquad \xi_{-m}(t) \to \xi_{-m}(t)e^{i\omega t}, \tag{32}$$

is obtained on the basis of a rotating wave approximation, or, in mathematical terms, on the basis of a transformation to Birkhoff normal form. We mention that the transformations hold for modes with arbitrary values of m.

As is well known there are two different types of solutions. The first type is a standing wave:

$$\xi_1(t) = r_s e^{i\omega_s t}, \qquad \xi_{-1}(t) = r_s e^{i\omega_s t + \phi_0}. \tag{33}$$

The corresponding pattern has the form

$$V(r, \phi, t) = \sin(\omega_s t)\sin(\phi + \alpha)\mathbf{v}(r). \tag{34}$$

The orientation of the pattern is not fixed due to symmetry. The second type of solution is given by the following expressions

$$\begin{aligned} \xi_1(t) &= re^{i\omega_r t}, \qquad \xi_{-1}(t) = 0, \\ \xi_{-1}(t) &= re^{i\omega_r t}, \qquad \xi_1(t) = 0. \end{aligned} \tag{35}$$

They describe rotating waves which may rotate in either the positive or negative ϕ-direction:

$$V(r, \phi, t) \sim \sin(\phi \mp \omega_r t)\mathbf{v}(r) \tag{36}$$

In the case that the real part of the coefficient A is positive, which we assume for simplicity, the stability of the two types of waves is determined by the real part of B. For $B_r > 0$ rotating waves are stable, for $B_r < 0$ the standing wave is stable. Thus this simple model already yields a description of the experimentally observed rotating wave and standing wave behaviour. However, it cannot account for the transitions between left and right rotating waves and standing waves.

The question arises how the complex transitional behaviour of the α-waves is generated. It is tempting to look for the origin of this complex dynamical behaviour in the interaction of the modes with $m = \pm 1$ with the modes with $m = 0, m = \pm 2$ since they contribute significantly to the total pattern (see the results of the Karhunen-Loeve expansion). In a previous paper [5] we have demonstrated that a deterministic dynamical system can be established on the basis of the assumption of O(2)-symmetry which qualitatively reproduces the characteristic behaviour of the phases. The assumption that the patterns are generated by an oscillatory instability of a basic state in the presence of O(2)-symmetry induced by modes with $m = 0, m = \pm 1, m = \pm 2$ already fixes the structure of the order parameter equations. They read (up to third order terms in the nonlinearity):

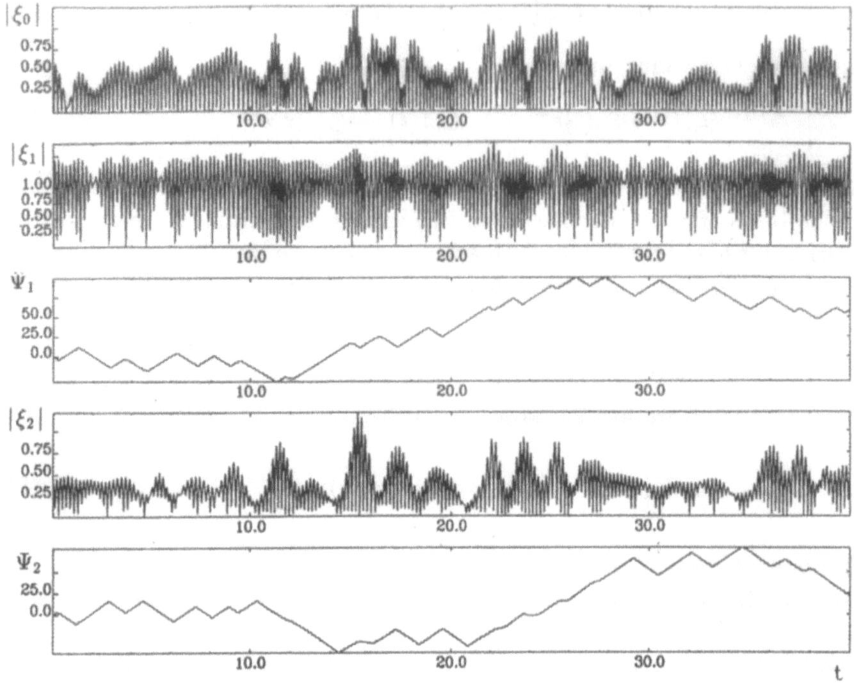

Fig. 11: Moduli and phases of the amplitudes $\xi_i(t)$ obtained from the O(2) model of α-waves.

$$
\begin{aligned}
\dot{\xi}_0 &= \xi_0 \left\{\lambda_0 + \alpha_0(|\xi_0|^2, |\xi_1|^2, |\xi_{-1}|^2, |\xi_2|^2, |\xi_{-2}|^2)\right\} + C_1 \left\{\xi_2 \xi_{-1} \xi_1^* + \xi_{-2} \xi_1 \xi_{-1}^*\right\} \\
&\quad + C_2 \left\{\xi_{-1} \xi_{-1} \xi_{-2}^* + \xi_1 \xi_1 \xi_2^*\right\} + C_3\, \xi_1 \xi_{-1} \xi_0^* + C_4\, \xi_2 \xi_{-2} \xi_0^*, \\
\dot{\xi}_1 &= \xi_1 \left\{\lambda_1 + \alpha_1(|\xi_0|^2, |\xi_1|^2, |\xi_{-1}|^2, |\xi_2|^2, |\xi_{-2}|^2)\right\} + A_1\, \xi_0 \xi_2 \xi_1^* + A_2\, \xi_{-1} \xi_2 \xi_0^* \\
&\quad + A_3\, \xi_0 \xi_{-1} \xi_{-2}^* + A_4\, \xi_0 \xi_0 \xi_{-1}^* + A_5\, \xi_2 \xi_{-2} \xi_{-1}^*, \\
\dot{\xi}_2 &= \xi_2 \left\{\lambda_2 + \alpha_2(|\xi_0|^2, |\xi_1|^2, |\xi_{-1}|^2, |\xi_2|^2, |\xi_{-2}|^2)\right\} + B_1\, \xi_0 \xi_1 \xi_{-1}^* + B_2\, \xi_1 \xi_1 \xi_0^* \\
&\quad + B_3\, \xi_0 \xi_0 \xi_{-2}^* + B_4\, \xi_1 \xi_{-1} \xi_{-2}^* \quad .
\end{aligned}
\tag{37}
$$

(The differential equations for the amplitudes $\xi_{-m}(t)$ are obtained from the above equations by replacing m with $-m$.) The evolution equation is obtained by applying the same invariance conditions as discussed in the case of the instability induced by modes with $m = 1$. The values of the various coefficients in the equations are not yet fixed. In a complete macroscopic treatment they should be fitted to the experimental data. This task is not done at present. However, a suitable choice of the values of the coefficients allows one to generate a dynamical behaviour which closely resembles the actually observed one. A numerical solution for a suitable choice of coefficients is exhibited in fig. 11. The reader may note that especially the 'random walk' of the phases is reproduced.

4.2 The O(3) model of α-waves

The previous model has basically used the assumption of O(2)-symmetry of the basic state. Since the EEG waves are detected on the surface of a half-sphere it is legimitate to consider a model with spherical symmetry. For a justification we refer to the following facts:

a) Assuming that the α-waves arise as an instability of a spherically symmetric state the normal modes are the spherical harmonics $Y_l^m(\theta, \phi)$. Figure 12 shows that the Karhunen-Loève basis may be interpreted as spherical harmonics with l=2 with different values of m. This interpretation is rather appealing since it answers the question why all modes of this group have to be considered as order parameters. It also determines that there are no modes of higher order in the radial direction.

b) There are time segments where seemingly complicated patterns represented on the circular disk have a simple interpretation in a spherical geometry. An example is exhibited in fig. 13.

Therefore, we shall establish a model based on the assumption that a spherically symmetric state becomes oscillatory unstable with respect to a group of modes with $l = 2$. Furthermore we mention that the results below obtained for a complete sphere may, under certain circumstances, extend to the case of a half-sphere.

Due to symmetry the order parameter equations have to be equivariant with respect to the transformations (we denote the positive frequency part of the amplitudes of the modes $Y_2^m(\theta, \phi)$ by $\xi_m(t)$):

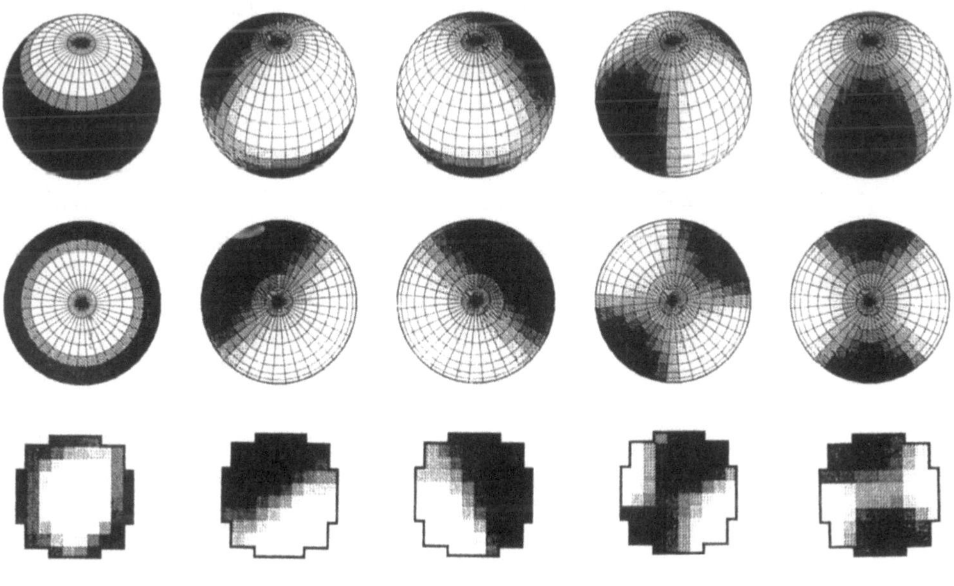

Y_2^0 $Re\{Y_2^1\}$ $Im\{Y_2^1\}$ $Re\{Y_2^2\}$ $Im\{Y_2^2\}$

Fig. 12: Interpretation of the Karhunen-Loève basis of α-waves in terms of spherical harmonics $Y_l^m(\theta, \phi)$.

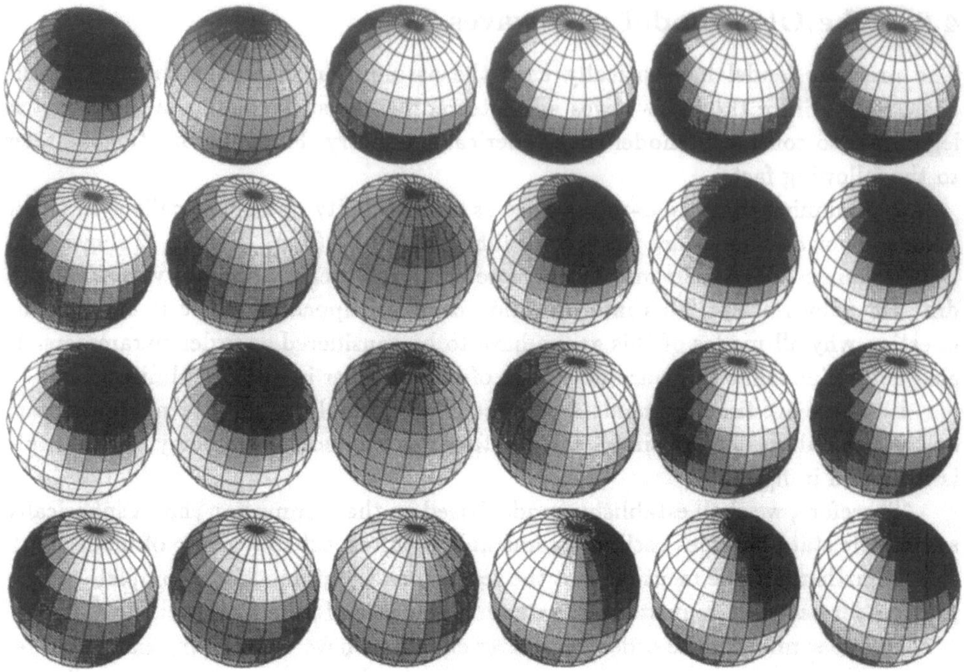

Fig. 13: Pattern with inclined axis: Standing wave with m=0.

$$\xi_m \to \sum_{m'=-2}^{2} [\exp[i(\alpha_+ L_+ + \alpha_- L_- + \alpha_3 L_3)]]_{mm'} \xi_{m'}. \tag{38}$$

The operators L_+, L_-, L_3 are the well-known representations of the generators of the SO(3)-group with respect to the irreducible representation with $l = 2$:

$$\begin{aligned} L_\pm \xi_m &= \sqrt{(2 \mp m)(2 \pm m + 1)} \xi_{m\pm1}, \\ L_3 \xi_m &= m \xi_m. \end{aligned} \tag{39}$$

The equivariance condition already fixes the structure of the order parameter equations and only a few coefficients remain undetermined

$$\begin{aligned} \dot{\xi}_m(t) = (\lambda + \alpha E)\xi_m(t) \quad &+ \quad \beta(-1)^m I \xi^*_{-m}(t) + \gamma \frac{\partial}{\partial \xi^*_m} H \\ m \quad &= -2, ..., m, ..., 2 \end{aligned} \tag{40}$$

There are only four complex coefficients $\lambda, \alpha, \beta, \gamma$, which remain unspecified. The quantities E, I are quadratic polynomials which are invariant under the symmetry operations of the O(3) group (we shall come back to the meaning of these invariants later on):

$$E = \sum_{m=-2}^{m=2} \xi_m \xi^*_m, \tag{41}$$

$$I = \sum_{m=-2}^{m=2} (-1)^m \xi_m \xi_{-m}. \tag{42}$$

H is an invariant fourth order polynomial:

$$\begin{aligned} H &= \sqrt{6}[\xi_2^*\xi_{-1}^*\xi_0\xi_1 + \xi_{-2}^*\xi_1^*\xi_0\xi_{-1} + \xi_2^*\xi_0^*\xi_1\xi_1 + \xi_{-2}^*\xi_0^*\xi_{-1}\xi_{-1} \\ &+ \xi_1^*\xi_0^*\xi_2\xi_{-1} + \xi_{-1}^*\xi_0^*\xi_{-2}\xi_1 + \xi_1^*\xi_1^*\xi_2\xi_0 + \xi_{-1}^*\xi_{-1}^*\xi_{-2}\xi_0] \\ &+ \frac{3}{2}[|\xi_2|^4 + |\xi_{-2}|^2] - |\xi_{-2}|^2|\xi_0|^2 - |\xi_2|^2|\xi_0|^2] \\ &+ 2[|\xi_1|^2|\xi_0|^2 + |\xi_{-1}|^2|\xi_0|^2] + 3[|\xi_2|^2|\xi_1|^2 + |\xi_{-2}|^2|\xi_{-1}|^2] \\ &- 3[|\xi_1|^2|\xi_{-2}|^2 + |\xi_{-1}|^2|\xi_2|^2 + |\xi_2|^2|\xi_{-2}|^2] + \xi_2^*\xi_{-2}^*\xi_0\xi_0 + \xi_0^*\xi_0^*\xi_2\xi_{-2} \\ &+ 2\xi_1^*\xi_{-1}^*\xi_0\xi_0 + 2\xi_0^*\xi_0^*\xi_1\xi_{-1}. \end{aligned} \tag{43}$$

Additionally, there are two invariant cubic polynomials:

$$J_1 = \xi_0^3 - 3\xi_0\xi_1\xi_{-1} - 6\xi_0\xi_2\xi_{-2} + 3\sqrt{\frac{3}{2}}[\xi_{-1}^2\xi_2 + \xi_{-2}\xi_1^2], \tag{44}$$

$$\begin{aligned} J_2 &= \xi_0^*[\xi_0^2 - \xi_1\xi_{-1} - 2\xi_2\xi_{-2}] \\ &+ \xi_0[|\xi_1|^2 + |\xi_{-1}|^2] - 2\xi_0[|\xi_2|^2 + |\xi_{-2}|^2] \\ &- \sqrt{6}[\xi_2\xi_{-1}\xi_1^* + \xi_{-2}\xi_1\xi_{-1}^*] + \sqrt{\frac{3}{2}}[\xi_1^2\xi_2^* + \xi_{-1}^2\xi_{-2}^*]. \end{aligned} \tag{45}$$

Starting from the order parameter equations it is possible to obtain a closed system of evolution equations for the invariants E, I, J_1, J_2, H:

$$\begin{aligned} \frac{d}{dt}I &= 2I[\lambda + (\alpha+\beta)E], \\ \frac{d}{dt}E &= 2[(\lambda_r + \alpha_r)E + \beta_r|I|^2 + 2\gamma_r H], \\ \frac{d}{dt}H &= 4[(\lambda_r + \alpha_r)E + 3\gamma_r E]H + 4\gamma_r[|J_1|^2 - |J_2|^2], \\ \frac{d}{dt}J_1 &= 3[(\lambda + \alpha E)J_1 - \gamma E J_1 + (\beta+\gamma)I J_2], \\ \frac{d}{dt}J_2 &= 3(\lambda_r + \alpha_r E)J_2 + i(\lambda_i + \alpha_i E)J_2 + 2(\beta - \gamma^*)I J_2^* \\ &+ (\beta^* - 2\gamma - \gamma^*)J_1 I^* + (5\gamma_r - i\gamma_i)E J_2. \end{aligned} \tag{46}$$

(The subscripts r and i denote real and imaginary parts.) Although we have no general proof it seems that the invariants form a generating set since the set of evolution equations is closed. (A set of invariants is called a generating set if each invariant polynomial can be expressed as a polynomial function of the set.)

The dynamical system possesses a variety of standing and travelling waves, exhibited in fig. 14. Their stability depends on the values of the coefficients α, β, γ. There are two different types of solutions representing rotating waves:

$$\begin{aligned} V(\theta,\phi,t) &= \xi_1(t)Y_2^1(\theta,\phi) + \text{c.c.} \\ V(\theta,\phi,t) &= \xi_2(t)Y_2^2(\theta,\phi) + \text{c.c.} \quad . \end{aligned} \tag{47}$$

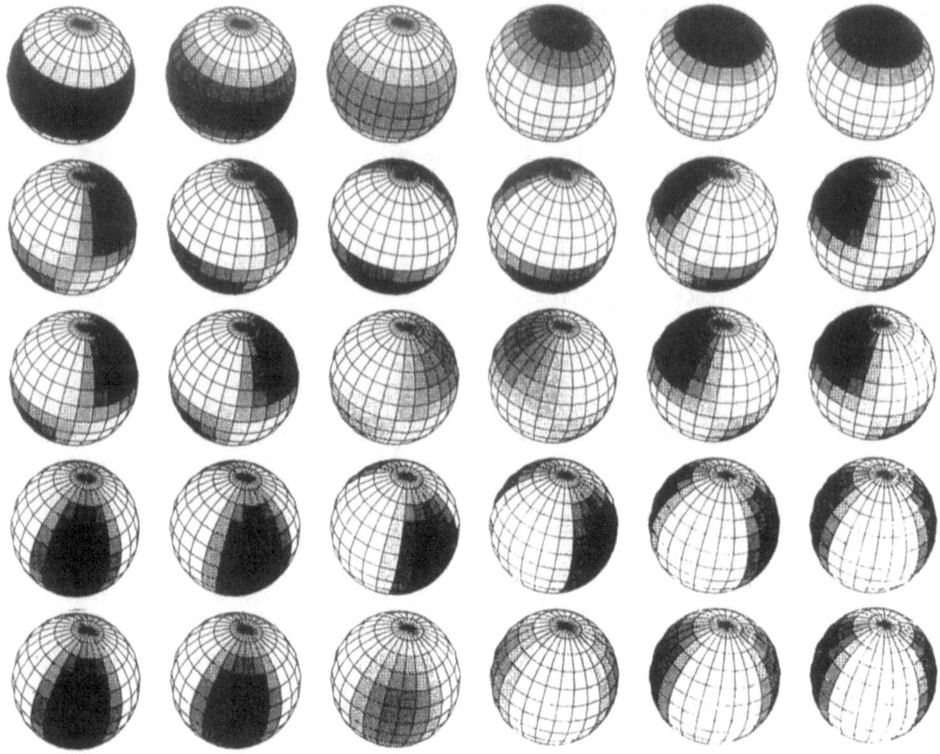

a

b

c

d

e

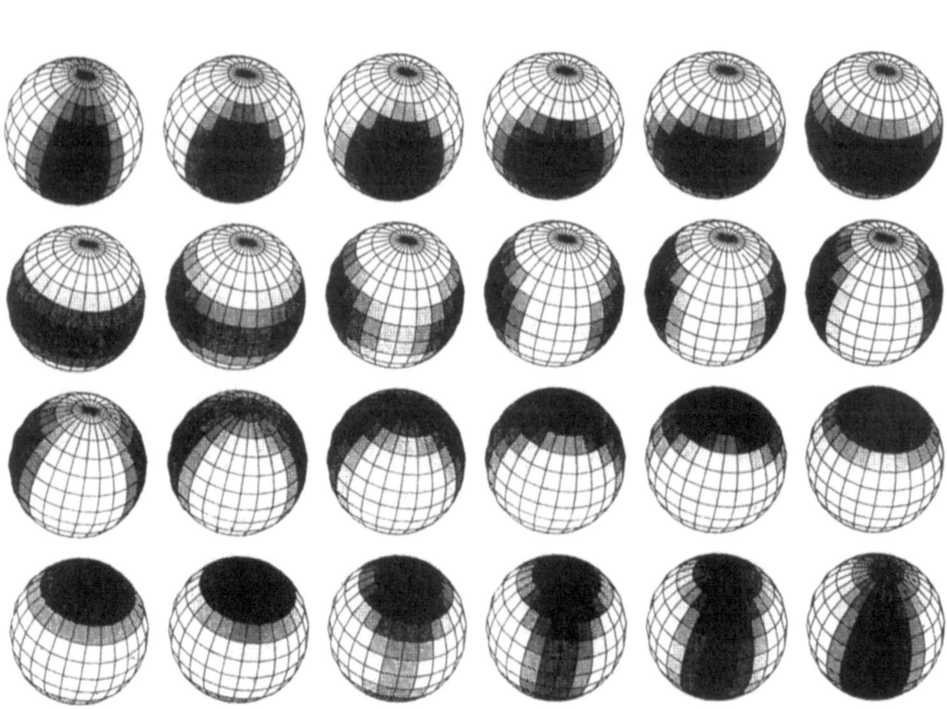

f

Furthermore, there are two different types of standing waves:

$$\begin{aligned} V(\theta,\phi,t) &= \xi_1(t)[Y_2^1(\theta,\phi)+Y_2^{-1}(\theta,\phi)e^{i\alpha}]+c.c. \\ V(\theta,\phi,t) &= \xi_0(t)Y_2^0(\theta,\phi)+c.c. \end{aligned} \quad (48)$$

Additionally, there is a 'pulsating wave' defined by

$$V(\theta,\phi,t) = \xi_0(t)[Y_2^0(\theta,\phi)+\frac{i}{2}(Y_2^2(\theta,\phi)+Y_2^{-2}(\theta,\phi))]. \quad (49)$$

The time dependence of the amplitudes $\xi_m(t)$ is proportional to $\exp(i\omega t)$. Due to symmetry the patterns can be arbitrarily orientated on the sphere. The representation of a rotated pattern is not as simple since its mode decomposition yields contributions of all modes with $L = 2$. Therefore it is difficult to identify the underlying patterns from an inspection of the time series of the amplitudes. However, the different types of motion can be distinguished on the basis of the invariants E, I, J_1, J_2, H. The rotating wave motions are classified by the following sets of invariants:

$m = 1$:	$I = 0$	$J_1 = 0$	$J_2 = 0$	$H = 0$
$m = 2$:	$I = 0$	$J_1 = 0$	$J_2 = 0$	$H = \frac{3}{2}E^2$

The standing waves are classified by the following sets of invariants:

$m = 0$:	$\|I\| = E$	$\|J_1\| = \sqrt{E}^3$	$\|J_2\| = \sqrt{E}^3$	$H = 0$
$m = 1$:	$\|I\| = E$	$J_1 = 0$	$J_2 = 0$	$H = 0$

The 'pulsating wave' is characterized by

puls. wave:	$\|I\| = 0$	$\|J_1\| = \sqrt{2}\sqrt{E}^3$	$\|J_2\| = 0$	$H = -\frac{1}{2}E^2$

We mention that in order for these relations to hold it is not necessary that the amplitudes $\xi_m(t)$ yield oscillatory behaviour of the form $\xi_m(t) = \text{const.}e^{i\omega t}$. In fact, these relations hold for a more complicated time dependence. An inspection of the time series of these invariants allows us to detect segments during which one of the above kinds of motion exists, irrespective of the orientation of the pattern on the sphere. Figure 15 presents the time series of these invariants. The first observation

Fig. 14: Standing and traveling wave solutions emerging due to an oscillatory instability of a spherically symmetric state.
(a) Standing wave (m=0) ;
(b) standing wave (m=1) ;
(c) rotating wave (m=1) ;
(d) standing wave (m=2) (equivalent to standing wave with m=1) ;
(e) rotating wave (m=2) ;
(f) pulsating wave .

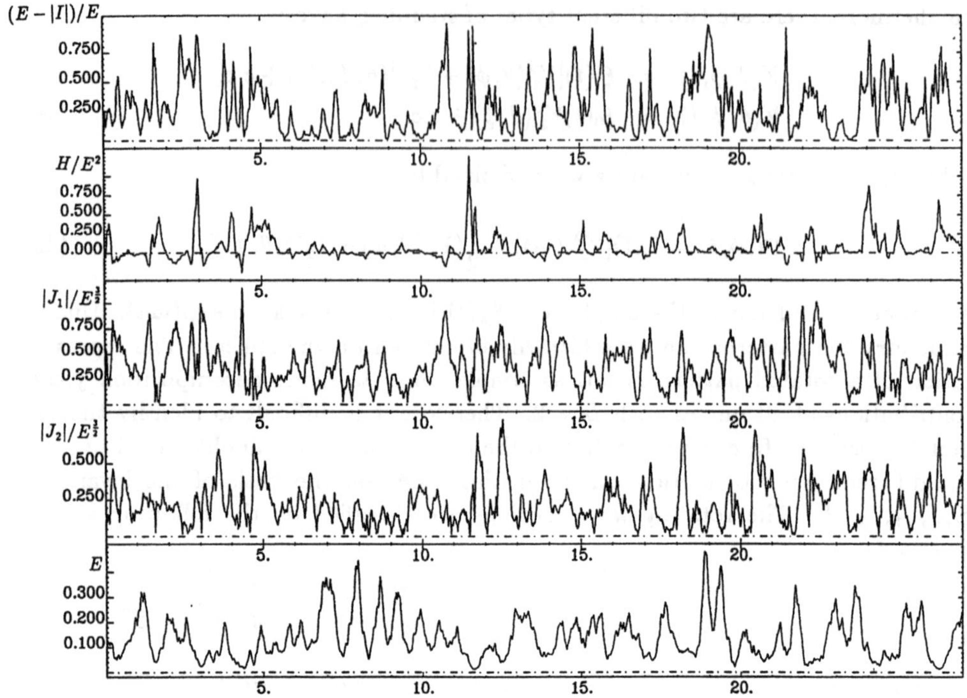

Fig. 15: Time series of the invariants $(E-|I|)/E$, H/E^2, $|J_1|/E^{\frac{3}{2}}$, $|J_2|/E^{\frac{3}{2}}$, E.

is that the quantity H/E^2 is nearly zero during a substantial part of the time. This excludes the possibility of rotating waves with m=2 as well as the 'pulsating wave', however, admits the existence of standing waves with $m=0$, $m=1$ and rotating waves with $m=1$. The axisymmetric standing wave can be separated from the standing and rotating waves by the quantities $|J_1|/\sqrt{E}^3, |J_2|/\sqrt{E}^3$. The additional measure to distinguish between standing and rotating waves is the quantity $(E-|I|)/E$, which equals one for a pure rotating wave with m=1.

The time series of the invariant E shows that this quantity exhibits some kind of oscillatory behaviour. We remind the reader that the quantity E is related to the total intensity of the brain waves. In the spherically symmetric model we have not found such a behaviour. This model can therefore not deal with the complex dynamics. Several ways to extend the model seem to be plausible. One way consists in explaining the complex dynamics as a result of degrees of freedom which change the parameters $\lambda, \alpha, \beta, \gamma$. A second explanation, which assumes a deterministic chaotic process generated by the set of modes with $l=2$ itself, is based on the idea that the spherical symmetry is broken in such a way that chaotic motion becomes possible (the above model based on O(2)-symmetry shows that this is possible). A third explanation would allow for an additional oscillatory behaviour of the amplitudes in the way that higher time derivatives (at least a second time derivative) of the amplitudes have to be added and could allow for the oscillating behaviour of E. These ideas will be pursued in the future.

5 Summary

The spatio-temporal analysis of EEG patterns based on the synergetic approach to complex systems yields results which already allow for a systematic mathematical modelling of the dynamics. We have shown that in the case of epileptic EEG patterns the underlying attractor can be reconstructed on the basis of a dynamical system which naturally arises as a normal form in nonequilibrium systems. Additionally, we have indicated how these spatial modes may be identified with modes obtained from a Karhunen-Loève expansion. In the case of α-EEG waves a similar analysis yields considerable insight into the spatio-temporal dynamics of the brain waves. As an important point the modes obtained from the Karhunen-Loève expansion exhibit spatial features which point out the existence of spatial symmetries. We have exploited these symmetries in order to establish models of the spatio-temporal behaviour of the α-waves. The first model is based on the assumption of O(2)-symmetry. It shows that an oscillatory instability induced by the same modes as obtained from the Karhunen-Loève expansion of the α-waves exhibits similar spatio-temporal phenomena. On the basis of the interpretation of modes obtained by the Karhunen-Loève expansion as spherical harmonics of order two we have established a model exhibiting O(3)-symmetry. This model has allowed us to introduce quantities (polynomial invariants) which enable us to identify certain types of standing and rotating waves in the time sequence of the α-waves. In summary, the present results underline the applicability of the macroscopic approach to complex systems outlined in reference [3] to the analysis of spatio-temporal EEG patterns.

References

[1] H. Haken: *Synergetics. An Introduction*, 3rd edition, Springer, Berlin (1983)

[2] H. Haken: *Advanced Synergetics*, 2nd ed., Springer, Berlin (1987)

[3] H. Haken: *Information and Selforganization*, Springer, Berlin (1988)

[4] A. Fuchs, R. Friedrich, H. Haken, and D. Lehmann: 'Spatio-Temporal Analysis of Multichannel α-EEG Map Series', in *Computational Systems - Natural and Artificial*, ed. by H. Haken, Springer, Berlin (1987)

[5] R. Friedrich, A. Fuchs, and H. Haken: 'Synergetic Analysis of Spatio-Temporal EEG Patterns', Proceedings of the workshop *Nonlinear Phenomena in Excitable Media*, Leeds (1989)

[6] R. Friedrich, A. Fuchs, and H. Haken: 'Modelling of Spatio-Temporal EEG Patterns', Proceedings of the workshop *Mathematical Approaches to Brain Functioning Diagnostics*, ed. by A. Holden, to appear in: Manchester University Press Series *Proceedings in Nonlinear Science*

[7] A. Babloyantz: 'Strange Attractors in the Dynamics of Brain Activity', in: *Complex Systems - Operational Approaches*, ed. by H. Haken, Springer, Berlin (1985)

A. Babloyantz, C. Nicolis, and M. Salazar: 'Evidence of chaotic dynamics of brain activity during the sleep cycle', Phys. Lett. **A 111**, 152 (1985)

A. Babloyantz and A. Destexhe: 'Strange Attractors in the Human Cortex', in: *Temporal Disorder in Human Oscillatory Systems*, ed. by L. Rensing, U. an der Heiden, M.C. Mackey, Springer, Berlin (1986)

A. Babloyantz and A. Destexhe: 'Low-dimensional chaos in an instance of epilepsy', Proc. Natl. Acad. Sci. **83**, 3513-3517 (1986)

[8] M. Schanz, unpublished

[9] G. Mayer-Kress and J. Holzfuss: 'Analysis of the Human Electroencephalogram with Methods from Nonlinear Dynamics', in: *Temporal Disorder in Human Oscillatory Systems*, ed. by L. Rensing, U. an der Heiden, M.C. Mackey, Springer, Berlin (1986)

S.P. Layne, G. Mayer-Kress, and J. Holzfuss: 'Problems Associated with Dimensional Analysis of Electroencephalogram Data', in: *Dimensions and Entropies in Chaotic Systems*, ed. by G. Mayer-Kress, Springer, Berlin (1986)

[10] I. Dvorak, J. Siska: 'On some problems encountered in the estimation of the correlation dimension of the EEG', Phys. Lett. **A 118**, 63-66 (1986)

[11] D. Lehmann:' Multichannel Topography of Human Alpha EEG Fields', Electroenceph. clin. Neurophysiol., **31**, 439-449 (1971)
D. Lehmann: 'Human Scalp EEG Fields: Evoked, Alpha, Sleep, and Spike-Wave Patterns', in: *Synchronization of EEG Activity in Epilepsies*, ed. by H. Petsche and M.A.B. Brazier, Springer, Berlin (1972)

[12] M. Golubitsky, I.N. Stewart, and D.G. Schaeffer: *Singularities and Groups in Bifurcation Theory: Vol.II*, Springer, Berlin (1988)

[13] V. A. Arnold: *Geometrical Methods in the Theory of Ordinary Differential Equations*, Springer, Berlin (1983)

[14] See e.g.: J. Guckenheimer, P.J. Holmes: *Nonlinear Oscillations, Dynamical Systems, and Bifurcations of Vector Fields*, Springer, Berlin (1983)

[15] A. Arnéodo, P.H. Coullet and E.A. Spiegel: 'Chaos in a finite macroscopic system', Phys. Lett. **A 92**, 369 (1982)

[16] J.P. Eckmann, D. Ruelle: 'Fundamental Limitations for Estimating Dimensions and Lyapunov Exponents in Dynamical Systems', Preprint IHES/P/90/11

Information Processing by Systems with Chemical Communication

A.S. Mikhailov

Institute for Theoretical Physics and Synergetics,
University of Stuttgart, Pfaffenwaldring 57/IV,
W-7000 Stuttgart 80, Fed. Rep. of Germany

Abstract. The aim of this paper is to show that large populations of biological cells or macro-organisms which communicate only chemically are capable of performing the basic functions of analog information processing. Specifically, I construct two hypothetical models of systems with chemical communications which possess the property of associative memory.

1. Introduction

Any living cell can be viewed as a complicated "chemical machine" within which thousands of different chemical reactions are simultaneously going on. Therefore, cells are sensitive to the chemical changes in their environment. They release chemicals into the intercellular medium and have receptors for different kinds of molecules in their membranes. The action of such receptors is usually based on the molecular "key-to-lock" principle. The total number of different specialized receptors can be enormous. Moreover, the experimental data on immune system provides evidence that living cells can even develop new chemical receptors for molecules that have not been met before. It should also be emphasized that cells may be extremely sensitive to the presence of some substances in the medium. A single molecule can sometimes be sufficient to trigger a complete change of the functional regime of a cell. High selectivity of chemical interactions between the cells provides the basis for *chemical communication*. By releasing and absorbing various transmitter molecules, the cells can exchange information about their current states.

Chemical communication is also common in populations of macro-organisms. Insects and higher animals often produce *smells* which transfer information to other individuals.

Chemical communication can be used to send some simple distinct messages such as alarm signals. When the communication is permanent, it may be sufficient to support very diverse forms of collective behavior which can resemble the dynamics of activity patterns in neural networks.

In the simplest formal mathematical models of neural networks each neuron i is considered as an element with only two possible states $S_i = \pm 1$. When $S_i = + 1$, the neuron is in the "active" state, while $S_i = - 1$ corresponds to the "passive" state of a neuron.

Each neuron is assumed to be physically connected to all other neurons. The connections are characterized by "synaptic strengths" J_{ij} which can take both positive and negative values (in the latter case the synapse is called inhibitory).

At randomly chosen time moments each neuron updates its current state. The new state S_i' is determined by the rule

Eds.: H. Haken and H.P. Koepchen

$$S_i' = \text{sign } h_i, \tag{1}$$

where the acting field h_i consists of the sum of the contributions from all other neurons,

$$h_i = \sum_j J_{ij} S_j . \tag{2}$$

Despite the apparent simplicity of this mathematical model of neural networks (proposed by McCulloch and Pitts [1] in 1943) it can support very complicated forms of dynamics. It is used in the majority of modern works on parallel analog information processing. When the synaptic matrix J_{ij} is constructed in an appropriate way, such a system can operate, for example, as a system with associative memory (see [2,3]) or to store and to retrieve the temporal sequences of patterns [3-5].

In Section 2 I show how *any* model of a neural network described by (1) and (2) can be effectively realized by systems with only chemical communications.

2. "Distant" synapses

About twenty years ago G.M.Shepherd [6] mentioned that *directed* chemical communications between insects can effectively play the role of distant "synaptic" connections. In my first hypothetical model (see also [7]) I want to explore further this idea, putting it into mathematical terms.

Suppose that we have a population of insects. As a gross simplification, I will describe the relevant state of each insect by a single binary variable $S_i = \pm 1$ (for instance, this can specify one of the two possible functional regimes of its behavior).

I will assume that the insects can communicate by sending and receiving some directed messages, which are transmitted by the *messenger* molecules. These molecules are released into the air and diffuse there, so that their uniform distribution is established in the medium. However, a messenger molecule can exhibit some action only when it reaches the insect to which it was addressed.

Each messenger bears the *address* i and the *message* $\sigma = \pm 1$ (the meaning of this property will be explained below).

Any insect j permanently produces messengers with all possible addresses i. However, the messages which are attached to the produced molecules depend on the current state of the sending animal (thus providing a way for the chemical communication between them). When the state S_j of an insect is changed, all generated messages σ are inverted.

Therefore, the generation of messengers with an address i by an insect j is specified by two parameters. The first of them, I_{ij}, gives the generation rate of such messengers, which is independent of their message. The second parameter is the binary variable w_{ij} which can take only values +1 and -1. It indicates which message σ will be sent by the insect j when it is in the state $S_j = +1$.

To prevent accumulation of messenger molecules in the medium, we should also include some mechanism of their decay. It will be assumed that all such molecules have the same decay rate γ.

Now we should specify how the messengers influence the states of receiving insects. It will be assumed that, when a message $\sigma = +1$ is delivered to the insect i, this would be an *order* for this insect to go to the state with $S_i = +1$. On the other hand, if the message is $\sigma = -1$, it would be interpreted as an order to go to the state $S_i = -1$.

Since many messengers with the same address may be simultaneously produced by different insects, within a small interval of time an insect can receive a large number of *contradictory* messages. In this situation it can make a decision about its next state by applying a sort of a voting procedure. Namely, it can count the number of messages with $\sigma = +1$ and $\sigma = -1$ and then determine its next state to obey the dominating message. Explicitly that would mean that the next state S_i' will be determined as

$$S_i' = \text{sign } h_i, \tag{3}$$

where

$$h_i = m_{+,i} - m_{-,i} \tag{4}$$

and $m_{+,i}$, $m_{-,i}$ are the local concentrations of molecules with the address i and the messages $\sigma = +1$ and $\sigma = -1$, respectively.

Thus the dynamics of a system of insects that communicate by exchange of directed messages is entirely specified. To investigate it, we first write down the kinetic equations for the local concentrations of the messenger molecules. It will be assumed that diffusion is fast and these concentrations are therefore the same in all volume elements of the medium occupied by the populations of insects. Under our assumptions the kinetic equations have the form

$$\dot{m}_{\sigma,i} = -\gamma m_{\sigma,i} + (1/V) \sum_j I_{ij} (1 + \sigma w_{ij} S_j) . \tag{5}$$

The last term in these equations is constructed in such way that the message σ which is sent by an insect j in the state $S_j = +1$ is +1 if $w_{ij} = +1$ and -1 if $w_{ij} = -1$. This message σ is inverted when the state of the insect is changed to $S_j = -1$.

The next step consists in the demonstration that the dynamics of the described model can be effectively characterized by a model of a neural network (1) and (2). We show this correspondence in the limit of short lifetimes $\tau = 1/\gamma$ of messengers. If the lifetimes are short in comparison with the characteristic time intervals between the transitions in the states of the insects, concentrations of the messenger molecules would adjust adiabatically to the momentary activity states of insects. In other words, in this limit S_j could be treated as the frozen variables in the equations (5) and the established messenger concentrations could be found to be

$$m_{\sigma,i} \cong (1/\gamma V) \sum_j I_{ij} (1 + \sigma w_{ij} S_j) . \tag{6}$$

We can use (6) to calculate the effective acting fields, which yields

$$h_i = m_{+,i} - m_{-,i} = (1/\gamma V) \sum_j I_{ij} w_{ij} S_j . \tag{7}$$

It is instructive to compare (7) and (3) with (1) and (2). We see that effectively our system of communicating insects is described by a model of a neural network. Moreover, by choosing the appropriate values of the parameters I_{ij} and w_{ij} we can emulate *any* neural network with an arbitrary synaptic matrix J_{ij}. To do this, we should simply take

$$I_{ij} = (\gamma V/2)|J_{ij}|, \quad w_{ij} = \mathrm{sign}\, J_{ij} \; . \tag{8}$$

Thus, we have demonstrated that a population of animals which employs the described system of chemical communication can perform any task possible for an abstract neural network. This conclusion can be viewed as a mathematical formulation of the original idea of a "distant synapse" which was proposed by G.M.Shepherd.

The established correspondence is of high theoretical importance. However, it can be noted that its practical realization would have required a very large number of different messenger molecules (twice the total number of insects in the population). It seems hard to imagine how this could occur under realistic biological conditions, even for the populations consisting of mere hundreds of species. Therefore the applicability of the described model in biology is highly questionable.

However, it can be easily realized in artificial information processing devices which represent the lattices of primitive information processors with local information exchange. We explore this possibility further in the final section. In the next Section we propose a different realization of a chemical communication network.

3. Associative memory with chemical mediators

The above scheme was universal in the sense that it allowed *any* neural network to be emulated. The price that had to be paid for such great universality consisted of the necessity to have a great number of different messenger molecules, which might be a serious problem for a biological system. It turns out that some simpler schemes are possible if we limit our investigation to the special networks used as systems with *associative memory*.

The task of associative memory can be formulated as follows (see, e.g.[2]). Suppose we have a set of prototype patterns $\{S_j^{(\alpha)}\}$, $\alpha = 1,2,\ldots,p$. We need to construct a dynamical system which would classify any applied pattern $\{S_j\}$ according to its similarity with the recorded prototypes. Moreover, it is desirable that the nearest prototype is automatically produced by a system in response to the applied pattern.

In neural networks this can be achieved in a variety of ways, by choosing the appropriate synaptic matrix. For instance, if the prototypes are *orthogonal*, i.e. if they satisfy the conditions

$$\sum_j S_j^{(\alpha)} S_j^{(\beta)} = N\, \delta_{\alpha\beta} \; , \tag{9}$$

it is sufficient to choose the synaptic matrix by following the *Hebb rule*, as

$$J_{ij} = (1/N) \sum_\alpha S_i^{(\alpha)} S_j^{(\alpha)} . \tag{10}$$

In my second hypothetical model (see [3,8]) of systems with chemical communications I will speak in terms of biological cells which can communicate by release and absorption of some *mediator* molecules. The principal difference from the first model is that now each kind of mediator molecule will be associated *with a particular prototype pattern*. We again assume that the state of any individual cell i is specified by a binary variable $S_i = \pm 1$.

Suppose that any cell j releases the mediator molecules of kind α only when its current state S_j coincides with the state $S_j^{(\alpha)}$ required by the respective prototype pattern. To prevent their accumulation, the mediator molecules are assumed to decay at a constant rate γ, the same for all kinds of mediators. Diffusion is fast and ensures permanent ideal mixing of reagents.

The kinetic equations for the concentrations m_α of the mediator molecules are

$$\dot{m}_\alpha = -\gamma m_\alpha + (\gamma/N) \sum_j (S_j^{(\alpha)} S_j + 1) \ . \tag{11}$$

It is convenient to take the generation rate of the mediator molecules which are the same as the decay rate.

We must also specify how the mediator molecules act on the cells.

We will assume that each kind of mediator tends to establish that state of a cell i which is required by the respective prototype pattern. In other words, each mediator molecule is again assumed to bear some *order* for the cells. The difference from the first model is that this order itself is not directed now, i.e. the same kind of mediator molecules will act on all cells precent. However, each cell interprets the received order in an individual way, since it keeps in its memory how it should respond to the reception of such a signal.

Since within a small time interval a cell can experience collisions with many different kinds of mediator molecules and thus receive a large number of contradictory orders, some "voting" procedure should be employed by it. The next state of a cell S_i' will be determined by the dominant order. Namely, we can assume that

$$S_i' = \operatorname{sign} h_i \tag{12}$$

where

$$h_i = \sum_\alpha S_i^{(\alpha)} (m_\alpha - 1) \ . \tag{13}$$

We begin the discussion of the model described by equations (11)-(13) with the analysis of the limit of short lifetimes of the mediator molecules (namely, we will assume that γ is much larger than the characteristic frequency of state transitions in the cells). Under this condition the mediator concentrations m_α adiabatically adjust to the momentary states of the cells and we have

$$m_\alpha = (1/N) \sum_j (S_j^{(\alpha)} S_j + 1) \ . \tag{14}$$

Substitution of these expressions into (13) yields

$$h_i = (1/N) \sum_j S_i^{(\alpha)} S_j^{(\alpha)} S_j \; . \tag{15}$$

But this is precisely what is found for a neural network (1)-(2) if the synaptic matrix J_{ij} is constructed by the Hebb rule (10)!

Hence, we can conclude that in the limit of short mediator lifetimes our system is effectively equivalent to a neural network with some *special* synaptic matrix given by the Hebb rule. Such neural networks were shown by Hopfield [9] to possess the property of associative memory. If we start from some initial pattern, it will be attracted eventually to one of the prototype patterns $\{S_j^{(\alpha)}\}$, provided that the patterns are orthogonal. (Some special initial conditions may result in convergence to the inverted prototype patterns that are also attractive.)

When we look at the mediator concentrations, we see that the concentration of the mediator which corresponds to the surviving pattern will approach the value of 2, if this is one of the memorized prototype patterns, or 0 if it is obtained by inversion of one of the prototype patterns. The concentrations of all other mediator molecules will saturate at the level of 1.

This means that the variables defined as

$$\eta_\alpha = m_\alpha - 1 \tag{16}$$

can be interpreted as the *order parameters* of our model. Each of them is associated with a particular prototype pattern. When this pattern is recognized, the respective order parameter approaches the value of $\eta_\alpha = +1$. Then all other order parameters vanish. Note that $\eta_\alpha = -1$ would correspond to recognition of the inverted prototype pattern α which is also stored in the memory.

In the considered limit of short lifetimes of mediators these order parameters are actually the overlaps between the current activity configuration of the network and the activity configurations corresponding to various prototype patterns. Indeed, as follows from (14), we have then

$$\eta_\alpha = (1/N) \sum_j S_j^{(\alpha)} S_j \; . \tag{17}$$

4. Slow mediators

It is interesting to examine what the properties of our model (11)-(13) will be in the *opposite* limit when the lifetimes of all mediators are *large* in comparison with the characteristic time intervals between the transitions in the cells. In this limit the activity pattern $\{S_j\}$ of the cell will adjust adiabatically to the momentary mediator concentrations m_α.

We consider in this section a slightly more general model where the deterministic transition algorithm (12) is replaced by the *probabilistic* transition law. Namely, we will assume that the next state S_i' of a cell is chosen with the probability

$$p(S_j') = \frac{\exp(h_i S_i'/\vartheta)}{\exp(-h_i/\vartheta) + \exp(h_i/\vartheta)} \; , \tag{18}$$

where h_i is again given by the expressions (13) and ϑ plays the role of an effective temperature.

In a given field h_i (which is determined by slowly varying concentrations of the mediators and thus is "frozen" from the point of view of the cell variables) the mean values of the cell variables S_i would be

$$\langle S_i \rangle = \sum_{S_i=\pm 1} S_i\, p(S_i) = \tanh(h_i/\vartheta). \tag{19}$$

We should substitute these mean values into the kinetic equations for the mediator concentrations:

$$\dot{m}_\alpha = -\gamma m_\alpha + (\gamma/N) \sum_i (S_i^{(\alpha)} \langle S_i \rangle + 1). \tag{20}$$

This yields

$$\dot{m}_\alpha = -\gamma m_\alpha + (\gamma/N) \sum_j \{S_j^{(\alpha)} \tanh(h_j/\vartheta) + 1\} \tag{21}$$

where

$$h_j = \sum_\beta S_j^{(\beta)} (m_\beta - 1). \tag{22}$$

It is convenient to change now to the order parameters $\eta_\alpha = m_\alpha - 1$. This yields

$$\dot{\eta}_\alpha = -\gamma \eta_\alpha + (\gamma/N) \sum_j S_j^{(\alpha)} \tanh[(1/\vartheta) \sum_\beta S_j^{(\beta)} \eta_\beta] . \tag{23}$$

Hence, we have derived the closed set of equations for evolution of the order parameters of our system.

These evolution equations are formally the same as the equations for the order parameters in the Hopfield model proposed by Rieger et al. in [10]. However, in the case of the Hopfield model of a neural network the physical meaning of these order parameters remains vague. Actually, only at equilibrium do they coincide with the overlaps (17) and thus receive a clear physical interpretation.

Contrary to this, in our model such variables have a definite physical interpretation because they are trivially related to the mean concentrations of the mediator molecules and thus are directly observable. Moreover, in our case we can systematically *derive* their evolution equations (23), whereas in the case of the Hopfield model these equations are proposed in a phenomenological manner.

The order parameter equations (23) can be written in the gradient form, i.e. as

$$\dot{\eta}_\alpha = -\gamma \frac{\partial F}{\partial \eta_\alpha} , \tag{24}$$

where the "free energy" is given by the expression

$$F = \frac{1}{2}\sum_\beta \eta_\beta^2 - \frac{\theta}{N}\sum_j \ln\{\frac{1}{2} + [\sinh(\frac{1}{2\theta}\sum_\beta S_j^{(\beta)}\eta_\beta)]^2\} \ . \quad (25)$$

Therefore, evolution of this system can consist only of approaching some steady stationary states which correspond to the minima of F.

Suppose we have only one nonvanishing order parameter, i.e. $\eta_1 \neq 0$ while $\eta_\alpha = 0$ for all $\alpha \neq 1$. Then (25) takes the form

$$F = \frac{1}{2}\eta_1^2 - \theta \ln\{\frac{1}{2} + [\sinh(\frac{1}{2\theta}\eta_1)]^2\} \ . \quad (26)$$

When $|\eta_1| << \theta$, we have approximately

$$F \cong \frac{1}{2}(1 - \frac{1}{\theta})\eta_1^2 + \frac{1}{12\ \theta^3}\eta_1^4 \ . \quad (27)$$

On the other hand, for $|\eta_1| >> \theta$ we have

$$F \cong \frac{1}{2}\eta_1^2 - |\eta_1| \ . \quad (28)$$

Thus we see that $\theta = 1$ is a bifurcation point. At $\theta > 1$ the "free energy" F possesses only one minimum at $\eta_1 = 0$. When the effective temperature θ is decreased below 1, two minima at $\eta_1 \cong \pm 3^{1/2}(1 - \theta)^{1/2}$ appear while $\eta_1 = 0$ becomes a maximum of the function F.

The behavior which was found above is typical for the second-order phase transitions. Note however that we observe it only if we assume that all order parameters except a single one are equal to zero. When there are many small but nonvanishing order parameters ($|\eta_\alpha| << \theta$), (27) is replaced by

$$F \cong (1 - \frac{1}{\theta})\sum_\alpha \eta_\alpha^2 + \frac{1}{12\ \theta^3}\sum_\alpha \eta_\alpha^4 + \frac{1}{4\ \theta^3}\sum_{\alpha_1 \neq \alpha_2} \eta_{\alpha_1}^2 \eta_{\alpha_2}^2$$

$$+ \frac{1}{12\ \theta^3} \sum_{\alpha_1 \neq \alpha_2 \neq \alpha_3 \neq \alpha_4} T_{\alpha_1\alpha_2\alpha_3\alpha_4}\, \eta_{\alpha_1}\eta_{\alpha_2}\eta_{\alpha_3}\eta_{\alpha_4} \ . \quad (29)$$

Here

$$T_{\alpha_1\alpha_2\alpha_3\alpha_4} = \frac{1}{N}\sum_j S_j^{(\alpha_1)} S_j^{(\alpha_2)} S_j^{(\alpha_3)} S_j^{(\alpha_4)} \quad (30)$$

are the fourth-order overlaps between the prototype patterns (which are assumed to be orthogonal).

Thus we see that in general, although $\theta = 1$ is still a critical effective temperature, the behavior of the system below the critical point is much more complicated and we cannot guarantee that the minima of F will correspond to pure prototype patterns. Instead they can correspond to some mixtures of our prototypes. It should also be noted that the coefficients $T_{\alpha_1\alpha_2\alpha_3\alpha_4}$ in (29) can even be negative, so that further terms in the decomposition of F must be taken into account.

Only in the limit of low temperatures $\theta \to 0$ do the pure patterns $\{S_j^{(\alpha)}\}$ become the minima of the "free energy" F. To demonstrate this, we can derive the approximation for F when at least one of the order parameters is large:

$$F \cong \frac{1}{2}\sum_{\alpha}\eta_{\alpha}^{2} - \frac{1}{N}\sum_{j} \left| \sum_{\beta} S_{j}^{(\beta)}\eta_{\beta} \right| . \tag{31}$$

Now we can consider a case when one of the order parameters is equal to 1, while all other order parameters are small (i.e. $\eta_1 = 1$ and $|\eta_\beta| << 1$). Then we have

$$\sum_{j} \left| \sum_{\beta} S_{j}^{(\beta)}\eta_{\beta} \right| \cong 1 + O(\eta_{\beta}^{3}) \tag{32}$$

and hence

$$F \cong \frac{1}{2}\sum_{\beta\neq 1}\eta_{\beta}^{2} + O(\eta_{\beta}^{3}) . \tag{33}$$

This shows that the pure prototype state with $\eta_1 = 1$ and $\eta_\beta = 0$ for $\beta \neq 1$ is a local minimum of F, provided that $\theta << 1$. Indeed, for small enough values of perturbations η_β the quadratic terms in (33) dominate over the cubic corrections denoted by $O({\eta_\beta}^3)$ and these terms are positive.

The above arguments reveal that, when the probabilistic transition algorithm (18) is used, we can have the behavior required for the systems with associative memory only in the limit of weak noise. Moreover, since the derived equations (23) for the order parameters in our model coincide with the equations proposed for the Hopfield model, we can make certain conclusions concerning it. Namely, we conclude that the dynamics of the Hopfield model defined in terms of the respective order parameters associated with different prototype patterns is very complicated (despite the fact that this model is very simply defined in terms of the "spin" variables S_j). The above analysis shows why the dynamics of the order parameters of the Hopfield model cannot be even approximately described near the critical points by standard Ginzburg-Landau equations.

5. Nonorthogonal patterns

Until now it has been assumed that the prototype patterns are orthogonal, i.e. they satisfy the condition (9). However, the model of associative memory with chemical mediators that was introduced in Section 3 can be easily generalized to treat the case of nonorthogonal prototypes. The only required modification consists of a slight change of the generation rule for the mediator molecules.

In the initial model it was proposed that the mediator molecules corresponding to the prototype α were generated by the cell j when its current state S_j coincided with the the state ${S_j}^{(\alpha)}$ in the respective prototype. Now we determine for each prototype $\{{S_j}^{(\alpha)}\}$ its complementary pattern $\{{\sigma_j}^{(\alpha)}\}$ and assume that the generation of mediators by a cell j takes place only when its current state coincides with ${\sigma_j}^{(\alpha)}$.

The complementary patterns are defined as follows. First we construct the overlap matrix

$$\Omega_{\alpha\beta} = \frac{1}{N}\sum_j s_j^{(\alpha)} s_j^{(\beta)} \qquad (34)$$

whose elements represent the overlaps between all prototypes (it degenerates to a unit matrix for the orthogonal prototypes). Then we find its inverse $(\Omega^{-1})_{\alpha\beta}$ and construct $\sigma_j^{(\alpha)}$ as

$$\sigma_j^{(\alpha)} = \sum_\beta (\Omega^{-1})_{\alpha\beta} s_j^{(\beta)} \ . \qquad (35)$$

After the modification, the kinetic equations (11) of the original model are replaced by

$$\dot{m}_\alpha = - \gamma m_\alpha + \frac{\gamma}{N}\sum_j (\sigma_j^{(\alpha)} s_j + 1), \qquad (36)$$

while equations (12) and (13) remain unchanged. It can be shown that this deterministic model will have the prototype patterns as its final steady states even if they are not orthogonal.

6. Discussion

In Sections 2,3 and 5 two basic hypothetical models were described which allow the function of associative memory to be realized for biological populations with chemical communications between the species. However, the proposed mathematical models can also find a potential application in the field of *artificial* distributed information processing.

Today great hopes are attached to the use of artificial neural networks as the devices of distributed information processing. Such devices should represent the networks of primitive information processors with many connections between the individual units. Already there are many examples which demonstrate the great advantages of the neural network approach [11]. Nonetheless, serious difficulties arise when we try to realize it at the microminiature level of molecular electronics.

It is extremely difficult to realize multiple connections between the primitive processing units at the molecular level. Now each "wire" would occupy the same volume as a processing element. Since the neural network models usually require each "neuron" to be connected with all the others, this difficulty may seriously limit their application in molecular electronic devices.

It is much easier to practically realize the *cellular automata networks* which represent regular lattices of simple processing units with connections only between local neighbors. I want to emphasize that both the models proposed above can be easily implemented using such networks.

Instead of some chemical molecules we should speak then in terms of "signals" which wander at random in the lattice. To obtain the diffusion-like wandering, we assume that each unit, occupying a definite site in the lattice, can receive signals from any of its nearest neighbors. The signal is then retransmitted to a randomly chosen nearest neighbor of the unit that received it. The signals are generated by the units according to certain rules which depend on a particular model. When a signal arrives at an unit, there is some probability that it will be "killed" by it (this will result in finite lifetimes of signals in the lattice).

A "signal" represents a short information string. If we want to implement the first model outlined in Section 2, the string should contain the address of the unit where it will act and the binary message variable σ. In our second model the signals bear the names of the patterns with which they are associated.

In addition to the ability to perform simple operations with signals, each unit in the lattice can change its "internal" binary state variable, as is required by the corresponding model. These changes are controlled by the numbers of signals of different kinds that are present at a given (discrete) time moment in a particular site of the lattice.

The principal difference between the two models lies in the fact that, in the first model, only the signals with the address which coincides with the address of a given unit can influence its internal dynamics. In the second model any signal can execute action on each unit, independent of its address, if it is found there.

Applications of these models in analog distributed information processing will be further investigated in a forthcoming paper. To end the present contribution, I will briefly discuss the possibility that the behavior described by such or similar models may indeed be observed in real biological populations.

From a biological point of view, a serious disadvantage of the first model is that it requires too many different kinds of messenger molecules. In fact, to implement it in a system with chemical communication the number of different messenger molecules must be twice the total number of communicating biological organisms. Therefore, although this model demonstrates the theoretical possibility of realizing the idea of "distant synapses" which was initially proposed in [6], it is not very realistic, at least for sufficiently large biological populations.

On the other hand, in the second model the number of different mediator molecules is the same as the number of prototype patterns that can be stored in the memory of a given biological population. It may well happen that, although the population itself is large, it needs to keep in its memory only a few distinct prototype patterns. Then just a few kinds of chemical mediators would be sufficient for operation of the associative memory. These conditions can be easily met in many natural biological populations. It would be interesting to undertake an experimental search for the described type of behavior in populations of macro- or microorganisms. To an observer, it would appear as a complex *mimicry* ability.

I want to acknowledge very stimulative discussions with Prof.H.Haken and to thank him for the excellent work opportunities during my long stay at the Institute for Theoretical Physics and Synergetics in Stuttgart.

References

1. W.C. McCulloch, W. Pitts: Bull. Math. Biophys. **5**, 115-133 (1943)
2. H. Haken: *Synergetic Computers and Cognition* (Springer, Berlin, Heidelberg 1990)
3. A.S. Mikhailov: *Foundations of Synergetics I. Distributed Active Systems* (Springer, Berlin, Heidelberg 1990)
4. L. Personnaz, I. Guyon, G. Dreyfus: Phys. Rev. A**34**, 4217 (1986)
5. H. Sompolinsky, I. Kanter: Phys. Rev. Lett. **57**, 2861-2864 (1986)
6. G.M. Shepherd: *Neurobiology* (Oxford Univ. Press, 1983)
7. A.S. Mikhailov, I.V. Mitkov, N.A. Sveshnikov: BioSystems **23**, 291-295 (1990)

8. A.S. Mikhailov, I.V. Mitkov, N.A. Sveshnikov: J. Nonlinear Biology, in press
9. J.J. Hopfield: Proc. Nat. Acad. Sci. USA **79**, 2554-2558 (1982)
10. H. Rieger, M. Scheckenberg, J. Zittarz: Z. Phys. **B72**, 523 (1988)
11. H. Haken (ed.): *Neural and Synergetic Computers* (Springer, Berlin, Heidelberg 1988)

Criteria for the Relative Degree of Order in Self-Organization Processes

Yu.L. Klimontovich

Department of Physics, Moscow State University, SU-117234 Moscow, USSR

Abstract. The aim of this paper is the consideration of the criteria for the relative degree of chaos or order in open systems on the basis of experimental data: K-entropy and Lyapunove exponents, Boltzmann-Gibbs-Shannon entropy, renormalized to the given value of the mean effective energy (S-theorem) and entropy production. The principle of minimum entropy production in self-organization processes is formulated.

1. Introduction

The answers to the questions: "Is It Healthy to be Chaotic? [1], "What is Self-Organization?" and others are absent up till now. But according to the author, the answers to these questions are not as essential as the establishment of criteria for the relative degree of order of the various nonequilibrium states describing open systems. These criteria will be useful for the objectivity of control in developing new treatments for a number of deseases.

2. The Evolution in Time. H-Theorem for Open Systems. Lyapunov Functionals

Boltzmann's H-Theorem. For the formulation of criteria of self-organization the following form of the H-theorem is useful. We denote by $S_0 - S(t)$ the difference between the entropies in equilibrium and in other arbitrary states. Using the definition of the Boltzmann entropy, we get

$$\Lambda_S = S_0 - S(t) = \varkappa n \int \ln\frac{f(R,v,t)}{f_0(v)}\, f(R,v,t) \geq 0 \,. \tag{2.1}$$

Springer Series in Synergetics, Vol. 55 **Rhythms in Physiological Systems**
Eds.: H. Haken and H.P. Koepchen

From the H-theorem it follows that

$$\frac{d\Lambda_s}{dt} = \frac{d}{dt}(S_0 - S(t) \leq 0. \qquad (2.2)$$

We see that Λ_s is a Lyapunov functional.
The result (2.1) relies on the fact that in a process of the evolution towards the equilibrium state, the mean energy remains constant

$$\langle E \rangle = \langle mv^2/2 \rangle = \text{const.} \qquad (2.3)$$

It is only then that the Lyapunov functional will be the difference in the entropies, and not in some other thermodynamic potentials. Important is the following.

The conservation of the mean energy $\langle E \rangle$ is a natural property of the Boltzmann equation rather than an additional restriction imposed on it. The fact that it is not the energy E that is conserved in the process of evolution, but only its mean value $\langle E \rangle$, points to the "**internal openness**" of the Boltzmann gas, the system under consideration. So, as we see, the statment that Boltzmann's H-theorem holds for a closed system lacks a more precise definition.

We shall distinguish two kinds of evolution: **Degradation and Self-Organization**". As the evolution in time proceeds towards the equilibrium state the entropy increases. So, we see that this process, when properly described on the basis of the Boltzmann equation, is an example of **degradation** [2].

Brownian motion. H-Theorem. Among open systems there is an important class of systems exhibiting the phenomenon of Brownian motion. As an example we will consider the Brownian motion in the Van der Pol generator, using the Fokker-Planck equation for the distribution function of energy of oscillations

$$\frac{\partial f(E,t)}{\partial t} = D\frac{\partial}{\partial E}(E\frac{\partial f}{\partial t}) + \frac{\partial}{\partial E}[(-a + bE)Ef]. \qquad (2.4)$$

Here D denotes the intensity of noise, $a = a_p - \gamma$, a_p is the parameter describing the positive feedback (pumping), γ and b are the coefficients of the linear and nonlinear parts of friction. The stationary solution has the form

$$f_0 = \exp\frac{F_0 - H(E)}{D}, \quad H(E) = -aE + \frac{1}{2}bE, \quad \int f dE = 1. \qquad (2.5)$$

Here we have used the symbols H(E), which stands for the effective energy (effective Hamiltonian), and D, which plays the role of the temperature. The nonequilibrium free energy is defined by

$$F(t) = \langle H(E)\rangle - DS(t), \qquad S(t) = -\int \ln f\, f dE. \tag{2.6}$$

A Lyapunov functional is introduced by the expression

$$\Lambda_F = F(t) - F_0 = D\int \ln\frac{f}{f_0} f dE \geq 0, \qquad \frac{d\Lambda_F}{dt} \leq 0. \tag{2.7}$$

Let us now show that H-theorem can be formulated in terms of entropy. To do this, we carry out renormalization to the given value of the mean effective Hamiltonian function ⟨H(E)⟩. The renormalized intensity of noise $\tilde{D}$ is therefore a functional of the corresponding distribution function $\tilde{f}$ and can be found from the equation

$$\int H(E)\tilde{f}(E,t)dE = \int H(E)f_0(E)dE. \tag{2.8}$$

Hence the Lyapunov functional will be determined by the entropies $S_0, S(t)$ [4,5]

$$\Lambda_S = S_0 - \tilde{S}(t) = \int \ln\frac{\tilde{f}(E,t)}{f_0}\tilde{f}(E,t)dE \geq 0. \tag{2.9}$$

For the evolution towards the stationary state the following holds:

$$\frac{d\Lambda_S}{dt} = \frac{d}{dt}(S_0 - \tilde{S}(t)) \leq 0 \quad \text{at} \quad \langle H(E)\rangle = \text{const}. \tag{2.10}$$

This result is the H-theorem for an open active system.

The main advantage of using the Lyapunov functional defined in terms of the entropy is the following: the entropy can be defined through the distribution function for an arbitrary nonequilibrium state. The free energy, unlike the entropy, does not possess all the properties which would enable us to use it as a measure of indeterminacy in the statistical description.

3. Evolution in Space of Controling Parameters. S-Theorem [3-10]

In [6-9] the criterion for depicting the relative degree of order of states belonging to open systems has been formulated. This statment was termed in [6,7] the S-theorem, "S" standing for "Self-Organization" to emphasize the fact that the S-theorem is a criteri self-organization.

Now assume that we have a choice of control parameters and consider the evolution of stationary states of an open system corresponding to different values of these control parameters **a**.
We single out a state at the value $a = a_0$ and another one with the value $a = a_0 + \Delta a$. Assuming that the distribution $f_0(x_0, a)$ corresponds to the state of physical chaos (the correctness of this assumption will be verified) we also introduce the distribution function $f(x, a_0 + \Delta a)$. In order to define the Hamilton function $H_{eff}(x, a_0)$ for the state of physical chaos, we use the distribution $\tilde{f}_0$ of the Gibbs canonical type. As an additional condition (see (12)) we shall use the equation $\langle H_{eff} \rangle = \text{const}$. From the solution of this equation we find the effective temperature as a function of Δa

$$\tilde{D} = \tilde{D}(\Delta a); \qquad \tilde{D}(\Delta a)|_{\Delta a = 0} = D. \tag{3.1}$$

By $\tilde{S}_0$ we denote the entropy for the state of the physical chaos. Hence the difference between $\tilde{S}_0$ and S is given by

$$\Lambda_S = \tilde{S}_0 - S = \int \ln \frac{f}{\tilde{f}_0} f dx \geq 0, \qquad \langle H_{eff} \rangle = \text{const}. \tag{3.2}$$

Here we have used the symbol Λ_S to denote the Lyapunov functional.

Thus, we have obtained two results (3.1),(3.2). The solution (3.1) can then be used for the definition of the relative change of the degree of order due to the transition $a_0 \rightarrow a_0 + \Delta a$. If the inequality

$$\tilde{D}(\Delta a) > D \tag{3.3}$$

holds (that is, if the effective temperature in the state with $a = a_0$ fulfilling the condition $\langle H_{eff} \rangle = \text{const}$. increases), then the process $a_0 \rightarrow a_0 + \Delta a$ is the transition from the less ordered state (from a phycal chaos) to a more ordered state. The difference in the entropy is a quantitative measure of the increase in the degree of order.

If the inequality (3.3) is not satisfied, then the change Δa is not a relevant one, thus a is not an effective control parameter. This indicates that we have to look for new control parameters.

Thus, the S-theorem serves as a tool for checking the correctness of the choice of the control parameters. If we have several parameters then the search for the most ordered state can be optimized [10].

4. Criteria for Assesing the Relative Degree of Order in Open Systems on the Basis of Experimental Data, Using the S-Theorem.

In many different cases an adequate mathematical model for the system under considerationin is too difficult to construct. It is necessary to have certain criteria directly based on experimental data.

Such criteria can be based on concepts such as K-entropy, Lyapunov indices, and fractal dimensions, which can be obtained from the exrimental realizations [11,12]. Let us show that the relative degree of order can be evaluated directly through the experimental data, using the criterion based on the S-theorem [4,5].

Assume that a set of internal parameters X has been chosen. We consider the two time realizations $X(t,a_0)$, $X(t,a_0+\Delta a)$. The observation time must be sufficiently long in order to construct distribution functions

$$f_0(X,a_0),\ f(X,a_0+\Delta a);\qquad \int f_0\,dX = \int f dX = 1. \tag{4.1}$$

Using the distribution function f_0, which describes, according to our assumption, the state of the physical chaos, we find the function

$$H_{eff} = -\ln f_0(X,a_0). \tag{4.2}$$

After renormalizing the distribution function to the given value $\langle H_{eff}\rangle$, $\langle H_{eff}\rangle$ serves as the effective Hamilton function.

The mean value of the effective energy in general will depend on Δa. Let us carry out renormalization to the fixed value of $\langle H_{eff}\rangle$. To do this we introduce the renormalized function $\tilde{f}_0$, which represents a canonical Gibbs distribution with the effective Hamilton function H_{eff} and the "effective temperature" $\tilde{D}(\Delta a)$. The effective temperature as a function of Δa is defined by solving the equation $\langle H_{eff}\rangle =$ const, As a result we have

$$\tilde{D} = \tilde{D}(\Delta a),\qquad \text{but now } \tilde{D}|_{\Delta a=0} = 1. \tag{4.3}$$

Now, as before in sect. 2,3 we use the distribution functions $\tilde{f}_0$, f to evaluate the difference in the entropies (3.2). If the solution of the equation $\langle H_{eff} \rangle$ = const is such that

$$\tilde{D}(\Delta a) > 1, \tag{4.4}$$

then the state with $a = a_0 + \Delta a$ is more ordered than the state with $a = a_0$, which denoted the state of physical chaos.

Instead of time realisations $X(t,a_0)$, $X(t,a_0 + \Delta a)$ it is possible to use the corresponding spectral intensities

$$I_0(\omega,a_0), \quad I(\omega,a_0 + \Delta a) \tag{4.5}$$

and the corresponding distribution functions

$$f_0(I,a_0), \quad f(I,a_0 + \Delta a). \tag{4.6}$$

The effective Hamilton function is now defined by the expression

$$H_{eff}(I,a_0) = - \ln f_0(I,a_0) \tag{4.7}$$

and the relative degree of order can be assessed by the method described above.

It is possible to use, instead of (4.6), the frequancy distribution function

$$f(\omega,a) = \frac{I(\omega,a)}{\int I(\omega',a)d\omega'}, \qquad \int f(\omega',a)d\omega' = 1. \tag{4.8}$$

The effective Hamilton function is defined in this case by the expression $H_{eff} = - \ln f(\omega,a_0)$ and the relative degree of order can be described by the method already introduced in sections 3,4.

5 K-Entropy and Entropy Production in the Statistical Theory

Let us consider two trajectories in the phase space of the system, corresponding to two different sets of the initial conditions, and introduce the relevant dynamic distribution $f^{(d)}$, and the distribution function smoothed over a certain time interval. We assume that the condition of local ergodicity (see [4,5]) is satisfied, so that we can introduce by averaging the distribution $f^{(d)}$, a statistical distribution $f(x_1,x_2,t)$ for the corresponding Gibbs ensemble. If the time interval of smoothing is shorter than the relaxation time, then the distribution function $f(x_1,x_2,t)$ will be time-independent. Also time-independent will be distribution function of distance between the tragectories

$$f(D,t,a) = \int \delta\left(D - \sqrt{(x_1 - x_2)^2}\right) f(x_1,x_2,t,a)dx_1 dx_2 . \tag{5.1}$$

We can now introduce the corresponding Shannon entropy

$$S(t) = - \int \ln f(D,t)\, f(D,t)dD . \tag{5.2}$$

If we can write the kinetic equation for the function $f(D,t)$ with the "collision integral" $I(D,t)$, then for the entropy (5.2) we can write the equation

$$\frac{dS}{dt} = \int \left(- \ln f(D,t)\right) I(D,t) \equiv \sigma(t) \geq 0. \tag{5.3}$$

Since, however, kinetic equations of this kind have not yet been considered, we have to use a less defined description.

Using the destribution $f(D,t)$ we introduce two characteristics: the mean distance at the time t, and the effective distance $\Delta D(t)$,

$$\bar{D}(t) = \int Df(D,t)dD; \qquad \Delta D(t) = \frac{1}{f(\bar{D},t)} . \tag{5,4}$$

For small deviations $\bar{D} - D$ we can go from the expression (4.2) for the Shannon entropy to the approximative "Boltzmann formula"

$$S(t) = \ln \Delta D(t) \tag{5.5}$$

Using the latter, we find

$$S(t) - S(t_0) = \ln \frac{\Delta D(t)}{\Delta D(t_0)} . \tag{5.6}$$

We see that with the increasing entropy, which characterizes the indeterminacy of the distance between the trajectories, the effective distance $\Delta D(t)$ also increases.

With the aid of (5.6) we now find the mean rate of change of entropy over the time interval $t - t_0$ - the statistical counterpart of the K-entropy

$$K_{stat} = \frac{S(t) - S(t_0)}{t - t_0} = \frac{1}{t - t_0} \ln\frac{\Delta D(t)}{\Delta D(t_0)} . \qquad (5.7)$$

It follows that K_{stat} defines the average across the finite time interval. The local change in the entropy is given by the equation

$$\frac{dS}{dt} = \frac{d}{dt} \ln\Delta D(t) \equiv \sigma(t). \qquad (5.8)$$

which follows from (5.5).

We see that the entropy production is not sign-constant: it will increase if the mean effective distance between the trajectories increases, and decreases otherwise. The change in the entropy vanishes as soon as the stationary state sets in, so that the distribution f(D,t) ceases to depend on time. The expression (5.5) for the entropy then takes the form

$$S(a) = \ln\Delta D(a). \qquad (5.9)$$

To compare the relative degree of order characterizing the different stationary states corresponding to the different values of the control parameters, we may once again use the criterion based on the S-Theorem. As before, there are two ways of tackling this problem.

The first one is based on use of the stationary solution of the kinetic equation for the distribution function f(D.t), but equations of this kind have not yet been considered. The second one is based on direct implementation of the experimental data. The calculation may follow the scheme outlined in sections 2 and 3.

Namely, we can use two time realizations $D(t,a_0)$, $D(t,a_0+\Delta a)$ in time. As before, the observation time must be sufficiently long, so that they can be used to obtain the distributions

$$f_0(D,a_0), \quad f(D,a_0+\Delta a); \qquad \int f_0 dD = \int f dD = 1. \qquad (5.10)$$

The effective Hamiltonian is defined now by the expression

$$H_{eff} = -\ln f_0(D,a_0). \qquad (5.11)$$

It will be useful to compare the results obtained in this fashion with the results derived in some other way, for instance, on the basis of the K-entropy.

6. Evolution of Entropy and Entropy Production in Transition from Laminar to Turbulent Motion.

The notion of relative degree of order is very important for the analysis of laminar and turbulent motion. The criterion proposed by the author (see [7]) is based on the evaluation of the Boltzmann-Gibbs-Shannon entropy renormalized to the given value of the mean effective energy using the S-theorem.

According to this criterion, the averaged turbulent flow has a higher degree of order then the laminar flow. This is just one of the arguments favoring the standpoint that regards the transition from the steady laminar flow to the steady turbulent flow as a non-equilibrium phase transition into a more ordered state - in other words, as a process of self-organization [4,5,8].

Using the S-theorem we obtained the difference of temperature of laminar and turbulent flows at a given value of the mean effective energy

$$T_L - T_T = \frac{1}{2} m \langle (\delta u)^2 \rangle \geq 0 , \tag{6.1}$$

and the corresponding difference in the entropies

$$T[S_L - S_T] = \frac{mn}{2} \langle (\delta u)^2 \rangle . \tag{6.2}$$

The difference of temperatures and entropies in (6.1,2) is determined by the sum of diagonal elements in the Reynolds stress tensor.

The higher organisation of turbulent motion is manifested, for one thing, in the way the momentum is being transferred. Namely, the transfer of momentum from layer to layer, which results in the viscosity in the laminar flow, is due to the independent changes in the momenta of individual particles. On the other hand, in the case of turbulent flow the transfer of momentum between the layers is a collective process. In other words, the individual, unorganized resistance to the laminar flow upon transition to the turbulent flow is replaced by collective (and thus more highly organized) resistance. This results in a considerable increase in the viscosity.

Thus, upon the transition to the turbulent flow pattern, new macroscopic (collective) links are established between the regions of the flow. This renders the structure of the flow much more complicated and, possibly, more organized.

Entropy production criterion. Entropy production can also serve as a comparative measure of the degree of order in open systems. We shall illustrate this by considering the transition from the laminar to the stationary turbulent flow.

Let us compare the entropy production in a stationary averaged turbulent flow that arises for Reynolds numbers $Re > (Re)_{cr}$ with an imaginary flow (that is unstable for this Reynolds numbers). It is shown (Klimontovich and Engel-Herbert [13], and [4,5]) that, subject to the additional condition that the stress on the channel walls should be equal for these flows, i.e. the dynamic velocities v_* are equal, the entropy production is less for the turbulent flow (σ_T) than for the laminar flow, i.e.

$$\sigma_T < \sigma_L. \tag{6.3}$$

The real process thus proceeds along the path with lower entropy production. This can be taken as an indication of a higher degree of order in the turbulent flow, and the transition from laminar to average turbulent flow can be regarded as a self-organization process.

As an illustration, we reproduce the result obtained by calculating the entropy production for Couette flow in a plane channel,

$$\frac{\sigma_T}{\sigma_L} = \frac{Re}{2R_*^2} \leq 1, \tag{6.4}$$

in which Re and R_* are the characteristic Reynolds numbers . The equal sign refers to the laminar flow in which the law of resistance takes the form $Re = 2R_*^2$.

7. The Principle of Minimum Entropy Production in Self-Organization Processes

The particular result (6.3) can be used for formulating a more general statment: "The principle of minimum entropy production in the processes of self-organization".

We consider the process of self-organization as a result of a nonequilibrium phase transition.

Let the point a_{cr} be the critical point for the nonequilibrium phase transition which leads to a new stable stationary solution. By σ_{st} we denote the entropy production in the stationary state at $a > a_{cr}$, Let us also extend, tentatively, the "old" stationary state into the same region which was stable when $a < a_{cr}$. The entropy production in the imaginary state (unstable when $a > a_{cr}$) will be denoted by σ_{inst}.

Using these definitions, the principle can be expressed by the inequality

$$\sigma_{st} \leq \sigma_{inst}. \tag{7.1}$$

Thus, the entropy production in the new stationary state, resulting from the phase transition, is less than the fictitious entropy production in the old state, interpolated above the threshold region. The evaluation of σ_{st} and σ_{inst} is carried out under an additional constraint which is chosen according to the structure of the expression of entropy production.

We have formulated the principle on the basis of a particular example. The general proof does not exist so far.

8. Entropy as Measure of Diversity in the Processes of Biological Evolution

Let us recall that the Boltzmann-Gibbs-Shannon entropy possesses a combination of properties which allows it to be used as a measure of inderminacy in the statistical theory and in the theory of information.

The role of the entropy, as we have seen, broadens considerably when we investigate the problem of the relative order of the nonequilibrium states in open systems.

Let us now use a very simple example to illustrate that the entropy can play yet another role: The role of a measure of the diversity of states, "species" in the processes of natural selection, which. according to Darvin, are the heart of the process of biological evolution.

As before, we characterize the system by the distribution func- $f(X,t)$. By Ω we denote the volume of the region which contains the values of X. If the distribution is uniform, then $f(X) = 1/\Omega$.

For the measure of **diversity** we use the local entropy $-\ln(\Omega f)$. The distribution function $f(X,t)$ is governed by the following equation [4,5]:

$$\frac{\partial f}{\partial t} = \alpha(\ln f - \langle \ln f \rangle) f + D\frac{\partial^2 f}{\partial X^2} . \tag{8.1}$$

Now we can write the relevant equation of entropy balance

$$\frac{dS}{dt} = -\alpha(\langle(\ln f)^2\rangle - \langle \ln f \rangle^2) + D\int \frac{1}{f}\left(\frac{\partial f}{\partial X}\right)^2 dX. \tag{8.2}$$

We introduce in these equations the **evolution parameter** α.

For values $\alpha < 0$ the distribution for the stationary state is uniform and the diversity is zero. At $\alpha = 0$ an evolutionary phase transition occurs, the phase transition towards greater diversity. When $\alpha > 0$, evolution is possible. It leads to a stationary (nonequilibrium) distribution provided there exists a stable solution with nonzero diversity, which is proportional to the intensity of the "mutation factor" D. So, the random processes play a constructive role.

References

1. Pool R.: Is It Healthy to be Chaotic? Science **243**, 604 (1989).
2. Nicolis G., Prigogine I.: Self-Organization in Nonequilibrium Systems. New York: Wiley 1977; Moscow: Mir 1979.
3. Haken H.: Information and Self-Organization. Berlin, Heidelberg, New York: Springer 1988; Moscow; Mir 1990.
4. Klimontovich Yu.L.: Criteria for the Relative Degree of Order in Self-Organization Processes. Sov.Phys.Usp 32(5), May 1989.
5. Klimontovich Yu.L.: Turbulent Motion. The Structure of Chaos. Moscow: Nauka 1990; Dordrecht: Kluwer Academic Publ. 1990.
6. Klimontovich Yu.L.: S-theorem. Pis'ma Zh.Tech.Fiz. **9**, 1412 (1983).
7. Klimontovich Yu.l: Pis'ma Zh.Tech.Fiz. **10**,80 (1984).
8. Ebeling W., Klimontovich Yu.L: Self.Organization and Turbulence in Liquids. Berlin: Teubner 1984.
9. Klimontovich Yu.L: S-theorem.Z.Phys.B **66**, 125 (1987).
10. Klimontovich Yu.L., Bonitz M.: Z.Phys.B **70**, 241 (1988).
11. Shuster H.: Deterministic Chaos. Weinheim: Physik-Verlag 1984.
12. Haken.H.: Advanced Synergetics. Berlin ,Heidelberg:Springer 1983.
13. Klimontovich Yu.L.,Engel-Herbert H.: Zh.Tech.Fiz. **54**, 440 (1984).

Index of Contributors